A Basket of Weasels

The weasel family in Ireland and other furred Irish
beasts: bats, the rabbit, hares and some rodents

James Fairley

With portraits of the animals by Roy Gaston
and text figures by the author

Published privately
Belfast, Northern Ireland

First published in 2001 by the author

15, Luxor Gardens, Belfast BT5 5NB, Northern Ireland

© James Fairley 2001

Edition of 1,000 copies

Printed by Clódóirí Lurgan Teo., Inverin, Co. Galway

Set in Times New Roman

ISBN 0 9502626 2 5

British Library Cataloguing in Publication Data.
A catalogue record for this book is available from the British
Library.

Contents

"As cranky as a basket of weasels"

Adage favoured by the late Mamie Poynton, sometime landlady
to the author, who is affectionately remembered.

Cover illustration:

A basket of Irish "weasels"

(There are no weasels in Ireland, but Irish
stoats are commonly referred to as such.)

Preface

This book is devoted solely to the wild mammals of Ireland and has been written with the layman very much in mind; it is intended both to entertain and inform. You will encounter here a blend of ancient and modern research, local history, many casual observations by naturalists and others, and additional ingredients of multifarious origin. Among other things, I have aimed to provide a digest of some recent scientific studies on Irish mammals which, because they have been published in professional journals, are not readily available to the general public. The sole criteria for inclusion of any information has been that it is interesting, reliable - I have always plainly indicated where I have doubts about anything - and pertains directly to Ireland, in other words that it stems from observations actually made here. Wherever it has been necessary to rely upon sources from outside Ireland, this is made clear. Whereas such an approach may be considered insular, I have a horror of the uncritical application of findings from Great Britain to Ireland.

In order to render the text of greater value to the professional naturalist, a list of just under 400 references has been appended, which are indicated by inconspicuous superscript in the text. To the same purpose, I have also included the scientific (Latin) name of each species of non-domestic animal at first mention; the scientific names of all of the plants are included at the back. Note that some English names (e.g. mullet, laurel) are too imprecise to permit provision of a scientific name. Partly to make the information of more value to full-time zoologists, there is somewhat more tabular matter than is usual in a book intended for the popular market. I hope that none of this will detract from the enjoyment of other readers.

A Basket of Weasels is a sequel, or companion, to an earlier volume, *An Irish Beast Book,* originally published in 1975, with a second edition in 1984. When the *Beast Book* was first written, mammalogists in Ireland were a rarity. Since then many people have entered this fascinating area of research, if only for a few years in order to secure a postgraduate degree, and there has thus been a remarkable proliferation of knowledge in the field. It must, however, be added that unfortunately many sound pieces of work remain unpublished and languish sequestered either in theses in university libraries or in internal government reports.

I must emphasise that this book deals with only a selection of species, the choice being determined by certain animals having already been featured at length in the *Beast Book,* the limited extent of our knowledge on others in Ireland and, it must be confessed, by personal preference. For instance, you will look in vain for chapters on Irish deer, which have been studied extensively over the past three decades - but rarely by my students or by me.

The more perceptive reader may be puzzled by my choice of title, because there are no (definitely no) weasels in Ireland. Nevertheless, we have five species which belong to the weasel family, each of which has been given a complete chapter. Even though both the stoat (often incorrectly known as the weasel in Ireland) and otter were each afforded chapters in the *Beast Book,* readers will find that the corresponding chapters here are largely of fresh material.

I am most grateful to Mr Roy Gaston, a much underrated Irish wildlife artist, for his fine portraits of the mammals, including those in the cover illustration.

I wish to express my gratitude to the following for help in gathering data for this book, usually by responding to queries - although relevant information sometimes arrived gratuitously - or by pointing me towards fruitful printed or manuscript sources: Professor Nicholas Canny (History Department, National University of Ireland [NUI] Galway), Mr Bill Clarke (Zoology Department, University College Dublin), Randall Crawford Ltd., Mr Sherwin Curran (Rentokil, Northern Ireland), Dr Owen Denny (Veterinary Service, Department of Agriculture of Northern Ireland), Mr J.L. Desmond (Secretary, The Irish Coursing Club), Dr Karina Dingerkus, Dr L.A. Dolan (Tuberculosis Investigation Unit, University College Dublin), Mr George Douglas (Pest Control Unit, Belfast City Council), Sir Josslyn Gore-Booth, Messrs Paul Hackney (Ulster Museum) and John Higgins (Dúchas), Dr Robert Hunter (History Department, University of Ulster), The *Irish Independent* and *Irish Times* (for printing letters of enquiry regarding martens and mink and the many correspondents, two numerous to mention individually, who responded to them), Drs Noel Kirby (Dúchas), Graeme Kirkham, Kate McAney (Vincent Wildlife Trust) and Kieran McCarthy (Zoology Department, NUI Galway), Professor Ian Montgomery (School of Biology and Biochemistry, Queen's University Belfast), Mr Peter Mooney (Dublin Port Company), Ms Sharon Murphy (Community Health Division, Dublin), Mr Terry Nevin (Port Health Division, Belfast City Council), Dr Jim O'Connor (Natural History Museum, Dublin), Messrs Ger O'Donnell and Paddy

O'Sullivan (both Dúchas), Mrs Angela Ross (Ulster Museum), Royal Sun Alliance Insurance, Mr Stephen Sharpe (Regional Information Branch, Department of Health and Social Security, Northern Ireland), Dr Caroline Shiel, Sorex Ltd., Dr Paddy Sleeman (Zoology Department, University College Cork), Dr Chris Smal, Mr Pat Smiddy (Dúchas), Mr Mark Smyth, Mrs Jo Whatmough and Professor Frank Winder (Department of Biochemistry, Trinity College Dublin). I must also acknowledge the help of Professor Michael Mitchell (formerly of the Botany Department, NUI Galway) at the publishing stage.

I am indebted to Dr Tony Mitchell-Jones and Mr T. McOwatt (English Nature) for authorisation to adapt an original illustration for Fig. 22, to the Deputy Keeper, Public Record Office of Northern Ireland, for allowing me to publish archival material, and Miss Rebeccah Cogan and Drs Rebecca Jeffrey, Paddy Sleeman and Alan Wolfe for consenting to my utilising some unpublished material from their theses. The first verse of the poem *The Badger* is reproduced by kind permission of Punch Ltd; the passage from "The shadow on the lake" by Joseph Lyster (*Journal of the Old Athlone Society* 1974-75) by kind permission of his nephew, also Mr Joseph Lyster; and that from *Muskrats and Marsh Management* courtesy of the Wildlife Management Institute, Washington DC. I would also acknowledge, with thanks, the Royal Irish Academy as the source of the material from the Ordnance Survey Memoirs. The chapter on the muskrat is closely based on an article I published in the *Irish Naturalists' Journal* in 1982.

I am most grateful for the facilities of the the the Belfast Central & Newspaper Libraries, the Public Record Office of Northern Ireland, the Linenhall Library and the libraries of the Department of Agriculture of Northern Ireland, the Northern Ireland Veterinary Service, Queen's University Belfast, the Ulster Museum, NUI Galway, the Royal Irish Academy, Trinity College Dublin, University College Dublin and the Zoological Society of London.

Finally, I am greatly obliged to those who have read the draft typescript: to Drs Kate McAney, Chris Smal and Alan Wolfe, who scrutinised individual chapters, to Mr Pat Smiddy, who looked over most of it, and Dr Caroline Shiel, who worked her way through the lot. They are responsible for very many improvements. Any mistakes are my own.

James Fairley

Introduction to the Weasel Family

The weasel is, as it were, a long mouse.
De Proprietatibus Rerum. Bartholomaeus Anglius. (1495?)
(English translation by J. Trevissa. 1535.)

Although, as emphasised in the Preface, the weasel itself does not occur in Ireland, the first five chapters of this book deal with the members of the weasel *family* living in the wild here: the mink *Mustela vison,* Irish stoat *Mustela erminea hibernica,* badger *Meles meles,* otter *Lutra lutra* and pine marten *Martes martes.* It therefore seems reasonable to give readers some sort of general introduction to the family, or the Mustelidae as zoologists have named it; its members are collectively known as mustelids.

The Mustelidae comprises some 65 species worldwide and includes the smallest and fiercest of all flesh-eating mammals. Moreover, the smaller a mustelid is the fiercer it tends to be.

Why so?

Mammals are warm-blooded: they maintain a constant internal body temperature and, like birds, this has allowed them to live in the coldest parts of the world and to go abroad elsewhere in temperatures that would either kill most other animals, or at the very least render them torpid. Regulation of body temperature therefore confers a continuity of existence on mammals and birds denied to lower animals. The disadvantage of a constant body temperature is that much more energy in the form of food is needed to fuel it. For example, a foraging mouse uses 20 to 30 times as much energy as a foraging lizard of equal weight, and most rodents burn off 80-90% of their energy simply in keeping warm.

Heat is dissipated over the surface of the body and one of the most important characteristics in determining surface area is size. Of course larger mammals have a greater surface area in absolute terms, but relative to their volume, larger animals have *less* surface area than smaller ones. A cube of 1 cm side has a volume of 1 cubic cm and a surface area of 6 square cm, a ratio of 1:6. But for a cube of 10 cm side the ratio is down to 1:0.6. Mammals are never cubical. Nevertheless, surface area is approximately proportional to the square of an animal's length and volume to its cube. So as an animal increases in size its volume goes up disproportionately more than surface area. A small mammal therefore needs to eat relatively more to keep warm. The speed at which fuel, in the form of food, is burned per unit body weight is called *metabolic rate*. A mouse has a metabolic rate ten times that of a cow, and a shrew a metabolic rate one hundred times greater than that of an elephant!

➠ Weasels, stoats and mink are the smallest of all flesh-eating mammals. Therefore they must kill and eat more in proportion to their body weights than larger carnivores. Moreover mustelids usually have long, slender, sinuous bodies - although there are exceptions, like the badger, which is a fairly stocky beast - and this further increases surface area. For the rounder and dumpier an animal, the smaller will be its surface area. Small size and elongate body therefore maximise relative surface area. A captive weasel *Mustela nivalis,* the smallest of all mustelids, will consume from a quarter to a third of its own weight each day and for wild ones the requirement is probably greater.[179] In contrast, the average daily consumption of meat by a lion *Panthera leo* in the wild is only about 3-4% of its body weight. Provided a kill is big enough, a lion is able to devour enough at a single sitting to last it at least 5 days.[296] On the other hand, a mink or stoat almost certainly hunts every day and the curiosity of these animals, born of an almost incessant quest for prey, is well known. It is the greater relative surface area of a small mustelid which necessitates its bloodthirsty ferocity. The least of the Irish mustelids - the stoat and mink - are by far the most savage.

➥ Mustelids have medium to short tails and the legs are nearly always short. At lower speeds they scuttle along with the back more or less level; but at a gallop they undulate, the elongated body alternately humping, as the hind feet land together behind the fore feet, and then extending as the fore feet are thrown forward. Interestingly, the Latin word *Mustela,* meaning a weasel, from which the name of the family is derived, can be divided into *mus* = a mouse, and *tela* = a web or something woven. So, paraphrasing, the Romans by "*Mustela*" probably meant a small furry creature that weaves up and down as it progresses: an apt description.

➥ Besides its increased surface area, the typical mustelid body plan has additional disadvantages, although it more than compensates in other ways. Small stature means that, even in moderate ground vegetation, the animals cannot see very far. In order to spy out the land, many are in the habit of sitting up on their hind legs, the lengthy body then working to their advantage. Longer legs would have meant greater breadth of stride and thus increase in speed with little extra effort. So for locomotion short legs are inefficient in terms of energy expended. On the other hand, combined with a body of small cross section, they allow the pursuit of small quarry, such as a rodent, through confined spaces: along the rodent's run and in some cases even down its burrow. But a mustelid cannot afford to sacrifice too much weight when it closes with its victim. The necessary extra mass is supplied by the long body.

As we have already seen with surface area, the scaling of the various parts of the animal body as it gets progressively bigger or smaller is not a simple relationship. We have another example in the stubby legs of mustelids. The volume of an animal, and therefore its weight, rises with the cube of its length (x^3), whereas the strength of a leg increases only with its cross section (x^2), so very big animals, like elephants and rhinos, have to have disproportionately stout, pillar-like legs to carry their weight. Conversely, if a large animal were to be gradually reduced in size, and its parts kept more or less in proportion, then the cross section of many muscles and bones, and therefore their strength, would decrease more slowly than body weight. So a small animal with a roughly similar body shape as a large one is relatively stronger. Mustelids are thus comparatively powerful animals, which explains why a lion cannot run while bearing a carcass of even half its own weight, whereas a stoat can easily carry away a dead rabbit *Oryctolagus cuniculus* over twice its size.[179] Neither have the short, robust legs of mustelids needed much modification to serve as powerful and efficient spades, as in the badger, or as paddles, as in the otter.

Mustelids have well developed scent glands around the anus and are noticeably smelly creatures. The polecat *Mustela putorius* and its domesticated version, the ferret, are proverbially malodorous. Indeed, the Latin name literally translates as the weasel with the putrid odour. Nevertheless, it is skunks (also mustelids) that are the most accomplished in this respect.

Mustelids are generally solitary animals; although the badger is a social mammal. In almost all other mustelids, encounters between adults of the same sex are aggressive. The male of most species is noticeably larger than the female, size giving an edge in fights with other males during the mating season. The female produces only one litter each year. In most species, apart from the breeding season, the sexes live apart and, at best, no more than tolerate each other. The act of mating is anything but tender. The male invariably seizes the female by the scruff of her neck and may drag her about roughly before mounting. The violence, nevertheless, is not gratuitous. It is actually essential to stimulate *ovulation,* the release of eggs from the ovaries, and thus for successful procreation. Copulation is repeated and prolonged, for hours in some species, but this makes fertilisation more likely. Such an extended performance is facilitated by a bone in the penis, the *baculum,* which is, incidentally, a feature in many other mammals and by no means exclusive to the Mustelidae.

Chapter 1. The Mink - a Semi-aquatic Weasel

In the entire weasel family, there are few that are more bloodthirsty and cruel than the Mink, and few that are more tireless and active.
Fur, a Practical Treatise. Max Bachrach.

As members of the weasel family are generally small and skinny, it should come as no surprise that their fur coats are often particularly dense and warm, especially as several species inhabit high latitudes. So, besides being persecuted by man for their depredations on game and poultry, many have also been relentlessly trapped for their pelts, among them the ermine (the stoat in its white winter pelage), marten, sable *Martes zibellina* and, of course, the mink, a mink coat being contended by some to be the very *ne plus ultra* in the decadent trappings of wealth. A warm coat is of further advantage to the mink because the beast lives beside water and is often in and out of it; essentially it is a semi-aquatic weasel. It is a paradox that "mink" is thus a byword both for sophisticated luxury and unparalleled savagery.

The mink that now occurs in Ireland was originally a native of North America and has been extensively bred in captivity for its pelt, not only in its homeland, but also in Europe, where escapees have established free-living populations. Moreover, the former Soviet Union pursued a policy of widespread deliberate releases to found populations in the wild for commercial exploitation. There is a second extant species of mink, the European mink *Mustela lutreola,* which is also semi-aquatic. Its numbers are falling as a result of loss of habit, through hydroelectric developments and pollution, and in unsuccessful competition with its introduced American relative. A third species, the sea mink *Mustela macrodon,* also native to north America, became extinct in the nineteenth century.

The skull of the mink is of the typical mustelid pattern (Fig. 1). There is a short, squashed-up facial region at the front and a long low cranium. The canines are long, for stabbing and gripping, and the teeth behind are longitudinal ridges, the upper and lower sets working like scissor blades as they shear past one another when the jaws close. At the back of the lower jaw bone, on either side, is a horizontal rod of bone which fits neatly into a corresponding transverse groove in the skull and thus acts as a hinge - so neatly, in fact, that even on a cleaned museum skeleton it is often awkward to separate lower jaw from skull. As a result, all of the jaw movement is up and down, the action ideal for slicing flesh.

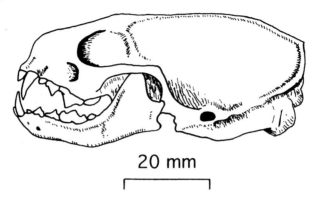

20 mm

Fig. 1 Skull of mink.

The mink has the typical elongate weasel shape. The head is pointed, unlike the broad flattish head of the otter, and the tail much the same diameter along its length. The otter's tail, in contrast, is markedly tapered and thick at the base. Because the mink is so bold, and also possibly because its evil reputation precedes it, people tend to overestimate its size, although it must be admitted that this is also true of many other wild mammals, from whales to rats. (Rats which put in an appearance on Irish housing estates are usually reported in the media to be of exceptional dimensions.) The biggest Irish mink carcass that I have weighed was only 1.56 kg, and the average weight of adult males (32) 1.23 kg and females(13) 0.74 kg.[106] My former postgraduate student, Dr Chris Smal, has been involved in most of the scientific work on Irish mink, and has live-trapped, marked and recaptured them in a study which extended from autumn 1983 to spring 1985, mainly in the Irish midlands around Mullingar - more of which later. He found that adult males generally weighed 0.9-1.3 kg and females 0.5-0.8 kg. However, the biggest one he ever captured was 1.9 kg - a monster![330] One of the ways of distinguishing the mink from the otter is that the latter is much bigger - about five times bigger. So mink are relatively small, which, as emphasised in the Introduction, in the weasel family equates with ferocity.

The fur of the wild mink is nearly always predominantly very dark chocolate brown: from any distance it appears black. There are also patches of white on the underside, which may occur almost anywhere on the throat, chest, belly and the inside of the legs (Fig. 2). The chin is usually white and occasionally also the tail tip. Isolated white hairs can appear anywhere else on the body. White hairs on the scruff of the neck are a common feature on females, and are caused by scarring as a result of the biting by the male during mating.[323]

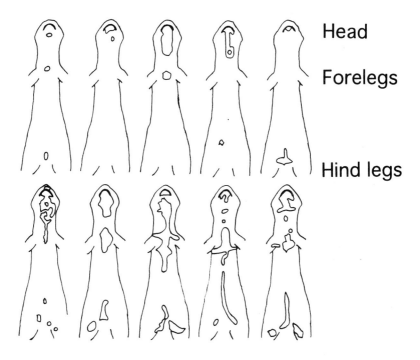

Head

Forelegs

Hind legs

Fig. 2 The undersides of some Irish mink showing the extent of variation in the patches of white. Skins are predominantly dark brown but shading has been omitted in order to show the white patches clearly. In addition there are sometimes individual white hairs scattered over the body.

A great variety of other colours has also been produced by selective breeding on fur farms of unusually coloured animals, most of which occur naturally in very low numbers in the wild. Even though a population of wild mink may be derived from fugitives from fur farms, of which a large proportion is usually exotically coloured, the coat reverts almost exclusively to the wild-type colouration in succeeding generations. This is partly because many of the genes producing the unusual hues are what geneticists term "recessive", whereas those for wild-type coats are dominant. In animals and plants which reproduce sexually, every gene is paired - one coming from the father and one from the mother - and a dominant gene from one parent will override a recessive from the other. This effectively means that a fancy coloured mink is only likely to breed true when crossed with another of the same shade. Indeed most exotic coat colours in mink are controlled by several genes working together; for the colour to show, all the pairs of genes have to be recessive, which will occur as a matter by chance in the wild only extremely rarely.

Of some significance also is the fact that, just as fur farmers select the exotics to breed from, the environment naturally selects the wild-type, which presumably has some sort of advantage over the other shades. Perhaps the latter are more easily seen by predators or potential prey, or maybe they are less attractive to potential mates. At any rate, like most wild mammals, only a small proportion of mink survive to breed, and the advantage enjoyed by those with wild-type coats swamps the fancy colours within a few generations. Selection for wild-type by nature is, after all, only to be expected. If this were not so, then fancy coloured mink would be more common in native populations in North America.

In the late 1970s I obtained and dissected the bodies of 108 trapped Irish mink, mainly from the north-west, 98 of which with coats sufficiently unbloodied to allow of an accurate assessment of colour. Only six were fancy coloured: three "silverblu", a light grey colour also known as "platinum" (a term no doubt coined to extol the luxurious nature of the garments fashioned from the pelts) and two "pastel" (a lightish brown). The colours were verified by showing prepared skins to a fur farmer. A further mink was white but lacked the pink eyes of an albino. Unfortunately, the skin was too damaged to preserve, but it may have been a "regal white". White skins are the result of several pairs of recessive genes and complex selective breeding, and are unlikely to turn up outside a fur farm by chance, even from stock only a few generations away from escapees. So the possibility that this animal had itself absconded cannot be precluded. Even if the 6% of the skins that I examined corresponded to the proportion of fancy colours in the population of Irish wild mink at the time, the percentage may have fallen since, for mink are now widespread and therefore the genes from any exotic escaping today would probably be more quickly diluted in matings with the wild population.

Probably the earliest mink farm in Ireland was set up at Killybegs, Co. Donegal, in 1951.[327] Mink ranching continued unregulated for another 15 years, although the usual quarantine restrictions had to be met. It is difficult to determine just how many people kept the animals before the trade was licensed, but the allure of a quick and substantial profit dazzled some, and the early years of the industry included many small, back-yard concerns[327] in which security was inevitably lax, or lax enough to enable such slim, sinuous, intelligent, resourceful, and undoubtedly still undomesticated mammals to break free. On enquiry, many fur farms in the 1960s acknowledged the escape of "one or two" but were reluctant to be precise.[75] To be fair, some ranch managers offered cash for mink trapped in their districts.[327] Although there is, understandably, now no available evidence, it is possible that some small-time amateur fur farmers, realising that they had bitten off more than they could chew, and being too tender-hearted or lazy to bump off their stock, disposed of it by the simple expedient of opening the cage doors. In addition there was at least one malicious mass release, of 350 animals at Swords, Co. Dublin, in 1964.[75] There have been similar widely-reported, mindless, multiple "liberations" in Britain in recent years by animal welfare fanatics, although it is difficult to comprehend how such criminal meddling would have benefited either the mink themselves or the local wildlife. Populations derived from escapees are usually referred to as *feral* mink.

A questionnaire in 1960 revealed that there were then at least 40 mink breeders in Ireland with an average of 70 animals per ranch.[75] The Department of Agriculture in the Republic was notified of 58 farms in existence prior to 1966, when licensing was brought in through an extension of the Musk Rats Act of 1933 (more of which in Chapter 9).[327] Similar regulation was enforced in Northern Ireland in 1968. But such legislative shutting of the stable door can have made little difference when the animal had already bolted. Anyway, a number of small unlicensed fur ranches persisted in the Republic into the late 1960s, one continuing in business to 1986.[327] As the license was then £5 for 5 years, such flouting of the law can hardly have been because the owner pleaded poverty. At least one operation north of the Border was still going 5 months after licensing became obligatory.[75] Gradually the small units went to the wall, for an efficient operation usually involves stocks of hundreds, if not thousands of mink. In the Republic the number of farms fell to 23 in 1969-71; in 1986 only seven licenses were granted and in 1998 there were eight. In Northern Ireland only five licenses were issued from 1968 to 1986, when there was only one farm in operation.[327] There has been no mink farming at all in Northern Ireland in recent years.

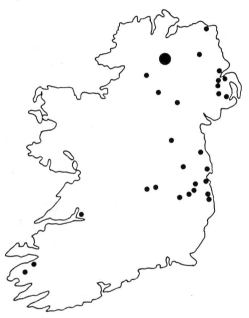

Fig. 3 Recorded sightings of feral mink in Ireland up to 1969. These almost certainly do not reveal the complete distribution of feral mink at the time. The large dot indicates the area where they were then known to be breeding.

The earliest survey of the distribution of feral mink in Ireland, published in 1969,[75] was simply a collation of assorted "confirmed records" collected by its authors (Fig. 3). It will come as no surprise that the sightings of the beasts corresponded approximately with the then known distribution of fur farms. Records of feral mink in 1969 were in the north, on the eastern seaboard inland for around 100 km, with scattered sightings in the south west, particularly on the Iveragh

Peninsula in Co. Kerry, where there was a fur farm. Although this survey did not apparently include Co. Donegal, it is elsewhere reported that mink had settled in the wild there as early as 1956,[327] although no further information is available.

The first breeding population for which there is any detail was near Omagh, Co. Tyrone, apparently as the result of the escape of 30 animals from a ranch in 1961. Of these, 15 were trapped or otherwise accounted for; the rest were believed to have made their way to the neighbouring River Strule. In 1965 the ranch was moved into Omagh, but apparently with little better security, for 20 more of the wily creatures were able to decamp, again probably to the Strule. A number were subsequently shot in hen runs. It may be that newly escaped mink are less reluctant to approach poultry than those born in the wild, the former having been accustomed to dining in the vicinity of homesteads. Moreover, adult mink in new surroundings, and with no previous experience of hunting, may be forced to such rash measures. Also relevant is the fact that numbers in the wild, and their consequent depredations, tend to peak shortly after initial escapes from captivity of the founders. Thereafter populations stabilise at a lower level as some individuals settle into territories and exclude the others.[323] Although a proportion of the Omagh animals were poisoned, trapped and hunted with dogs, in 1967 mink, some of them mothers with kits, were reported along a wide stretch of the Strule, Mourne and Glenelly Rivers and, in the following year, some were killed in back gardens in Omagh.[75]

Between 1965 and 1969 there were a number of other sightings of mink at large in Ireland.[75] Several individuals were apprehended in hen runs near Ballynahinch, Co. Down. One was killed in Belfast on the Connswater River, where it flows beneath Newtownards Road. A fisherman saw mink twice in pools on the River Finn in Co. Fermanagh. A female was shot in a hen house near Dunshaughlin, Co. Meath, but only after it had polished off eight of the residents. In Co. Dublin one was killed in the act of attacking rabbits actually in the Agricultural Institute's animal house at Dunsinea! Another was dispatched in a garden at Mount Merrion, in the Dublin suburbs. A mink was trapped under a bakery in a tunnel communicating with the River Dodder, and another, shot at Rathangan, was sent to a taxidermist. In Co. Kerry in 1968, an individual was shot in a farmyard on Valencia Island, where there had never been fur farms. This animal presumably swam across the sound from the mainland - there being no bridge at the time - a minimum of half a kilometre.[75]

In the 1970s the short-lived, government-run and much-lamented Irish Biological Records Centre collected records of a variety of species of wildlife and compiled distribution maps from them. Unfortunately the Centre was abolished during internal reorganisation of the civil service. The Centre published maps of mink distribution in 1974 and 1979[243] (Fig. 4) in which the known range of the mink had extended, and in which its presence over a large part of Donegal was acknowledged. Most records were again from the north and mid and south-east (although there were a few from the extreme north-east) with some sightings in the south-west. The mid-west was still practically free of them.

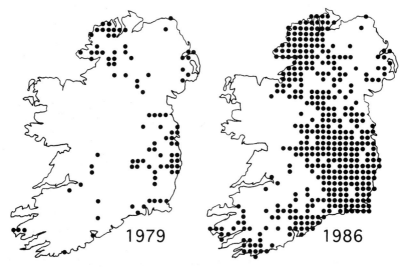

Fig. 4 The recorded distribution of feral mink in 10 km squares of the Irish National Grid in 1979 and 1986.

The most complete survey of all, and the most recent,[327] was in the middle 1980s, and included all the earlier data, plus new records obtained mainly by the personnel of Dúchas. (Dúchas is the current name for the government body in the Republic incorporating the Wildlife Service, the National Parks etc., which I use throughout this book for simplicity, even when referring to former times when any one of several others names was in vogue.) Such records were in the form of sightings, road casualties, trapped animals and from searches specially undertaken for droppings and tracks beside streams, rivers and lakes (Fig. 4). The searches not only determined the presence of mink but also many places where there were certainly none of them. Whereas such negative records are not shown in Fig. 4, they covered most of the blank area in the mid-west. Mink were then widespread in the east, except for the extreme north-east, although there were a few records from the north of Cos Antrim and Londonderry, and in the north-west. Records for the south-west were more scattered. The mid-west - Cos Mayo, Galway and Clare - was effectively free of them. The large number of negative records from these three counties and the lack of reports from Regional Gun Clubs there, which are eager to notify the appearance of mink, confirm that the animal was then almost completely absent from the mid-west. While it appears as if the River Shannon then presented a barrier to spread, this is difficult to credit for a semi-aquatic animal like the mink. It is much more likely that, at this stage, feral populations, having originated in the north, south and east, had simply not had time to spread any farther.

By the late 1990s mink were well settled in Co. Galway and I had had numerous reports of them from around Lough Corrib. When in July 1996 18 hens were slaughtered at a farm at Kylemore, Mr Ger O'Donnell, of Connemara National Park, Letterfrack, set a trap in the woods nearby and within a few days caught a mink. Since then there have been reports as far west as Clifden.

In order to determine the situation in Clare and Mayo, I wrote letters to a number of local and national newspapers in 1999, to which there was a gratifyingly large and illuminating response. From this it was clear that the animals were then widespread in both of these counties and had reached the western seaboard, at least at Westport and Newport in Mayo, and Miltown Malbay and Quilty in Clare.

Several replies came from rod fishermen, some of whom remarked on the apparent lack of fear shown by the animals, one writer describing how a mink actually sniffed at his wader, which is rather too close for comfort. Rod fishermen, in particular, have both time and opportunity to observe wildlife around fresh water, although not enough time for some, one correspondent lamenting to me that he had been "born to fish, but forced to work". Another fisherman told me of a mink using the boats in his boathouse as dens, and of droppings and a dismembered coot *Fulica atra* left in them. My former colleague at National University of Ireland (NUI) Galway, Dr Kieran McCarthy, who has researched on eels *Anguilla anguilla* on the loughs on the River Shannon for years, confirmed that it is by no means unusual there to dislodge a mink from a boat which has been drawn up on the shore for some time.

The fearless character of the beast was particularly well drawn in incidents described by my correspondents in Clare and Mayo. Of domestic stock, ducks are understandably at greatest peril from mink because of the mutual fondness for water. A farmer at Miltown Malbay described how his flock of ducks mysteriously diminished daily until, one morning, he heard a commotion among the survivors. On investigation, he saw a mink in the process of slaughtering yet another. Although he shouted and threw everything to hand at it while his wife ran to fetch a shotgun, the animal continued to dart in and out of a drain, unrelenting in its efforts to finish off the half-dead bird. Eventually the shotgun arrived, with the inevitable consequence for a marauder unacquainted with the better part of valour.

A fisherman's ghillie at Pontoon, in Co. Mayo - one of Ireland's premier spots for game fishing - wrote of frequent sightings of mink and of seeing them attacking mallard *Anas platyrhynchos* more than once. One day a mink pursued by dogs ran into his house, much to his mother's alarm. The animal scuttled from one room to another and even tore at a packet of dog food. Eventually one of the dog's dispatched the intruder under the refrigerator.

Systematic data on the recorded distribution in Northern Ireland to date come from 5 km grid records in the Ulster Museum. Tyrone and Fermanagh have been the counties with most records of mink, with somewhat fewer for 'Derry, Armagh and Down. From this database it appears that Co. Antrim is still largely uncolonised.

The aspect of mink that interests most people is what they eat. Whereas I examined the stomach contents of all of the mink which I dissected, the findings have been superseded by a major study in which one of my students, Mr Dermot Ward, analysed 2,510 scats, gathered by Smal during the course of his work.[363] Incidentally, the word "scat" is used technically by wildlife biologists to mean "dropping" - usually of a carnivore; I have never seen the faeces of deer, mice or bats so referred to.

Mink defecate beside the water, often in the same sorts of places as otters (see Chapter 4), but their scats are readily differentiated by the use of one's nose. Whereas those of otters have a pleasant odour, mink scats have a smell which is indisputably nasty. Otter droppings are

MINK

highly variable both in size and shape. Mink scats are compact, cylindrical, 3-8 cm long, about 1 cm in diameter, often dark green or brown, and usually pointed at one or both ends.

Chris Smal studied mink both by collecting droppings and by trap-mark-recapture, mainly around Mullingar in two, fairly clean rivers: along 14 km of the River Inny, which is slow-flowing, 4-8 m wide, with a muddy bottom, and its tributary, the River Glore, 10 km long and 2-4 m wide, which along most of its length cuts deeply into its limestone bed. His study also included Lough Ennell, a shallow lake of 14 km[2] which at the time of study received effluent from Mullingar in the north via the River Brosna. Consequently Lough Ennell was, not to put too fine a point on it, thoroughly polluted, although, judging by its appearance, somewhat less so at its southern end, which is furthest from Mullingar. This Lough holds large populations of waterfowl.[162] Smal also collected scats around Lough Lene, also in the region, which is 4 km[2] in area and only slightly contaminated, although he did not trap mink there. The countryside around the lakes and watercourses is mixed, including pasture, rough grazing, bog and conifer forest. As Smal also often came across otter droppings in the course of his work, he collected them as well. Eventually these too were analysed (Chapter 4).

Each scat was placed in an envelope and dried; when the faeces of many mammals are thoroughly dehydrated in this way, they can be stored in a dry place more or less indefinitely. Because Smal's study was funded by Dúchas, the envelopes were government issue: the brown manila envelopes emblazoned with the Symbol of State, the Irish harp - the kind in which official documents, like those from the Revenue Commissioners, are sent. Seldom can these envelopes have contained anything to arouse such pleasant anticipation. A heap of them awaiting attention, bulging and stained by the scats within, presented a bizarre, even surreal picture.

Dermot Ward weighed each dried scat, crumbled it, and identified the contents using assorted literature and reference collections from potential prey of bones, scales, hair and feathers. With scats containing more than one type of prey, he also judged, subjectively by eye, what percentage of the scat was accounted for by each category of prey.

The presentation of results from scat analyses poses problems. For example, suppose one found that brown rat *Rattus norvegicus* occurred in 25 (= 25 occurrences in scats) out of a collection of 50 scats. Then we could say that rat made up 50% of the diet by *percentage occurrence.* Unfortunately a difficulty arises if you look at the complete list of all types of prey, each expressed as percentage occurrence, because they will not add up to 100% should any scat contain more than one kind of prey. So it is often difficult to see clearly what proportion of the whole diet is made up of each type of food. We can get over this by expressing the number of occurrences of rat (and every other individual prey category) in a collection of scats as a percentage of the total occurrences of all prey categories; this we shall call *percentage frequency*. This still has disadvantages, because some small preys, such as insects, could appear in a lot of scats but only in very small amounts. Percentage frequency might therefore tend to overestimate the importance of such items. A third approach is to work out the dry weight of each prey category in the faeces. As the dry weight of each scat was recorded and the percentage of each food category in it had been estimated, it was easy to calculate the *percentage weight* that any prey contributed to any collection of droppings. If, dear reader, you are now scratching you head, do not worry. Percentage frequency and percentage weight usually yielded similar results, as shown in Table 1, which gives the content of each prey category in all two and a half thousand scats. Nevertheless, some small prey items, such as stickleback and other invertebrates, evidently do indeed seem to be overestimated by percentage frequency.

Table 1 reveals a varied menu. Crayfish *Austropotamobius pallipes,* fish, birds and also mammals and frogs *Rana temporaria* were all important components of the diet, and the mink apparently utilises just about any animal, on land or in the water, which is unable to escape, put up sufficient resistance, or is found dead.

The mink is apparently a jack of all trades predator and, even though it lives beside water, it does not, like the otter, live predominantly on fish (Chapter 4). Otters have webbed feet but such webbing is negligible on a mink, and it is a poorer swimmer. Laboratory studies in Britain show that a mink normally locates fish *before* diving, often poking its head below the surface the better to pinpoint its intended victim. Then it slips into the water, pushing off from the bank with its hind feet to gain momentum. Pursuit, capture and emergence from the water take, on average, 10 seconds. The maximum recorded length of dive was only half a minute.[88] The animal's diving ability is therefore surprisingly limited and it apparently operates efficiently by locating its prey before submerging.

While the remains of rabbits and hares were not distinguished from

Table 1 - Percentage of each food category found in 2,510 mink scats collected in the Irish midlands 1983-85. For explanations of percentage frequency and percentage weight see text. + = present.

Food category	Percentage frequency	Percentage weight
Total mammal	**4.8**	**4.7**
Rabbit or hare (Lagomorpha)	2.8	3.5
Brown rat	1.1	0.8
Field mouse	0.3	0.2
Pygmy shrew	0.3	0.1
Mustelid	0.1	0.1
Unidentified mammal	0.1	+
Total bird	**17.2**	**23.5**
Rails (Ralliformes)	6.3	9.6
Ducks or geese (Anseriformes)	3.7	4.8
Passerines (Passeriformes)	3.2	3.6
Waders or gulls (Charadriiformes)	1.5	2.5
Poultry or game birds (Galliformes)	0.2	0.2
Pigeons (Columbiformes)	0.3	0.5
Hawks (Falciformes)	0.1	0.2
Unidentified bird	2.0	2.1
Total fish	**17.0**	**12.3**
Perch	7.4	5.8
Pike	1.4	1.1
Trout family (Salmonidae)	1.8	1.3
Rudd	+	+
Roach	0.2	0.2
Unidentified carp family (Cyprinidae)	0.7	0.4
Eel	3.4	2.6
Stickleback (Gasterostidae)	1.3	0.3
Unidentified fish	0.8	0.6
Frog	7.4	6.3
Crayfish	44.6	50.1
Other invertebrates	4.6	1.4
Grooming fur	2.0	0.8
Other items	2.4	0.9
Unidentified	0.1	+

each other in the scats, only young hares are likely to have been at risk and most of this food category is probably rabbit. As a stoat can easily kill a rabbit, a mink should have even less problem, and yet is small enough to catch one in a burrow. Rats were almost as important, and along watercourses near farm buildings, they should be readily available.

Both field mice *Apodemus sylvaticus* and pygmy shrews *Sorex minutus* are common throughout rural Ireland wherever there is sufficient cover, though the former are absent from waterlogged land. But neither were eaten often. Unfortunately for the mink, both are small, presumably making them hardly worth catching if much effort is involved. The field mouse is nimble and is no mean jumper, fully capable of reaching a height of 0.5 m in a leap. The shrew is an expert in the use of cover, especially that at ground level. A mink would often be quite incapable of following one along its runs, through dense vegetation, or in the mat of vegetation below herbage, or in leaf litter, because there would be insufficient room. Shrews are also distasteful to many mammals.

One of the instances of a mustelid in a dropping (Table 1) was almost certainly of a stoat. Although the mink may have killed the animal, an adult stoat would be a formidable opponent, even for a mink. So carrion is the most likely explanation. There were three other scats with mustelid in them, in each case represented by mink hair actually attached to skin, which distinguished them from grooming fur: tangles of mink underfur, found in some scats. All three were collected at the same place on the same day. Whether they represented a kill (perhaps during a fight) or carrion feeding, it seems safe to assume that they stemmed from a single, and unusual, incident.

The birds the mink most often ate were rails. This has also been found to be the case in several studies of the prey of mink outside Ireland. The only common Irish rails are the moorhen *Gallinula chloropus,* often called the "water hen" in Ireland, and the coot, the latter being particularly common on Lough Ennell. As avian prey, these seem almost designed for the mink's convenience. Both are aquatic, roost near the water's edge, make nests nearby at ground level, are smaller than ducks (the second most common avian prey: Table 1), and are therefore easier to subdue, have a poorer ability to fly off, and are flightless for a time when moulting. Rails spend much more time in reed beds than ducks, where they would be more prone to ambush. In

North America wildlife biologists consider that the sounds made by young American coots *Fulica americana,* and a tendency for their parents to leave them unattended, as factors in their vulnerability to mink.

Although it was impossible to distinguish geese from ducks, in the scats, the former are only a theoretical possibility, adult geese probably being too big a proposition for a mink. One of my correspondents claimed to have several times seen a mink dive at around 20 m from a duck on the surface and attack it from below, and Dr Simon Berrow, in Co. Cork, has published a description of a submerged mink popping up and seizing a juvenile shag *Phalacrocorax aristotelis* by the neck.[27]

Small birds, such as thrushes, sparrows, finches, tits - what some ornithologists call "LBJs" or "little brown jobs" - and also crows are classified as passerines (Table 1). These were nearly as important as ducks. They are probably taken unawares on the ground.

Feathers of galliform birds - poultry and game birds - occurred in only five scats and none of them showed the dark and light bands characteristic of all species of game bird in Ireland. It appears therefore that all were from poultry. Several of my correspondents have recounted how mink had slaughtered hens, and also pheasant *Phasianus colchinus* poults in pens. Indeed, one of the mink which I dissected, whose gut contained feathers of game birds, had been shot after having been found asleep in such a pen surrounded by the corpses of its former occupants. This is an example of surplus killing, a phenomenon common in mammalian carnivores when presented with an excess of potential prey. This may merely represent instinctive exuberance in the act of slaughter, possibly aggravated by being surrounded by terrified birds fluttering and flapping about. On the other hand, on such occasions when surplus killing is feasible in the wild, it is prudent for a mink to dispatch everything available, and in so doing lay by provisions for the future. Many carnivores, among them the mink, cache excess prey to conceal it from scavengers. Notwithstanding the spectacular nature of such massacres, the data in our study show that game birds are unimportant to mink. One cannot even rule out the possibility of some of the domestic fowl in question being carrion, for instance the leftovers of a meal of a fox *Vulpes vulpes*. Besides, it is not unknown for a keeper of poultry occasionally to dump, rather than bury, the dead bodies of hens.

There appears to be no reliable account of mink attacking sheep or lambs. The size difference between predator and prey is probably too great and, in any case, such predation would usually have to take place near water. Assertions of this kind almost invariably describe the mink as sucking the blood of its victim, which is nonsense. No mammalian predator does this - not even vampire bats, which imbibe blood by lapping it. Of course, many mammalian carnivores will attack prey at the throat, where there are large blood vessels close to the skin, but only because this is usually where the victim is most vulnerable.

The hawk feathers may have been from a kestrel *Falco tinnunculus,* Ireland's commonest bird of prey, which takes its food from the ground. While a mink might surprise one there, the bird may have been carrion, perhaps shot, illegally, by an opportunistic hunter.

The carp family, the Cyprinidae, are represented in the Mullingar district by roach *Rutilus rutilus,* rudd *Scardinus erythropthalmus,* bream *Abramis brama* and possibly tench *Tinca tinca* and carp *Cyprinus carpio.* Unfortunately, the only hard parts of these fish which allow identification to species are the pharyngeal bones, bones in the throat of these fish which, strange as it may seem, bear teeth, the arrangement of which is diagnostic for species (Fig. 5). Ward only recovered pharyngeal bones in a minority of cases and so could usually only say that the fish concerned belonged to the carp family.

Roach   Rudd   Minnow   Bream

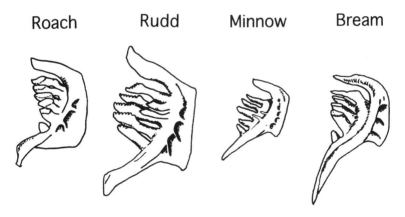

Fig. 5 The pharyngeal bones of some members of the carp family (Cyprinidae) to show the teeth. Not to scale.

When the mink went afishing they were evidently more likely to catch perch *Perca fluviatilis* and eel *Anguilla anguilla* than anything else. In view of the limited diving ability of mink, the predominance of these fish in the diet may be more related to their habits than to their relative abundance. Both are slow swimmers, or at least are not so easily disturbed as trout *Salmo trutta* and pike *Esox lucius*. It is usually impossible to distinguish the members of the trout family or "salmonids" (Family Salmonidae - which includes trout, salmon *Salmo salar* and a few other species in Ireland) in droppings. The only species in the Mullingar district is, in fact, the trout and its low consumption seems to suggest that mink are not, as some people suppose, a threat to wild populations of game fish.

Sticklebacks were of negligible importance in the diet. These fish can scarcely ever be worth bothering about when each one might require an entire dive and the spotting procedure preceding it.

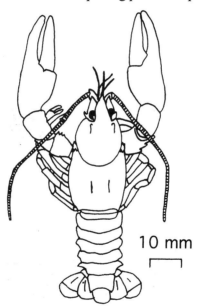

Fig. 6 A freshwater crayfish Austropotamobius pallipes.

All of the district around Mullingar lies on limestone, a fact conducive to an abundance of freshwater crayfish, which require lime for their shells, but also fairly clean water and, for preference, a firm bottom. Crayfish are crustaceans which look rather like lobsters (Fig. 6), but in Ireland they rarely grow longer than 15 cm. They can be surprisingly common. In parts of the Clare River catchment in Co.

Galway, where I have collected them, it seems almost as if every boulder shelters at least one. Furthermore, when water levels are low, I have had no trouble in catching them by hand - something which I could not do to any of the other prey mentioned above. Ease of capture explains their abundance in the scats. During the winter, the remains often included the eggs, which are red and easy to spot. At this time the females carry a bundle of them beneath the tail and are said to be *berried*. Berried females are reputed to be especially secretive, but this apparently does not prevent mink from catching them.

Fore foot   Hind foot

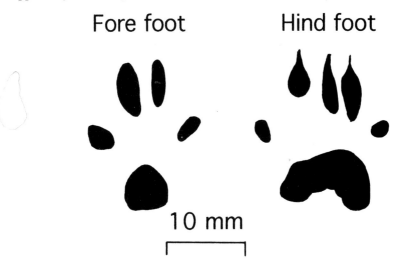

10 mm

Fig. 7 Tracks of mink.

A drawback to eating crayfish, of course, is the large proportion of indigestible shell. Mink eat the whole of the tail: the succulent muscle within and the skeleton encasing it. The latter is made up of a series of plates which would be awkward to strip off. However, the carapace - the continuous hard covering around the front end - is discarded together with the legs and sometimes the tail fan. Such leavings are characteristic, although not absolutely diagnostic, of the mink and their appearance beside a watercourse or lake is therefore at least a hint that mink may be about. By the way, the other sign to look for (besides leavings of crayfish and scats) is tracks, which are sometimes clearly imprinted in mud (Fig. 7). Of the tracks of other mammals that you are likely to see in mud along the edge of a lake or river, those of rats are much smaller and of otter much bigger. Despite the illustrations of animal tracks in field guides, most are rarely clear enough to allow of identification with certainty, except in snow or squelchy mud.

Other invertebrate prey included insects, snails, woodlice and a millipede. Almost all of these were small and were doubtless snapped up in passing rather than sought out deliberately. As some insects were fragmented, they may have originally been eaten by other prey. However, *Dytiscus* beetles were found in 38 scats and, at 24-38 mm long, are probably big enough to make them worth at least some minor effort in catching.

The relative importance of the different preys varied with locality. If we assume that the maximum distance that a mink will normally hunt away from water is more or less constant - it is often considered to be around 100 m[88] - it follows that a river-dwelling mink should have more land to hunt over than one living beside a lake, because a river has two banks, whereas a lake has but one shore. Besides, a lake provides a greater expanse of water. It would thus be reasonable to predict more terrestrial prey in the diet of a mink living on a river. This was indeed the case. A much greater percentage of the scats collected beside the rivers contained mammals than those from the lakes. Since frogs are, outside the breeding season, mainly terrestrial, it is also understandable why the mink also took them more often beside the rivers.

The mink caught far more birds at Lough Ennell than anywhere else (63% by percentage weight: 31% rails plus 12% ducks plus 20% others). There are, after all, large populations of wildfowl there. But there were hardly any crayfish in the Lough Ennell scats - less that 1% by percentage weight. This is explained by the highly polluted nature of this lake. The few scats that did contain crayfish were found at the southern, less contaminated, end. Crayfish were also eaten over three times more often on the River Glore than on the Inny, because they prefer a firm bottom, and the Inny is muddy.

While some prey categories varied seasonally at some localities, none of the variation was consistent, either between localities or sometimes even between years at the same locality. It is, nevertheless, worth noting that on Lough Ennell mink ate more rails in summer and autumn, when adult birds were breeding, and inexperienced, recently-fledged young most plentiful, and more ducks in winter, when numbers were augmented by migrants and by dead birds during the shooting season. Lead shot were actually present in one scat containing duck feathers.

Before leaving the subject of food, it seems worth raising a recurrent problem for those who investigate the diet of carnivorous mammals by analysing scats. The fact is that some prey have more hard parts than others, and it is the hard parts - bones, hair, feathers and shell - that resist digestion and appear in the faeces. So faeces may give a distorted picture of the proportions in which different items are actually consumed. This is probably not often of any great importance with an animal which takes only one sort of food, such as a bat eating mainly insects. However, one might reasonably expect that, in the diet of mink, the importance of crayfish, which have a massive shell, would be overestimated relative to some other items, especially fish which, because they are supported by the water, usually have less skeleton than amphibians, birds and mammals of comparable size.

To investigate this I carried out a feeding experiments on a captive male mink which Chris Smal trapped for me.[121] The theory behind this is as follows. A prey of a given species is weighed and fed to a mink, and the remains of it that turn up in the droppings are dried and weighed. Weight of prey is then divided by dry weight of its remains in faeces to give a correction factor for that species of prey. Correction factors are obtained for other species in the same way, and all are used to correct dry weights of prey in the droppings collected in the field to actual weights of prey eaten.

I kept the experimental animal for 2 months in a cage in an undisturbed, windowless outhouse at the university, but with a light controlled by a time switch adjusted to provide the prevailing daylength. The cage was 120 x 51 x 41 cm - the dimensions of the standard breeding cage on fur farms - and had a floor which could be removed and replaced. A detachable nest box with sliding door allowed the animal to be secured during operations in the cage. A week was allowed for acclimatisation.

At close quarters the mink had an awesome presence. It usually hissed - a warning - when I entered the room and then retired to its nest box as I moved toward the cage. It showed signs of distress only when it was twice locked out of its nest box to enable me to see whether this needed cleaning. Then it screamed repeatedly, the ultimate threat. At close quarters such screams are piercingly loud, and unnerving, and thus possibly of some practical value.

Each morning the animal was closed in its nest box, and the cage floor, with the faeces, removed and replaced with a fresh one, and fresh food left in the cage. The nest box and mink within were then weighed.

The food supply was adjusted so that a more or less constant body weight was maintained - in other words that sufficient, but not too many edibles, were available. Fresh water was also provided daily. Food consisted of whole dead animals, comprising: pygmy shrew, rabbit, rat, field mouse, a variety of birds (including passerines, rails and ducks), frog, several species of fish and crayfish. Each food item was weighed and one or more items of the same species were supplied together on a given day. When this was impractical, with small items such as shrews, contrasting foods were served, so that they could be easily separated in the droppings. Because food normally passes through the gut in under 3 hours, the food eaten on any day was always available as faeces on the next. Large items, like rabbits and ducks, were left until completely devoured or until it was obvious that any leavings were inedible. Remains in the faeces were dried and weighed.

Unfortunately the results of the work were of very little help in assessing what remains in faeces represented in terms of what was consumed. There was a great deal of difference in correction factors between the different species of fish, of mammal and of bird. However, it is usually only possible to identify feathers to family in scats of wild mink, so one cannot distinguish between large and small species of the same family, which had different correction factors. This was because, in general, larger species of bird and mammal tended to leave proportionately less remains, partly because, with relatively less surface area, they had comparatively less hair and feathers, but also because with large items, like rabbits, coots and ducks, the mink managed to avoid eating at least some of the hair, feathers and bones. The rabbit's stomach was rejected in its entirety as its contents, grass, were unfit for consumption by a carnivore. In contrast, every last bit of small prey was devoured, except for crayfish, which were shelled in the customary manner. The most serious complication was that large individuals of the *same* species also tended to leave relatively less material in the scats, and so a range of correction factors were obtained for the same species, rendering accurate correction unattainable.

A further possible complication, which we obviously could not deal with, is that, although in our experiment large food items were left until the mink had finished them, this may not always happen in the wild. It took 6 days for our mink to consume a rabbit, and 3 for a duck. In the wild a mink might cut through the skin of such prey, ingesting only a little feather and hair in the process, devour flesh and internal organs, and then sleep it off. Although mink sometimes drag large prey back to the den, this may not always be possible and, in the meantime, even if the partly eaten kill were cached, scavengers, such as foxes, rats or crows, might come upon it and complete the job. The proportion of hair or feathers that the mink would actually swallow might therefore bear no relation to that if it had had the time and opportunity to eat everything it wished. Besides, if prey were readily available, the less choice parts of a kill might often be rejected.

All that we could conclude from the experiment was that, in Table 1, the importance of fish is probably underestimated and crayfish overestimated. The prey which produced proportionately the most remains, and therefore the lowest correction factor was, surprisingly, the pygmy shrew which, being tiny, has a relatively large surface area for hair to grow upon.

The mink's habit of bringing half-eaten food items home is sometimes a help in spotting a den, which is also often surrounded by droppings. I particularly recall helping Smal to collect scats from an island in Lough Ennell, while, only a couple of metres away, a mink, half way out of its hole, screamed abuse at us. Outside the breeding season, and when a mink is not hunting, much of its time is spent snoozing in a den, especially if there is plenty of prey about. In the den it is safer and, in cold weather, warmer - an important factor if you are weasel sized and shaped. Pathways used by mink near dens are often particularly well marked.

Dens may be amongst boulders or between the roots of trees and are often burrows commandeered from rabbits. Mink may also set up house in outbuildings or even, as described above, in boats. Most accounts of dens state that the animals do not dig them themselves, but Smal found that in the soft, peaty soil of the Irish midlands the animals did sometimes excavate holes for themselves and he described four in river banks.[329] All were 8-11 cm in diameter. One had only one entrance, the others two, with one of these 0.6-1.2 m above the normal water level and the other 0.3-0.6 m higher. Despite this, when the rivers were in spate, they would all have been flooded.

And now to reproduction.

From information kindly supplied to me from two of Ireland's largest mink ranches, it is clear that the breeding season of farmed mink in Ireland is identical to that documented in Britain, and indeed in the USA. From the bodies I dissected, it was also plain that the breeding season for feral mink in Ireland corresponds to that on farms. As in most mammals which have a distinct mating season, the testes enlarge as it approaches. In the bodies which I dissected, I found that the testes, which together weighed around half a gram from July to early December, had by January, when sperm was first produced, rapidly swollen to over 5 g: a six to tenfold increase in size in a matter of weeks![106] Sperm disappeared from the male reproductive tracts in April and testis weight fell sharply. So males are fecund from January to the beginning of April. Smal, who examined the external appearance of the testes in the live animals he trapped, came to similar conclusions.[330]

As most of the animals that I dissected had been dead for some time, the ovaries were sufficiently fresh in only a few to allow microscopic examination, and thus to determine whether egg cells were about to be, or had recently been, shed. But there were sufficient fresh ovaries to indicate that ovulation (release of eggs) and therefore the period over which females mate, was consistent with that in farmed mink: from about the end of February to early April. Smal noted swelling around the reproductive opening of live females from January to March.[323] However, it is known in Britain that females show such swelling in January, but will not mate then.[44] Almost all of Smal's females bred each season.

I obtained only one female that was visibly pregnant, with five young. Fortunately the placenta of each unborn mink, like that of many other mammalian carnivores, leaves a marked scar on the uterus where it has been attached. Such scars persist for some time after birth. I dissected four females with such placental scars. Overall, therefore, I had five figures for litter size, ranging from 4-6, with a mean of 4.8.[106] The litter size on Irish mink farms is normally three or four, averaging 3.5,[323] which is surprising, since the mothers there, unlike their counterparts in the wild, are guaranteed an adequate food supply, and might therefore be expected to produce the larger litters. But there may be no contradiction between my estimate of young in the uterus and that of litter size. Some young may die before birth, and be reabsorbed in the uterus. This is known to occur in a number of species of mammal and I have evidence of it myself in foxes and field mice.[111]

It is interesting to note that work elsewhere has shown that ovulation in an individual female continues after her first copulation of the breeding season, and that more than one male will cover her, although it is often the last male that sires the young. Sometimes a litter of kits born to a mother may have more than one father. [88]

Smal noted that females were often noticeably pregnant in April and that all had given birth before the end of May.[323] This corresponds closely to the whelping season on Irish mink farms, where most mink are born in May.

Mink, several other mustelids and some other mammals exhibit a halt in the development of their embryos called *delayed implantation.* After fertilisation the egg divides up to give a ball of cells, the *blastocyst.* Growth then ceases and the blastocysts, which are minute, remain in suspended animation in the uterus for a time before attaching or *implanting* in it. The delay can amount to months in some species. Delayed implantation apparently functions in ensuring that mating and births occur at the best times in the year, regardless of the actual time that would be needed from fertilisation to birth if development were continuous. As a result of the delay, the gestation period for mink varies between 40 and 75 days. Although most matings in one Scottish population of feral mink were, as in Ireland, throughout March, most births occurred in the first week of May only.[88] Presumably for some reason this was the most advantageous time for the young to be dropped there.

The kits are born deaf, blind, hairless and helpless. They begin to take solid food at 3-4 weeks but are not weaned until 6-7 weeks and are cared for exclusively by their mother.[88] The male takes no part in the rearing of his children. Why should he when their paternity is in doubt? The young begin to appear above ground with their mothers in June. By September they are big enough to be indistinguishable from the adults at any distance.[330]

The largest family groups that Smal noticed were of mother plus three young. If weaned litters of more than three are indeed uncommon in Ireland, then this, compared to my average figure of 4.8 before birth - on a small enough sample to be sure - tends to indicate that either there are indeed casualties in the uterus, or that some young commonly die between birth and weaning. As one would expect, mothers show little fear where the welfare of their young is concerned. One lady correspondent described to me how, in an inlet on the shores of Lough Derg in Co. Tipperary, she watched a mother mink transferring her young, probably from one den to another. The mink repeatedly swam out to point on the opposite bank, disappeared into cover, reappeared holding a kit in her mouth, swam back to the jetty on which my correspondent - who supplied photographs of the procedure - was actually sitting, trotted along it, and disappeared into nearby vegetation.

Chris Smal investigated the populations of mink around Mullingar by laying live traps at suitable places, such as near dens, where drains entered watercourses, at bridges and beside logs and large rocks.[330] He usually set one trap at each site but, during the summer when there were family groups about, he laid additional ones. He continued trapping for 6-12 days every 10 weeks, visiting the traps daily either by boat or on foot. The length of each trapping session was influenced by available time and the weather, flooding in particular cutting operations short. The work continued for 2 years on the Rivers Inny and Glore, but for only 19 months on Lough Ennell. He ran similar, but shorter, less intensive, studies at two other localities: the Burn Daurnett, an acidic upland stream in Co. Donegal, and the River Nore, in Co. Kilkenny, which lies on limestone but which is 10-20 m wide. He also worked for a few months on the Ballysodare River, in Co. Sligo. In all localities he arranged with local landowners and gun clubs that no mink would be killed for the duration of the project. While it may seem elementary that such a condition be met in order to obtain valid results on populations, this wasn't the case in most previous live-trapping studies on mink populations, all of them, of course, outside Ireland.

When Smal first captured a mink, he anaesthetised it by injection, weighed it, noted its reproductive condition, marked it for future identification by punching a series of small holes in its ears and sketched the pattern of white patches on the underside as an additional aid to future recognition. When first caught on all subsequent sessions, it was again anaesthetised, weighed and the reproductive condition noted. If he caught it a further time during any particular session, it was identified by the ear punches and the colouration on the underside and immediately released. No mink ever died during anaesthesia. This capture-mark-recapture work enabled a picture to be built up of numbers, survival, reproduction, and much else besides.

Altogether Smal caught and recaught a total of 180 individuals in all his study areas. Generally there were about two captures of a male to every one of female, because the former range over greater areas, are more mobile and therefore encounter more traps. Indeed, in the great majority of mammal species, males operate over larger areas than females. A bias towards males is usual in most trapping studies of mammals for this reason. He also caught two moorhens, two cats, four pine martens (all in Sligo), and a trout in a flooded trap.

Mink were easiest to trap during the mating season, in March, and in late summer and autumn, when the juveniles are learning to fend for themselves. Numbers actually abroad and hunting on their own account therefore peak at the latter season and decline thereafter as a result of deaths and emigration. Although overall numbers are obviously greatest after the whelping season and fall thereafter. Nevertheless, locally a population can rise during the mating season, if males looking for mates move in.

Any young male that had not reached a weight of 0.9 kg by December disappeared from the population after February. Almost certainly such weaklings are weeded out by natural mortality, probably by competition with other mink, and hence by stress and ultimately starvation, disease or predation; both foxes and otters are known to kill mink.[330] Slaughter of one species of mammalian carnivore by a bigger one is by no means unknown: the act may provide food for the killer directly and very probably cuts down the competition.

Life's stresses on the adults were also revealed by body weight. Adult males tended to become lighter in late winter and spring, presumably as a result of increased activity and strife over females during the breeding season. One animal slimmed down from 1.66 to 1.02 kg, a reduction in body weight of 39%. Males on mink farms are, incidentally, most aggressive in March. The females, of course, also lost weight in spring as a result of giving birth to the kits, but the weight of most female adults continued to decline right up to July. This is explained by suckling, which is more demanding on many mammalian mothers than most of us appreciate. Human offspring may take 16 years or more to grow to adult size. Young mink do this in a matter of months; for some species of small mammal it takes only weeks.

Table 2 - Maximum, minimum and mean numbers of mink per kilometre of river or lake shore in trap-mark-recapture studies in Ireland.

Study area	Minimum	Maximum	Mean
River Glore	0.73	2.30	1.37
River Inny	0.54	1.01	0.79
Lough Ennell	0.17?	0.47	0.37
Burn Daurnett	0.10	0.30	0.25
River Nore	0.48	1.31	0.73

Table 2 gives the maximum and minimum numbers of mink per kilometre of river or lake shore that Smal recorded in his study areas. In view of the frequent propaganda on the "mink menace" in the press, these figures, representing a variety of rivers and lakes, strike one as amazingly low, but are by no means so when compared with those obtained in studies outside Ireland.[88] In fact the Irish populations were comparatively high.

As explained above, numbers fluctuate over the year, as they do in most populations of wild mammals, being highest when the young are born and lowest just beforehand. However, when such an annual pattern of variation in numbers is much the same from year to year, with about the same minimum and maximum numbers in the same seasons, the population is effectively stable long term. Wildlife biologists describe such regular fluctuation in numbers as a state of *dynamic equilibrium.* Although investigated over only 2 years, the mink populations on the River Glore and Lough Ennell appeared to be in dynamic equilibrium, while that on the Inny increased slightly overall. Again, despite occasional outcries to the contrary, this suggests that we are not eventually going to be overwhelmed with mink. Note also that, in dynamic equilibrium, if litter size is at least three and the sex ratio of adults is unity, then for each mother father and three kits in May - five mink in total - there must on average be only two at the next breeding season: three-fifths of the population must vanish. Mink have a harsh life.

Availability of prey is the most likely factor determining the different densities on the various study areas (Table 2). The most easily caught items are crayfish, which are therefore most likely to influence the abundance of mink. Crayfish were commonest on the River Glore and next commonest on the Inny. The two highest recorded densities of mink correspond. While the River Nore is on limestone and there are crayfish in it, we have no information either on their densities or on the diet of mink there. With a stronger current, there are possibly fewer crayfish than on the former two rivers, and mink might have more trouble catching them too, because of the deeper water. Densities of mink are, however, much the same as those on the Inny. The Burn Daurnett is acidic and holds no crayfish whatsoever, and mink were scarcest of all there. Densities of mink on Lough Ennell were higher, despite there being precious few crayfish. But there are compensations in the form of birds, particularly rails and ducks.

Smal divided the mink he studied into three categories: *transients,* which were only caught during one trapping session; *temporary residents,* which were caught on more than one session but disappeared in less than 3 months after they were first trapped; and *residents,* which were trapped over periods of more than 3 months. Some of the residents were around for much longer and, bearing in mind that the population study ran for only 2 years, maximum recorded residency may underestimate that possible for successful mink. Using the period of residency, and the fact that mink are born around May, Smal could also estimate minimum ages. The maximum recorded residences for females were 102 weeks and 90 weeks, on the River Glore. Both of these animals were juveniles when first trapped, in November. So both were over 2.5 years old. A male, which was around for 72 weeks on the Glore, but was first caught as an adult, was estimated to be over 3 years old. Another male stayed for 82 weeks on the Inny.

Unfortunately for mink in general, few are so lucky. On any given trapping session on the Glore, Inny and on Lough Ennell, usually one third to two thirds of the individuals trapped were residents. The rest were varying proportions of temporary residents and transients, the latter usually predominating. Because of the great turnover among the transients and temporary residents, this means that the majority of individuals trapped fell into these categories. The *average* length of residency on the Inny and Glore was only 16 weeks; less on Lough Ennell. Findings from the River Nore were more limited but were broadly similar. On the Burn Daurnett there were no residents at all. Absence of crayfish and a general shortage of suitable prey on this upland site made residency an uncertain proposition.

A resident mink is a privileged mink, a successful mink. One of the most important achievements for any individual wild mammal is to establish a *home range:* an area where it can live more or less permanently and where it goes about its normal, everyday activities of feeding, resting, mating and caring for its young. A home range confers great advantages. Familiarity with the terrain means that the animal knows the best places to find food, to shelter from the elements, to hide from enemies and to find members of the opposite sex for mating. A home range needs to be big enough to provide everything necessary for life. Different habitats will supply necessities in different measure, so individual members of the same species living in different habitats may have home ranges of different sizes. Resident mink have home ranges; transients do not. Temporary residents may have a home range, but for a shorter time. The difference between a home range and a *territory* is that the latter is a home range defended against other members of the same species. The home ranges of mink are indeed usually territories, although there is some overlapping of those of males and females. That mink are territorial is clear because, except for the mating season, they are intolerant of their fellows and will fight with them. If this were not so, it would be difficult to explain why such a large proportion of the population consists of transients or temporary residents, which would presumably stick around if unmolested. If mink were not territorial, theoretically all could live in harmony together, but, as resources are finite, all might eventually be fatally disadvantaged. As things actually stand, transients are condemned to death or to the hazards of a life of vagrancy. Salvation for a transient may come when a resident dies and a territory falls vacant, or if a resident is unable to defend its territory adequately.

Although the mink is a jack of all trades predator, even a resident individual labours under a great disadvantage. Because it lives close to water, either on a river, or on a lake shore, or on the coast, it is confined to a narrow strip of ground. In other words its territory is linear. The time and energy necessary to patrol the borders of a linear territory and chase off potential competitors is disproportionately great. You can see this clearly by drawing a square on graph paper and beside it a long thin rectangle of equal area. The distance around the edge of the rectangle, or even down one of its sides if it is narrow enough, is much greater than that around the square. This is the big drawback of hunting in a territory partly on land and partly in water: a much greater time must be spent in patrolling if the territory is to be guarded. It seems too that a resident mink does not even do a particularly efficient job in excluding temporary interlopers, or else transients would not be so common. For example, compared with the mink, the pygmy shrew, which is also territorial for much of the year but has a more compact territory, is a model of efficiency in shooing off competitors.[111] So even a successful mink, besides sharing the drawbacks in terms of energy loss of being small and skinny like stoats and weasels, has the added disadvantage of an inconveniently shaped territory, and therefore one which is energy-demanding to patrol. No wonder mink are so fierce.

How much of a pest is the mink, a versatile predator that operates both on land and in water? First of all, because it is an alien and still something of a novelty, it tends to attract more adverse publicity than our other wild carnivores. In addition, as it shows little fear of man and is abroad in the day as well as at night, a mink is more likely to be seen than Ireland's other predatory mammals, thus inclining many to overestimate its numbers. Smal's studies above indicate that it is not numerous, even under favourable conditions, or certainly not as numerous as many of our other wild mammals.

Many of the species eaten by mink breed rapidly and are well able to sustain predation: rats, mice, shrews, rabbits, frogs and fish. Of course the terrestrial prey also occur well away from water and the majority are therefore out of the mink's ambit anyway. Comparatively few game fish are taken; they are too fast. Locally mink can, however, be a thoroughgoing minor nuisance on a fish farm, where large numbers of fish are penned together. There control is simply a matter of regular trapping. Mink are easy to trap. There are no substantiated incidences of lamb killing and, while the mink will undoubtedly prey upon poultry and domesticated ducks, there is no reason to suppose that its activities in this direction are any more pernicious than those of foxes. In any event, mink kill poultry mainly in the vicinity of water; foxes may do it anywhere. Analysis of the diet shows that predation on free living pheasants is negligible. Notwithstanding, losses of pheasant poults in pens can be dramatic but, again, no more so than such havoc wrought by foxes. In any area where pheasants are penned for release, the destruction of mink should be part of a general programme of predator control.

The prey populations most likely to be affected are those of waterfowl and colonies of birds which nest on the ground near water. Therefore, rails and wild ducks are the obvious candidates inland. Locally mink could have reduced their numbers. One of the recommendations from Smal's research was that mink should be culled on Lough Ennell to protect the wildfowl there. Nevertheless, Ireland's numerous ornithologists have published little in the way of scientific evidence to indicate that populations of birds have declined as a result of mink predation, nor even seem to have expressed strong opinions to this effect.

The naturalist-journalist Mr Michael Viney, in a piece in the *Irish Times* of 21 February 1998, remarks that on the coast

> The terns at Our Lady's Island Lake in Co. Wexford have survived the mink and other predators only by the most determined electric fencing and wardening.

Moreover, in 1989 the late Clive Hutchinson, one of the most distinguished birdmen of his generation, recorded that numbers of terns at Lady's Island Lake undergo marked fluctuations

due in part at least to sudden changes in water level, to
predation by rats, mink and large gulls and to varying
disturbance at the site.[163]

In both cases mink do not appear in the dock alone. The most recent and
comprehensive account of the status of the various tern species in
Ireland, published in 1997,[150] does not even mention mink under the
heading of "Threats".

Whatever your feelings on the matter, mink are here to stay.

Finally there is evidence in Britain that mink numbers are declining
in some districts,[346] apparently corresponding to the recovery of otter
populations there, which crashed in England in the late 1950s and
1960s as a result of the use of toxic pesticides.[151] Chris Smal tells me
that, in the course of fieldwork in various districts in the late 1990s, he
has formed the opinion that numbers of mink have also fallen in parts of
Ireland too. He has seen no sign of them at all in some places where
they were present before. This can hardly be due to otters, for otters
have remained common over most of Ireland since the 1950s. We await
developments.

Chapter 2. New Data on the Irish Stoat

> One dark night - it was a *very* dark night, and blowing hard too....a company of skirmishing stoats, who stuck at nothing, occupied the conservatory and the billiard room, and held the French windows opening on to the lawn.
> *The Wind in the Willows.* Kenneth Graham.

Whereas both the weasel and stoat occur in Britain, only the stoat is found in Ireland, where it is often confusingly referred to as "the weasel". It has a light, reddish-brown coat with a creamy underside and a black tip to its tail. Like most of the small members of the weasel family, it is also possessed of a long sinuous body, short legs and a head that is both tapered and flattened; and the male is larger than the female.

The Irish stoat is a distinct subspecies *Mustela erminea hibernica,* confined to Ireland and the Isle of Man. Whereas the line dividing the chestnut of the upper parts from the cream of the lower is almost straight on stoats in Britain, on the Irish subspecies it is usually irregular, with the brown often encroaching onto the throat, chest and belly, and exceptionally forming a band right across the underside. The hair on the upper lip is white in Britain, whereas it is brown in Ireland, rarely with a white spot or two. The British form also has a white edging to the ear, which is lacking in the Irish. Although these differences readily distinguish the two forms, one still sees some illustrations purporting to be of Irish stoats which are nothing of the kind.

It is often stated that the Irish stoat is smaller than the British. Although this is true in the north of Ireland, size increases as one goes south. In the southernmost quarter of Ireland, males are nearly as large as those in Britain.[107] Representative measurements are given in Table 3.

Like the weasel, the stoat is a bloodthirsty and ferocious predator and there are many descriptions of the Irish stoat attacking, or being found in possession of, the bodies of rabbits, rats and birds.

Examination of the reproductive tracts of Irish stoats[99] has shown, not surprisingly, that the breeding season is similar to that in Great Britain.[151] Males are fecund from March well into the summer and mating takes place from mid April to mid June. As in the mink, there is delayed implantation. In fact the blastocyst does not become implanted in the uterus until the following March, with births in April and May. So gestation lasts 11 months. It has been observed, albeit outside Ireland, that paedophilia is an integral part of the sex life of stoats. Males are not fecund until the year following their birth. On the other hand, females are sexually mature when less than 3 weeks old, even though tiny, blind, deaf and helpless. They are mated by an adult male in the nest and successfully produce litters in the following year.[151] Since the male in question is most likely to be the dominant one in the area, this could in theory result in incest - the father copulating with his infant daughters. But turnover in the population is probably high, and consequently the dominant male in any year is unlikely to be in the same enviable position in the year following, when his daughters are born, if he is alive at all.

Table 3 - Weights and body measurements of some Irish stoats.

| | | Weight in grams | Length in millimetres of | |
			Head and body	Tail
County Down	Male	194-293	240-274	72-104
	Female	95-161	184-221	57-73
Counties Waterford	Male	302-420	259-288	88-114
& Limerick	Female	117-197	212-245	64-77

In 1975 Miss Eden Thomson, a zoology graduate of Edinburgh University, was living at Killarney and she and I agreed that it would be interesting to investigate stoat populations on the lowland part of Killarney National Park by trap-mark-recapture. Dúchas agreed to make the traps and pay any modest out-of-pocket expenses.

The best live traps for stoats are of wood and operate on the see-saw principle. A portion of dead rabbit is placed in the trap to attract the stoat's attention. The stoat walks up the see-saw, which swings when the beast passes the fulcrum, and the part behind the animal blocks the entrance and locks in place. All traps were weathered for some weeks to remove the smell of fresh wood and two were kindly tested by Mr Roger Forster, then gamekeeper at Adare Manor, Co. Limerick, at a site there known to be frequented by stoats, three of which were captured in 6 days. He had had long experience of trapping the animals and was good enough to come to Killarney to check and advise on Thomson's field-craft. So we knew that the traps worked and were being properly set in likely places.

Stoats are commonly seen by walls and hedges, in dry ditches and other pathways which provide cover. So the traps were laid in these, especially at the sides of gates or at gaps in walls and hedges, the trap entrances thus providing bolt-holes at the far sides of open spaces. The traps were covered with stones, sods, dead wood or moss so as to blend in with their surroundings.

Sad to say, although around 30 traps were operative on most nights for just over a year, only six stoats were caught: three in June, one in July and two in October. We were forced to conclude that there were few stoats in the Park, perhaps no resident ones at all. This was astonishing as the area is renowned for its wildlife. Irish stoats appeared a shaky prospect for research. It was therefore with some interest that in 1982 I learned that a PhD project on them was being contemplated by Mr Paddy Sleeman at University College Cork. As he is now Dr Sleeman, it will be clear that his efforts met with some success.

Sleeman obtained his data mainly in two ways. First, he advertised widely for dead bodies, together with information on where and when they had been found and on the sort of surrounding terrain. It is quite legal to kill the stoat in Northern Ireland, but in the Republic, although in no way endangered, it is a protected species. Nevertheless, corpses were still to be had south of the Border, mainly as road casualties. Several were also trapped under license, some by Sleeman himself using Fenn traps, which kill instantly and humanely. A few were confiscated from dogs - the latter had actually been seen to kill some of them - and one from a cat. Odd specimens were also recovered from deep-freezes by Irish zoologists. (Many of the latter squirrel away assorted corpses that come their way in the belief that they may prove useful at some future date. When I retired from Galway I disposed of several such items that were over 20 years old.) Altogether Sleeman accumulated almost 200 bodies. A great deal may be deduced from a collection of this kind, as we shall see.

Second, from May to November 1985 he caught stoats one at a time in see-saw traps in the Fota Estate on Fota Island in Cork Harbour and then radio-tracked them. Each individual was anaesthetised using ether, and a radio-tag attached to it by means of a collar. The animal was thereafter located using a receiver and hand-held, directional aerial.

Fota is some 3.2 km² in area but is not at present an island, being attached to the mainland by an isthmus and by railway embankments. It is well wooded in parts but there is also a lot of farmland and a wildlife park, which attracts many visitors.

Radio-tracking wildlife sometimes results in bizarre experiences. Sleeman claims that some of the visitors to Fota that he encountered began to comb their hair or otherwise attempt a brisk smartening up of their appearance. This he put down to his carrying a radio-receiver, a large directional aerial and notebook and to his wearing headphones. The audience apparently assumed that they were about to appear on television.[313,316]

The radio-tagged stoats were located both night and day and usually from one to three fixes (precise locations) were obtained every 24 hours.[316] The effective reception range for the radio signal was only 100 m in woodland, but rose to 300 m in the open. It rarely took more than 15 minutes to obtain a fix, but Sleeman deliberately restricted the number of fixes each day to avoid disturbing the animals unnecessarily. In the event, they did not seem to be easily unsettled and were apparently unconcerned by visitors, campers or estate workers in their vicinity. The life of the radio-tags varied. Three females and one male were tagged, but on three of these transmitter life was inconveniently short, from 6 to 20 days, periods probably insufficient to reveal the complete home ranges. The areas over which these animals were observed operating were 2, 10 and 11 ha. Fortunately the transmitter on the remaining female, christened Sally, continued to produce a signal from May to July and Sleeman achieved nearly 100 fixes on her. At this time she may have been caring for a litter of kits, for when first caught it was obvious from her nipples that she was producing milk, although no young were ever seen with her. As results were therefore mainly from one animal, they may be atypical due to possible idiosyncrasies of that individual, but they are certainly interesting and suggestive. Clearly there is room for further radio-tracking of Irish stoats for those with the enterprise, patience and stamina to catch enough of the beasts and to track them.

Altogether Sleeman obtained 131 radio fixes on positions of his stoats and the percentages of these in various habitats are shown in Table 4.[319] Clearly the animals preferred woodland to open country, and even in the latter the majority of fixes were in cover of some kind: up trees or in hedges. Woodland and the cover outside it provide protection from enemies, including man, and are also probably the best places to find prey. The expertise with which the animals used cover is demonstrated by the fact that in a total of 106 days of tracking, Sleeman actually spotted a radio-tagged stoat only 11 times.[315]

He obtained confirmatory evidence of the preference of Irish stoats for woodland from two other sources. He had information for the surrounding habitat where many of his corpses were picked up. There were also notes on habitat with many of the distribution records for stoats that had been accumulated by the Irish Biological Records Centre (Chapter 1). Out of a total of 590 records pooled from both sources, 54% were from open country, 37% from woodland, 2% from built-up areas and 6% from the seashore. Although stoats were therefore most often recorded in open country, only 6% of Ireland is wooded, so woodland is evidently most favoured.

Table 4 - Types of habitat in which 131 fixes were obtained on radio-tagged Irish stoats on Fota Island.

Habitat	Percentage of total fixes
In woods	
Up trees	8
Underground	60
In cover	17
Total percentage in woodland	85
In open country	
Up trees	6
Underground	2
Pasture	3
Hedges	2
Total percentage in open country	13
In reedbeds on seashore	1

Table 4 shows that for about two-thirds of fixes, the stoat was below ground. This may actually underestimate the proportion of subterranean time. For it was sometimes impossible to locate a radio-tagged stoat and this is most readily explained by the animal being too far below ground for the signal to be picked up on the radio receiver. There are limestone caves on Fota and the animals might have been lodging in them on such occasions. The time spent underground was almost certainly in dens throughout. In other words the animals were resting and, in Sally's case, possibly also caring for young. It may seem strange that such an apparently active mammal as a stoat should put in two-thirds of its time taking it easy but, once its stomach is full, there is little else to do until more food is needed, except mate in the appropriate season or chase intruding stoats away. The logical place to rest is below ground, away from predators and the cold. Even in an Irish summer cold may be a significant consideration for at least part of each day for a small, elongate mammal, and consequently one with a large surface area. And doing nothing also conserves energy.

Perhaps equally surprising is the amount of time that was spent in trees. But it is well known that stoats will climb them to kill adult birds or nestlings. In this context it is noteworthy that the Irish name for the stoat, easóg, has also been applied to squirrels, indicating confusion between the two, most readily accounted for by both being at least partially arboreal. Nevertheless, prey were not the only reason for ascending trees. One of the dens was up a lime in the cover provided by the dense mass of twigs that often develops in limes, known as a "witches broom", and which is a pathological effect of a parasitic fungus.

During radio-tracking, Sleeman was able to locate no fewer than 19 dens.[315] Most of these were in burrows, but three were in piles of sticks or stones and, of course, one was in a tree. It is common for mammalian carnivores to expropriate the burrows of other mammals which are not in a position to raise any objection. The reader will recall that mink habitually do this. Stoats are no exception. Sleeman identified the former owners of 14 of the 15 burrows serving as dens. Nine had belonged to rats, four to rabbits and one to field mice. All of the tracked stoats had more than one den. Sally had 12.

When the radio fixes were plotted on a map it was clear that the animals spent a disproportionate amount of their time at the edges of their ranges. On a few occasions it was also apparent that an animal was actually travelling along the periphery of its range. A resident stoat must pay special attention to the boundaries of its property if it is to keep intruders out. It is well know from studies outside Ireland that stoats are highly territorial, each individual usually defending its own patch, at least against members of the same sex. The borders of the territory are marked with scent from the anal glands, by dragging the backside across the ground.[151] While a scent mark may make a stranger think twice about entering an occupied territory, it is rather like a notice board bearing the inscription "Trespassers Prosecuted". Unless it is backed up by regular personal appearances, it tends to be ignored. So the stoat must patrol the edges of its territory if its ownership is to be taken seriously.

Although Sleeman's four animals were monitored successively, three of the stoats had territories which did not, in fact, overlap each other (Fig. 8). The fourth, that of a juvenile female tracked in October (Range B in Fig. 8), apparently lay completely within that which Sally occupied (Range A in Fig. 8), at least up to when her transmitter failed in July. Either Sally had disappeared and the youngster had taken over part of her range, or Sally was still around but tolerated her. As there is some evidence from Swedish studies that juvenile females may be allowed to stick around in their mother's range for a time, it is possible that the youngster was Sally's daughter of the season.

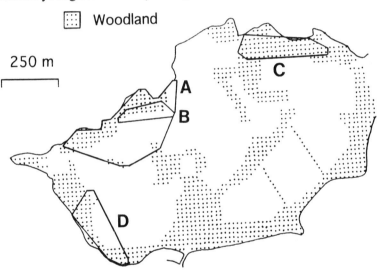

Fig. 8 Map of Fota Island, Co. Cork, with the home ranges of radio-tracked stoats indicated. Range A belonged to Sally.

Sally's territory apparently increased during the period over which she was tracked, from 15 ha in May to 17 ha in June to 22 ha in July. Of course, such expansion may be illusionary, the accumulating fixes simply giving a clearer and clearer picture of the true territory. However, it is at least as likely that the expansion was genuine. As mentioned above, Sally may have had a family to care for when she was being tracked. Readers may recall from Chapter 1 that suckling is extremely demanding on those mammalian mothers whose offspring, like those of the stoat, grow rapidly. If rearing young, Sally would have had to eat more. As the young grew, more and more milk would have been required, and, after weaning, she would have had to supply solid food until the kits were independent. This could have led to her gradually expanding her territory to boost the supply of available prey.

The few sightings of stoats which Sleeman had were interesting. He watched them stalking rabbits several times, usually at dusk, and once hunting rats in a haystack. He saw a stoat carry a dead mouse twice. He also spotted them raiding birds nests, including those of dunnock *Prunella modularis,* wren *Troglodytes troglodytes* and woodpigeon *Columba palumbus.* One managed to enter a cage in the wildlife park and killed all but one of the quaker parakeets *Myiopsitta monachus* there.[313]

Males predominated among the bodies Sleeman obtained as road casualties.[312] Readers may recall from Chapter 1 that Chris Smal caught male mink more often than females, and that there is a bias towards males in most trapping studies of mammals, because males generally have larger home ranges. So male stoats are more likely to have roads in their territories than females, and therefore to cross them. Adult males were knocked down far more often from March to May, with a peak in casualties in April. This overlaps the mating season of April-June. Radio-tracking studies of stoats in Sweden have shown that males expand their home ranges in spring, doubtless in quest of females at the onset of the mating season. Such an expansion in Ireland, with presumably a commensurate rise in the likelihood of roads being crossed, could account for the increased casualties. Interestingly, the number of males Sleeman obtained from roads in June, when the mating season would not have finished, was comparatively low. Perhaps this was because most of the males that had expanded their ranges onto roads and that were at risk of being knocked down, had succumbed by May.

There was a surge in road deaths of young stoats from June to August, peaking in July. In Britain juveniles usually become independent and disperse from July to September.[151] It may be recalled that four of only six stoats Eden Thomson caught in the lowland part of Killarney National Park were taken in June-July. These may have been dispersing through the Park. At least three of them were juveniles; the fourth escaped before it could be fully examined.

Sleeman investigated the diet of stoats by analysing the gut contents of his corpses. Unfortunately the insides of some of the road casualties had been eaten by scavengers. In addition many of the guts were empty because digestion, like that in the mink, is rapid. Since stoats evidently spend much of their time in dens, there is a good chance that any above ground are hunting, and therefore with the previous meal digested and defecated. Altogether he obtained 89 stoat corpses that had food remains in their guts, mainly from southern Ireland. The results of analysing these are given in Table 5.[318]

Whereas there had been many casual observations on Irish stoats killing birds and mammals prior to Sleeman's study, there was little in the way of systematic investigation of scats or gut contents to determine the relative importance of the different kinds of prey. Casual observations are likely to be biased. For example, one is much more likely to see a stoat killing large prey, such as rabbits and rats, than small, like mice and shrews. I had myself published two small series of analyses of gut contents, all but one of the animals being from Northern Ireland, but they totalled only 29 individuals. Moreover, one of my collections was made from 1963 to 1966, when rabbits were scarce because of myxomatosis, and the other, from Clandeboye Estate, Co. Down, from 1968 to 1970, where they were then common.[111]

Table 5 - Percentage frequency of various prey items in the guts of 89 Irish stoats.

Prey item	Percentage frequency
Rabbit	32
Rat	14
Field mouse	5
Bank vole	3
Pygmy shrew	26
Stoat	1
Bird	18

While Sleeman did not detail how he distinguished between hare and rabbit in the gut contents, he apparently used the fine structure of the fur. In any event he recorded no hares, and rabbits were the single most important prey, as they are in Britain. Indeed, after myxomatosis all but wiped out the rabbit population in Britain in the 1950s, stoat populations declined steeply. Every gamekeeper in Ireland that I talked to about stoats in the 1960s was of a similar opinion regarding the Irish stoat.[111]

Rats and birds were also important. It proved possible to identify the remains of birds in most of the stomachs to family: nine stomachs contained passerines (the small perching birds mentioned in Chapter 1), four pigeons and three waders or gulls. In four of the guts the condition of the feathers indicated that they had come either from nestlings, which would not have been able to fly, or from moulting adults, who might have been slower in taking wing than those with a complete coat of feathers.

The most surprising result was that pygmy shrews figured frequently on the stoat's menu. This emphasises the wisdom of a systematic analysis of scats or gut contents in investigating the diet of a predator. Because shrews are so small, stoats are unlikely to be seen killing them. Anyway, both dispatch and consumption of shrews are likely to be in cover. There are no recorded observations whatsoever of an Irish stoat catching a shrew. Furthermore, shrews were unexpected as they are distasteful to many mammalian predators; I found none at all in the stomach contents of over 500 foxes.[111] More surprising still is that stoats and weasels in Britain have rarely been recorded as eating shrews.[59]

The stoats consumed appreciably more shrew in summer and autumn (June-November) than from March to May. (Sleeman obtained few bodies from December to February.) This simply reflects the numbers of shrews available. Through the winter, populations of pygmy shrews are low and stable. Breeding starts in spring and numbers increase dramatically as the early litters enter the population so that, by June, the adults that have overwintered are outnumbered by young of the year.[111]

In mustelids, males are usually appreciably larger than females and in some of the smaller species it is known that this affects prey selection, the male taking more larger prey items and the female more of the smaller ones. This has been shown to be the case for weasels in Britain, although not, to date, for stoats.[59] Sleeman's results, however, demonstrate it to be true of Irish stoats. Shrews made up 14% of the diet (by percentage frequency in gut contents) in males and 44% of females, a result that is statistically significant.

The stoats, like mink in Chapter 1, ate relatively few field mice, which, although common, are agile and no doubt often leap to safety. Although the figures for bank voles *Clethrionomys glareolus* and field mice in Table 5 are similar, they are misleading. Only a third of the stoat bodies came from within the geographical range of the bank vole in Ireland. So voles, which cannot leap, are likely to be caught more often where they are available, and may well be as important as birds and rats there.

There was one instance of cannibalism, although it was, of course, impossible to say whether the dead stoat had been killed by the eater or had been found dead.

Sleeman, like any good scientist, was critical of his results and pointed out that, because all of the gut contents he examined were from road casualties, his results might be biased towards rats and shrews, which are sometimes common on the verges of Irish roads. He may be right for, in the 29 gut contents of Irish stoats which I looked at, there was only one instance of rat and no shrew remains at all. Unlike Sleeman's animals, all of mine were either trapped in a well wooded estate, or shot, and all therefore probably met their ends well away from a road. That Sleeman's material was from roadsides may also account for the contrast between his findings and those in the only major published study of gut contents of stoats in Britain.[59] In it rat and shrew turned up only occasionally despite there being not just one species of shrew available, as in Ireland, but three, two of which are common. All of the British stoats were obtained on estates. Further analyses of gut contents of Irish stoats from a variety of habitats would be instructive. Unfortunately obtaining corpses other than from road casualties would be difficult, as this would involve trapping and therefore a great deal of work. In any case in the Republic this could be done only under license. By the way, stoat scats are usually extremely difficult to find, so systematic investigation of the diet necessitates corpses.

Sleeman who, before his research on stoats, earned an MSc degree for work on the parasites of Irish deer, also examined 196 corpses for parasites.[314] About half were infested with small numbers of the louse *Trichodectes erminae,* which occurs only on stoats. Lice are parasites of birds and mammals. Most species are fastidious about just what they will feed on and can make a living from only a few species of host, often only from one. Unlike a flea, a louse spends its whole life on the body of its host. The eggs are cemented onto hair or feathers and are the familiar "nits" found on the heads of schoolchildren infested with the human head louse *Pediculus humanus capitis*. The egg hatches to give a nymph, which resembles the mature louse. This grows and moults twice, the third moult producing the adult. The process from egg to adult takes about a month. The intimacy of the association of a louse with its host and the time needed to complete its life cycle is well illustrated by the human body louse *Pediculus humanus corporis* - a more serious parasite than the head louse, as it spreads disease - which lives in clothes. If one changes one's clothes and has a bath at least once a month, the insect is unable to complete its life cycle and it is therefore impossible for an infestation, which may involve several thousand of the critters, to build up. Transfer of lice from one host to another can only occur by close contact.

Although some kinds of flea are confined to a single species of host and only occur on others by accident, many fleas can live happily on several host species and some, like the so-called human flea *Pulex irritans,* on many. Even fleas which do not normally parasitise a particular species can sometimes rub along on it for a while. Fleas lay their eggs either in the host's nest, or in its hair, from which they are later dislodged into the nest when the host settles down in it. The larvae hatch in the nest and feed *inter alia* on dried blood from the host, some of which has passed through the adult flea's alimentary tract undigested. Particular species of flea larvae may require particular nest conditions and particular blood. So, although an adult flea may survive on the wrong host, its offspring will not develop in the wrong nest. As the larvae of nearly all fleas spend their time in the hosts' nest, most mammals which do not have a nest, or some sort of regular bed, generally do not suffer much from fleas. Unlike lice, most fleas can survive for long periods away from a host.

In Ireland the stoat does not have a specific flea of its own, neither can any of the fleas so far found on it be regarded in any way as genuine stoat fleas. Those which do occur on Irish stoats are normally found on rodents, rabbits and birds. While feeding, predators are at risk of affording the fleas that have been living on their prey at least a temporary home, and it had been assumed that Irish stoats pick up most of their fleas in this way.[111] Sleeman's studies suggest otherwise. He compared the fleas he found on his road casualties, about a quarter of which produced fleas (a minimal figure, for these insects tend to leave the body when it cools), with those on the rats he accidentally caught in his Fenn traps, and a few others that he obtained on rats from other sources.[309,320] The data for stoats and rats are compared in Table 6. I have, incidentally, searched the literature and have found that all four of the species on the Irish stoat in the table had occasionally been recorded on it before with, just as in the table, the rodent fleas *Ctenophthalmus nobilis* and *Nosopsyllus fasciatus* predominating. Sleeman argued that, if the stoats were contracting fleas from their prey, then the numbers of the different kinds should have been roughly in proportion to the numbers of guts containing rabbits, birds and shrews. But nearly all were from rodents and there were none which specialise on shrews at all. Clearly fleas on stoats do not reflect the diet, but there is a strong similarity to the composition of fleas found on rats. This is most readily explained by stoats regularly expropriating rats' burrows and perhaps even their nests. It will be recalled that the majority of the dens used by stoats on Fota had formerly been the property of rats. The rat fleas thus exact a modicum of vicarious revenge on the stoat for the eviction of the rightful owner. Rats and stoats have approximately the same girth and the burrows of the former may be the most suitable for the latter. Mouse holes would often be too small and, as rabbit burrows are wider than a stoat, perhaps they tend to be more draughty.

Table 6 - Numbers of fleas collected from the bodies of Irish stoats killed on roads and from brown rats mostly caught in Fenn traps.

Species of flea	On stoats	On rats
Mouse/vole/shrew flea *Hystrichopsylla talpae*	-	8
Common rodent flea *Ctenophthalmus nobilis*	28	183
Rat flea *Nosopsyllus fasciatus*	17	56
Bird flea *Dasypsyllus gallinulae*	3	-
Rabbit flea *Spilopsyllus cuniculi*	1	-

The stoats also sometimes carried ticks including *Ixodes ricinus* and *Ixodes canisuga,* but mainly *Ixodes hexagonus.* All parasitise a range of mammalian hosts but *hexagonus* is the commonest tick on mustelids. Dog owners are usually familiar with ticks, all stages of which attach themselves by burying the mouthparts in the skin and then gradually swell as they engorge with blood. Once replete they drop off and digest the meal. Access to a host generally occurs when it brushes against the tick, which lies in wait on vegetation. Since an extended wait may be necessary - patience is the tick's long suit and many die before finding a host - the proportion of its life that the parasite spends feeding is relatively short. When a host arrives it must be like a rescue ship to a marooned mariner.

Ticks are related to spiders and scorpions and, like them, the adults have eight legs. The female lays her eggs in a secluded place. The egg hatches to give a six-legged larva. After it feeds, it grows and moults to produce an eight-legged nymph, which must again feed, grow and moult to become an adult. Both larva and nymph may overwinter before feeding. The larvae and nymphs of many ticks loiter close to the ground, on grass for instance, and thus end up on small mammals. The adults hang around higher up and therefore tend to have larger hosts, which is more satisfactory as the adults, being bigger, require more blood. As the smaller hosts are often rodents and the larger ones can be man or his domestic animals, ticks are important vectors of disease. A stoat, of course, being a low slung beast, tends to pick up the juvenile stages. Of the 332 ticks Sleeman recovered, 91% were larvae, 7% nymphs and less than 2% adults.

Since ticks are unevenly spread, even in areas where they occur, their distribution on hosts is clumped. Many hosts are free of them; some harbour a few; but a small minority, which have had the misfortune to brush against vegetation with many ticks on it, have a lot of them. Less than 20% of Sleeman's corpses were infested, but one stoat carried a burden of 266 larvae. Bad luck of this kind, of having a home range in the wrong place at the wrong time and contracting a load of parasites as a result, was also illustrated by a single stoat carrying no fewer than 1,819 larvae of the harvest mite *Neotrombicula autumnalis*. The adult mites live in soil and it is only the larvae that are parasitic. These animals are minute and are locally common in the countryside in late summer and autumn - hence the name - and will attack wild and domestic mammals, poultry and man.

Those who feel uncomfortable reading about parasites should skip this and the next two paragraphs, which deal with the nastiest of the stoat's tormentors, the sinister and horrifying skull worm, a roundworm which rejoices in the scientific name of *Skrjabingylus nasicola*. Even some zoologists experience a little initial difficulty in getting their tongue around this one. Unfortunately the skulls of most of the stoats Sleeman examined were smashed - hardly surprising as the animals were mainly road casualties. Of 36 intact skulls that he was able to

IRISH STOAT

prepare, half had been infected with skull worms,[311] which are apparently common parasites of stoats and weasels everywhere. Infected skulls show obvious damage behind the eyes: swellings which make the skull asymmetrical, with some thinning and perforation of the bone. If a fresh head is dissected, it will be seen to contain a mass of red roundworms, each up to 25 mm long, closely packed into a swelling in the nasal sinuses. As the skull of an Irish stoat is, depending on sex, only 35-45 mm long, the parasites are relatively enormous. One infected skull which Sleeman dissected out contained four females and a male.

Although we do not know for certain, it is difficult to believe that the swellings do not put pressure on the brain, and the wriggling of the worms alone is likely to be intensely annoying. Stoats and weasels are sometimes seen running back and forth, leaping about and turning somersaults and cartwheels. Among the youngsters this is usually believed to be play, or in an adult a stratagem to attract the attention of rabbits or birds, which, thus bemused, fall easy prey to the wily mustelid. However, a stoat may behave like this on its own and in the absence of prey. I observed this myself near Askeaton, Co. Limerick. The exhibition, which lasted several minutes, was at times too fast for the eye to follow, and included flying somersaults reminiscent of those by human gymnasts, but at lightning speed. Such solitary displays are also usually interpreted as play. But maybe the gyrations are abnormal behaviour brought on by skull worms: perhaps fits resulting from pressure on the brain or writhings in an attempt to relieve irritation caused by the movements of the parasites.

Roundworms begin life as tiny larvae, which grow and moult four times before becoming adult. Not all roundworms are parasitic, but the females of those which are concentrate on reproduction, because the chances of a larva eventually reaching a new host are slim. This, incidentally, is true of parasitic worms in general. A steady stream of the larvae of the skull worm passes out of the sinuses of its host through the nasal passages to the throat, and hence to the gut and outside with the faeces. To develop further, a larva must find a slug or snail, into which it bores. Transmission to a stoat could occur directly if it were to eat an infected mollusc, but slugs and snails apparently do not figure in the stoat's regimen. The next step therefore occurs when the mollusc is dined upon by a shrew or small rodent, in which the worm then encysts. If the shrew or rodent is eaten by a stoat, the worm emerges, passes through the wall of the gut, migrates along the spinal cord to the skull and settles down to a secure and comfortable existence.

In June 1936 two ladies, out walking on the seashore at Tawin, on Galway Bay, saw six or eight stoats - a family party is not impossible at this time of year - carrying something in their mouths and hurrying across a stretch of grass to a loose stone wall. Upon spotting the intruders, the animals disappeared into the wall, leaving behind their burden, which proved to be two young mullet. The ladies kept quiet to see what would happen and eventually a stoat emerged and retrieved one of the fish. The other was alive and unharmed, so they returned it to a pool on the shore, the tide being some way out. The stoats had apparently been returning from a fishing expedition, for their coats were soaked and the fish were freshly caught.[347]

This incident remained as a curiosity in the literature until recently, when there has been significant further evidence that the Irish stoat may turn fisherman.

In the summer of 1983 at Derrynane, Co. Kerry, an observer watched, from about 25 m, a pair of stoats make four excursions from a field to a rock pool. Each time one carried back a shanny (or common blenny) *Lipophrys pholis*. Whereas the stoats were not actually seen in the water, ripples were evident and the animals were obviously wet.[79]

Dr Simon Berrow, of University College Cork, watched a stoat for over an hour on Achillbeg Island, Co. Mayo, in June 1992. It commuted thrice over 200-300 m from a dry stone wall to a boulder-strewn rocky shore, each time returning with a fish. The first two fish were both about one third the length of the stoat and looked like members of the blenny family (Blenniidae). The third fish was thought to be an eel, being about as long as the stoat itself. Although the stoat was not actually seen to catch any of the fish, it was observed for some time apparently foraging among the boulders below high-tide mark.[29]

At the end of October 1992, Mr Ger O'Donnell, of Connemara National Park, was sitting in his land-cruiser at Renvyle Strand, Co. Galway, about 50 m from the sea. He was surprised to spot a stoat, carrying a small fish, emerge from some boulders about 20 m away. A child then passed, causing the animal to retreat into the boulders. Shortly afterwards, it reappeared without the fish and bounded down to the sea. O'Donnell then lost sight of it but 10 minutes later saw what was presumably the same stoat with possibly the same fish, come out of the boulders again and run off. He thought the fish was about 5 inches (13 cm) long and almost certainly a pollack *Pollachius pollachius*.[250]

Michael Viney has recently received additional correspondence on stoats fishing.[359] This comparatively large number of chance sightings of stoats with fish suggests that those dwelling by the seaside may often be competent anglers.

Chapter 3. The Badger - a Social Weasel

Last of the night's quaint clan
He goes his way -
A simple gentleman
In sober grey:
To match the lone paths of his
In woodlands dim,
The moons of centuries
Have silvered him.
The Badger. P.R. Chalmers.

Badgers in general - and there are several species worldwide - differ from most other mustelids in being stocky, chunky animals. The European badger, the only species found in Ireland, is also highly unusual, possibly even unique, in being a fully social mustelid, for badgers usually live in small groups, which zoologists often refer to as clans. The term is apt, for the clan comprises a number of more or less related individuals associated to their mutual benefit, although not always living in perfect harmony. Just as in a Scottish clan, there is usually a laird - the dominant boar, and lady - the dominant sow.

The animal is certainly robust, with a brawny, wedge-shaped body which slopes down from the hind to the fore quarters to a short, thick neck and relatively small head. The badger is the largest of Ireland's terrestrial carnivores, adult boars in Great Britain weighing 9.1-16.7 kg and sows 6.6-13.9 kg.[151] On the skulls of most mammalian carnivores a flange of bone, the *sagittal crest,* runs along the mid line on top of the brain case from front to rear and serves as an anchorage for the big muscles which pull up the lower jaw. The flange is strikingly developed on the badger, and therefore indicative of an especially potent bite, and may grow to a height of 15 mm on some old boars (Fig. 9). The tooth nearest the back on each of the upper rows of teeth is disproportionately big and broad and this, together with the powerful jaw muscles, means that once a badger has made up its mind to hold on to something, it is no easy task to dislodge it. The Irish mammal I would least like to be bitten by is the badger. At the other end of the body is a stumpy tail, appended almost as an afterthought. The legs are to the usual mustelid pattern, being short and stout. They are particularly strong, even for a mustelid, are equipped with stout claws, especially those on the fore feet, and have transformed the animal into a formidable digging machine.

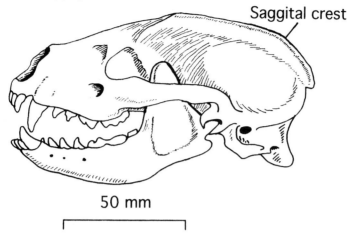

Fig. 9 Badger skull showing the sagittal crest.

At a distance the the badger appears to be grey but in fact the long outer hairs on its back and flanks are light at the base and tip but dark in between. The hairs on the legs and underside of the body are pretty much uniformly dark, but they are often sparse on the chest and belly. The head is white with a dark stripe down each side encompassing the eye, markings which are both distinguished and distinctive and which enable anyone to recognise the beast in an instant. Those who suggest that the stripes may act as a form of camouflage have never seen a badger emerge for its sett - the term given to its subterranean home - in the failing light of dusk, for the stripes are then its most conspicuous feature. Indeed they may be a form of what zoologists call *warning coloration*, other examples of which are the contrasting black and white markings of skunks or the black and yellow banding on the abdomen of wasps. Such colouring means, "Leave me alone. I have adequate means of defending myself". A skunk has stink glands, a wasp its sting and the badger its jaws and claws, but it is better for each that any would-be molester should be able to recognise this right away and leave well alone: hence the warning coloration. Recognition by a potential molester of such coloration as a warning may be the result of past experience or may even be instinctive.

Even from the outside, a *main sett* - the principal breeding sett and centre of a clan's activity - is often an impressive structure, sometimes with many entrances, the record being 44 and the average 6.9 for 1,378 setts investigated during the Badger and Habitat Survey of Ireland, undertaken in the Republic from 1989-93 under Chris Smal's supervision.[324] The elaboration of some setts is probably only achieved over a long period of time, each generation adding its pennyworth. One sett in Co. Cork mentioned in the literature in 1892 was still in use in 1993.[321]

Outside the sett are conspicuous spoil heaps, the visible evidence of the extent of the animals' excavations. A sett is often dug into a slope, with the resultant benefit that any earth brought to the mouth of a burrow tumbles down the incline and saves the badger effort in dispersing it further. Another possible advantage of such a location is that entrances at different levels may promote ventilation. This has been demonstrated by placing smouldering vegetation beside them, the smoke being drawn in at some and drifting outwards at others.[241]

Scattered over the spoil around an active main sett there is likely to be disused or left-over bedding, often dried grass, dead leaves or bracken. Bedding is apparently important in preventing heat loss, particularly so when there are cubs in the sett.[241] At any rate the animals will go to considerable pains to gather it. Bundles are carried clasped to the chest by the forelimbs and chin, during which the badger travels backwards propelling itself with its hind legs. Even though it is not looking where it is going, it still seems to head mainly in the right direction. The whole procedure is undertaken with great deliberation and, if not particularly elegant, is at least efficiently done. Bedding is replaced at intervals and may even be brought up for airing before being lugged underground again.

Near the sett there is often a tree with rough bark, often an elder, bearing numerous grooves and abrasions where the badgers are wont to scratch with their fore claws. The marks may extend as far as a metre up the trunk because the animal often stretches up to its full length before dragging its claws down the bark. The function of such a scratching tree is obscure. Perhaps badger claws need regular sharpening but the tree may also be some sort of territory marker. This would be likely if there were scent glands between the badger's toes, although this has yet to be demonstrated. There is, however, a scent gland beneath the tail and a pair of glands which open internally, close to the anus. All three produce oily secretions with a rancid, musky odour. Those from the anal glands are mingled with the dung.

There are usually some paths running between the entrances to the sett and one or more leads away from it for up to several hundred metres. Due to the appreciable body weight carried on each of the badger's feet, because of its claws, and as the paths are trampled night after night (for the badger is a creature of habit), they are well marked, often devoid of vegetation and easy to spot. They are readily distinguished from those made by humans, being much narrower, and may run through hedges, or under shrubs far too low to allow of human passage. Where the route passes beneath barbed wire, wisps of badger hair may be seen caught on the barbs. Paths may be used by generations of badgers and it is almost impossible to discourage the animals from traversing them. Thus if a mesh fence is to be erected across one, it is wise to leave a gap or a badger gate, which is rather like an extra large cat flap, to allow the animals free passage. Otherwise the fence, unless stoutly constructed, is likely to be damaged. When a new road is built across a badger path, casualties are likely.

Within a few metres of the sett there is often a latrine, consisting of a series of shallow pits, perhaps 10-15 mm deep, in which the animals defecate. When each pit is full, it is left uncovered and another prepared. Because the droppings are perfumed by the secretions of the anal glands, the latrine could serve an additional function in advertising the ownership of the property to other badgers, although it is unlikely that members from neighbouring clans would often intrude so far into an adjoining territory. But latrines are also strategically sited elsewhere in the clan's territory, as we shall see. Badgers, when undisturbed, do not usually defecate in the entrances to their homes, neither do they often bring prey back home. This is in marked contrast to the fox, the entrance to whose dwelling in May or June, when there are cubs to feed, is commonly promiscuously scattered about with scats and the putrefying remains of rabbits, rats and birds.

Besides the main sett, the clan generally has a number of auxiliary ones. There may be another close at hand, known as an *annex* sett, sometimes almost as large as the main one and nearly always connected to it by a path. Further away there can be a few more, smaller and somewhat less often occupied, called *subsidiary* setts. Out towards the edges of the territory there may be a single isolated burrow or two known as *outlier* setts. Because the badgers use outlier setts only sporadically, when uninhabited they may used by foxes or rabbits.

In the Badger and Habitat Survey,[324] Chris Smal, together with 50 conservation rangers of Dúchas and some volunteers, searched for setts in each 1 km grid square in the extreme south-west corner of each of the 10 km squares of the Irish National Grid in the Republic. This is equivalent to 1% of the total land area of the 26 counties and altogether 729 squares were checked. The workers recorded full details for each sett, including whether it was "active" (i.e. showed signs of recent tenancy, such as tracks, fresh digging, discarded bedding, badger hairs or fresh droppings in a latrine nearby). They also noted the habitat in which the sett was located and the general nature of the terrain in the 1 km grid square. The work was done only from late October to mid May, when cover by vegetation was low and the setts were therefore easiest to find. This study therefore provided valuable information, from a representative sample of the total land mass of Ireland, on the external appearance of the setts, on the sorts of places where they were dug, and on the types of habitat favoured by badgers. Furthermore, with data on numbers of badgers in setts obtained through the Department of Agriculture, it was also possible to estimate the total population of badgers in the Republic.

The Department of Agriculture arranges the removal of badgers under licence from areas in which tuberculosis (TB) in cattle has been attributed to infection from badgers. The word "removal" in this context appears frequently in the literature on the control of bovine TB and is a euphemism. In the Republic it is usually by snaring, normally followed either by shooting or occasionally by lethal injection. The snares used are "stopped" and fitted with a swivel, in other words they cannot close fully and do not tighten if the badger twists about, so it is not strangled. Snares are also regularly inspected for captures to minimise the periods that the captives are held in them. Nevertheless, the badger cannot enjoy the experience much. Unfortunately cage traps are heavy, cumbersome and impractical for local extermination. The carcasses thus obtained are routinely sent for testing for TB. Such snaring is intensive: the recommended level is 10 snares per sett entrance and 500 snare nights per clan (1 snare night = 1 snare set for 1 night). Incidental to such measures is the accumulation of figures for numbers of badgers in the different kinds of sett (i.e. main, annex, etc.). In this way Smal was able to obtain reliable data for 36 clans from 16 counties to use in the Badger and Habitat Survey. It would be misleading to leave the reader with the impression that this sort of operation eradicates every last badger from an area. Returns are dependent on the intensity of snaring and on how long it is continued. Smal evaluated trapping success statistically and was thus able to apply factors to correct the numbers of animals killed per sett upwards to an approximation of the true numbers.

A summary of the statistics from the Badger and Habitat Survey is provided in Table 7.[324] Evidently about half of the main setts had an annex sett and on average there was about one subsidiary and one outlier sett to each main sett. Judging by the average number of badgers per sett type and the fact that fewer subsidiary and outlier setts were actually in use, it is clear that the individuals of a clan reside for most of their time in the main sett. On average there were 5.9 adults in a clan up to a maximum of eight. The ratio of male to female adults was about one. From the overall data Small estimated that there were about 34,000 clans in the Republic with a total adult population of some 200,000 badgers. There was an average of one clan per 0.5 km² or 2.95 badgers/km².

Table 7 - Summary of some statistics on setts obtained during the Badger and Habitat Survey.

| | Type of sett | | | |
	Main	Annex	Subsidiary	Outlier
Percentage of active setts	84	81	71	59
Average number of entrances per sett	6.9	4.0	3.0	1.3
Numbers of other setts per main sett	-	0.50	1.32	1.08
Average number of adult badgers occupying active setts	4.26	0.92	0.93	0.21

In parallel and in cooperation with Smal's work in the Republic, Dr Sarah Feore ran a sett survey along similar lines in Northern Ireland as part of her doctoral research at Queen's University Belfast. She also surveyed a 1 km square in each of the 10 km squares of the National Grid in Northern Ireland along similar lines.[127] The total population of badgers in Northern Ireland, calculated in the same way, came to 52,000.[126] Thus the total for Ireland as a whole was 252,000.

Although the badger was scarcer in the earlier years of the twentieth century than it is today,[231] the frequency of the Irish word *broc* in the names of townlands attests to the wide distribution of the beast in historic time. I have uncovered 61 such townlands, of which 39 have names beginning "Broc" and three "Bruck". Many end in or embody "broc": e.g Altnabrocky, Ballybrocky, Castlebrock, Clonbrock (5), Cloonbrock, Coolbrock, Derrybrock (2), Derrynabrock, Dunbrock, Legnabrocky, Letterbrock and Meenabrock (3). In addition there are also Badgerfort, Badgerhill (2), Badgerisland and Badgerrock, making at least 66 townlands relating to badgers.

Spectacular as some main setts are from the outside, this pales into insignificance on the few occasions when it has been possible excavate one in its entirety. Normally such work does not bear contemplation: not only is a license required but the expense is prohibitive. The only realistic way of operating is to hire a mechanical digger complete with driver for a week or more. Fortunately there have been at least two studies in Ireland in which main setts have been thoroughly charted. In both instances they lay in the paths of projected new major roads. These excavations were funded by the local authorities and were in autumn, when any cubs that had been born in the previous spring would have been weaned, before the whelping season in January-February and indeed the harsher conditions of winter, during which the animals may lie up during the colder spells. In Scotland it has been shown that in winter the badger may spend much more time below ground, which reduces heat loss, and that the animal often drops its body temperature by several degrees, which further economise on energy. In other words it enters a winter lethargy, a sort of low grade hibernation.[139]

One of the setts excavated in Ireland was in woodland at Abbotstown, Co. Dublin, and lay on the route of the Northern Cross motorway. Mr Bill Clarke and Dr Gregory O'Corry-Crowe, of University College Dublin, together with various assistants, mapped it in 1992. Dublin County Council placed a Hitachi mechanical digger and driver at their disposal.[48] The other excavation was in 1998 at Carrigoran, near Newmarket on Fergus, Co. Clare, during a major road improvement scheme for the N18/N19. Chris Smal, with the help of several others, mapped this sett, which lay at the edge of a wood and extended into an adjoining field.[325] Clare County Council provided a JCB for the heavy work. I saw some of this myself and was amazed at the precision that could be achieved by such a monstrous machine when an experienced operative was at the controls. Such skill is essential for accurate mapping, especially where galleries run close alongside or underneath each other.

Apart from the actual digging itself, there is a great deal else to be done in investigations of this kind. All trees and bushes must first be removed leaving their roots intact, otherwise the tunnels may be damaged before they can be laid bare. The area then has to be marked out in a grid using pegs and cord or rope, rather in the manner of an archaeological dig, so that all of the entrances, and eventually the tunnels themselves, can be accurately charted. At Abbotstown, this was particularly tricky because the sett lay on the curved side of a hill. Then there are the measurements to be made on the dimensions of the galleries and chambers themselves and appropriate photographs to be taken. The word "chamber", by the way, is normally used in this context to indicate an enlarged cavity, usually with bedding present.

Not the least of the problems is the eviction of the badgers from their property. In both cases this was achieved by blocking the mouths of the burrows, a few each on successive nights, with earth splashed with animal repellents. This usually discouraged the beasts from reopening them, although the procedure had occasionally to be repeated, badgers being pertinacious creatures. Smal also used one-way devices on the final entrances to make sure that every last badger had left. In the event, all of the animals were successfully evacuated unharmed from both Abbotstown and Carrigoran. As an additional inducement to departure in both areas, a mixture of syrup and peanuts, of which badgers have been shown to be inordinately fond, was spread around subsidiary setts which did not lie in the paths of the projected roads. At Abbotstown this appeared to be successful in persuading the tenants to take up residence in the alternative accommodation, although not at Carrigoran.

Some general statistics on the setts are provided in Table 8 and there is a map of the one at Abbotstown in Fig. 10. The number of badgers living at Carrigoran was not determined, but watches at Abbotstown on 8 nights spread over the previous August to October revealed that the clan there was made up of only five individuals. Apart from a separate, small, blindly ending tunnel (Fig. 10), almost all of the passageways and chambers at Abbotstown were in use at the time. Clearly both clans liked their homes extensive and amply equipped with chambers, variously placed: some at the end of a gallery, others as an enlargement somewhere along its length, and still others thrust out from its side. Although the passageways were narrow at both sites, averaging about 0.25 m wide by 0.17 m high - surely a neat fit for a badger - the overall volume of both setts was considerable. At

Table 8 - Data from two excavated main setts in Ireland. Measurements are in metres.

Location	Length x width	Number of tunnel systems	Total length of tunnels	Average depth	Entrances	Chambers	Average chamber width and height
Abbotstown, Co. Dublin	34 x 12	2	267	0.84	15	25	0.75 x 0.78
Carrigoran, Co. Clare	27 x 19	5	213	0.91	20	14	0.65 x 0.37

Abbotstown it was estimated that some 18 m³ of earth had been mined and dragged to the surface in the course of construction. Clarke and O'Corry-Crowe reckoned that the sett there had been created over many years. The extent and complexity of both setts is none the less amazing, and badgers must be compulsive diggers. Incidentally, the size of these two main setts is by no means exceptional; several larger ones have been mapped in Great Britain. However, the chambers in the Abbotstown sett were peculiar in being generally somewhat higher than they were wide. Those at Carrigoran, which had much lower ceilings, are typical of setts that have been investigated elsewhere.

o   Entrance

c   Chamber

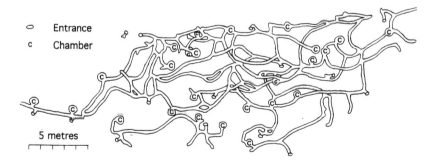

5 metres

Fig. 10 Plan of a sett excavated at Abbotstown, Co. Dublin.

The most complex part of the Abbotstown sett was towards its centre, where there were passages at several levels (Fig. 10). This was probably because the soil was deepest there. Elsewhere the underlying shale, which the animals would have found impossible to quarry, was closer to the surface and would have restricted such elaboration. The bedding in the chambers was generally of grass, most of it fresh, although sycamore and beech leaves served in one. Badger fleas *Paraceras melis* were common in the bedding. (The badger has a flea all its own, although it also occurs on foxes because they sometimes shelter in setts.) Intriguingly, there were badger droppings in three places, demonstrating that the animal does not always defecate above ground, away from the sett. There were also a few bones: two from badgers, and one each from a rabbit and a sheep. These hint that prey may occasionally be brought home, but such evidence is inconclusive: the bones might have been carried in by foxes.

The main sett at Carrigoran covered a somewhat larger area, was more convoluted, and proved to embrace no less than five separate tunnel systems. This was probably at least partially dictated by a wet ditch, under which it would have been impractical to burrow, and a raised bank, through which the animals had not attempted to drive any galleries. One end of the sett lay next to a patch of wet clay and silt, which, being soft, may have prevented further tunnelling in that direction. One tunnel system was disused at the time. Of the total of 20 entrances, seven were in active use, four inactive, seven disused and partially blocked, and two collapsed, as were a few of the tunnels. Such damage was probably the result of trampling by cattle or possibly the passage of agricultural machinery. Setts dug under open fields are naturally subject to such hazards. Of the 14 chambers, seven were active, all with fresh bedding, five inactive with old bedding, one collapsed and one so crammed with bedding that it had evidently been abandoned. Bedding throughout was mainly grass. Besides the main sett, Smal also dug out most of a subsidiary sett at Carrigoran. This had 64 m of tunnels, 10 entrances and three chambers.

A bizarre feature of both setts was the use of strips of polythene bag or sheeting as part of the bedding. Badgers may have taken to polythene more widely. When Dr Sarah Feore dug up a sett on a site scheduled for a housing estate outside Newtownards, Co. Down, and lying beside an existing estate, it contained innumerable plastic bags of the supermarket kind.

It will be apparent from the foregoing that new roadworks often cause problems for the badger. A range of measures has been devised to ameliorate the situation, including badger-proof fencing; one-way badger gates to allow animals which inadvertently wander onto major roads to escape; badger underpasses - pipes laid under new roads to allow the animals safe passage beneath along their traditional routes; and raised badger paths in the bridges and culverts used to reroute watercourses beneath the highway.

There have only been two studies of the badger's diet in Ireland. In 1963-64 I obtained badger stomachs from a number of sources, but mainly from a countryman in Co. Down, whose only apparent source of earned income appeared to be the trapping, snaring and shooting of wildlife, including foxes (for which a bounty was then paid), badgers and otters. In those days the only mammals with legal protection were those classified as game. It was a family business and I recollect being taken aback when his wife assured me of her participation and showed me her rifle, which was fitted with a telescopic sight.

Setting aside the stomachs which were empty and those containing only fragments of grass and bark, which might have consumed in an animal's efforts to escape a snare or trap, and pieces of unidentifiable flesh, altogether I obtained 15 with food in them.[96] These indicated a diet including rabbit, hare, birds, insects (especially beetles and caterpillars) and earthworms, which were the most important prey and turned up in eight of the stomachs, sometimes in quantity. During May-October 1964 I also collected seven samples of badger droppings from two latrines in the same county,[96] totalling 1,071 g after drying in an oven at 60°C for a couple of days. I detected some bird remains, beetles and, in autumn, oats: badgers sometimes feed on cereals, although rarely do any significant damage.[151] The bulk of the material, however, at first appeared to be soil, until I looked at samples of it under a microscope. Much of it was unquestionably soil, but in it were numerous *chaetae,* the tiny hard spines by which an earthworm gains a purchase on the sides of its burrows as it passes through them. As there are eight chaetae on each earthworm segment and an adult earthworm has 82-250 segments depending on species, this means rather a lot of chaetae per worm. Since most of an earthworm is soft, it is digested in the badger's gut, so that only the chaetae and the soil in its intestine are left. The worm's gizzard, which is thick walled, is also known to survive passage through the badger, but probably not the cooking which I gave the droppings.

Dr John Whelan and Mr K. Boyle did a more systematic study at

University College Dublin.[31] They collected scats from badger latrines in the Kilmashogue district at the foot of the Dublin Mountains every month from October 1982 to March 1983, washed them through fine sieves and identified all of the items which remained. From this they estimated the relative volumes of the various food items eaten. The calculation is complex but is based on extrapolating from remains in droppings to whole animals. For example, an earthworm gizzard was taken to mean that a whole earthworm had been eaten, and a few teeth and fur from a field mouse as indicating consumption of a complete field mouse. Such a procedure has obvious shortcomings, but at least the importance of earthworms, so much of which is digested in the gut, is unlikely to be underestimated.

Table 9 - Percentage volume of various food categories in badger faeces collected from one district in south Co. Dublin.

Prey category	Percentage volume
Vertebrates	3
Earthworms	46
Other invertebrates	9
Blackberries	5
Acorns	3
Oats	6
Scavenged	28

Altogether there were 100 samples of faeces, so the findings should provide a very good indication of the diet of the badgers in the region in autumn and winter. The results are shown in Table 9. The "scavenged" food represents a variety of mainly vegetable items that the badgers had consumed at a dump in the vicinity, including fragments of red paper serviettes, which the beasts, and obviously some of the people who threw them out with their other garbage, seemed to favour. Perhaps they had been used to mop up something tasty. Earthworms were the staple item, making up about half of the diet, or two-thirds if we exclude material from the dump. The "other invertebrates" were adult insects (mainly beetles) and larvae (predominantly caterpillars). There were also a few bumble-bees and there is an account in the older Irish literature of badgers regularly taking them in one district in Co. Wexford.[221] The "vertebrates" included passerine birds and two instances of field mouse. In view of the mouse's agility and the badger's digging propensities, it seems more likely that these rodents had been excavated from their burrows rather than caught in the open. There were no remains of rabbit whatsoever, but the situation might have been different in summer, during the bunnies' breeding season, when there would have been plenty of nests of young to dig out. As one might expect, the oats, acorns and blackberries, which would only have been available in quantity in autumn, only appeared in the scats then.

It seems unlikely that badgers are sprightly enough to catch birds very often. So at least some of those in the above analyses may have been found as carrion. Young birds that cannot fly properly are also potential prey as are nestlings. Mr Pat Smiddy, a conservation ranger operating in east Cork and west Waterford, has made the only relevant observations in Ireland.[334] (Smiddy is, incidentally, one of the most prolific authors of scientific papers and notes on Irish birds and mammals. In 2000 the National University of Ireland recognised his work with the award of an Honorary Degree.) One of the nests belonged to a kingfisher *Alcedo atthis* and had been built in a hole in a river bank, 45 cm in and 26 cm from the surface of the soil above. The badger had dug down vertically into the nest chamber and presumably consumed all of the brood, which was about a week old. The other instance was of a dipper *Cinclus cinclus* nest, build under an overhang on an earthen river bank. Again, the badger had driven a hole directly down, some 21 cm through the overhang, to the nest.

In Great Britain it has long been known that at least some badgers can unroll and eat a curled up hedgehog *Erinaceus europaeus*, leaving only its spiny coat behind. Recent research has also shown that, at least in some places in Britain, badgers are a regular scourge of hedgehogs and may even control their numbers locally.[206] Michael Viney writes that one of his correspondents was awakened by what sounded like a baby crying in the garden, and was more than surprised to shine a torch on a badger pinning a hedgehog to the ground. The correspondent struck the badger, which ran off, leaving the hedgehog badly mauled about the face.[359] It is perhaps here worth quoting in full a note published in the *Irish Naturalist* of 1917 by G.H. Pentland (1850-1932), a gentleman naturalist of Black Hall, Drogheda, Co. Louth.[274]

About fifteen years ago Badgers appeared in my woods and

soon formed a flourishing colony. As they increased, the Hedgehogs, which were then very plentiful gradually disappeared. For five or six years I never saw one at all. Then something happened to the Badgers. Their numbers dwindled till the tribe was reduced to one or two and they seem on the verge of extinction and the little Hedgehogs are reappearing! Cause and effect evidently.

The overall picture of the badger's diet in Ireland is very much the same as that in Great Britain: an assortment of food mostly found on or under the ground, including vegetable material, birds, mammals and insects, particularly beetles and caterpillars. Notwithstanding, the staple food is earthworms. We might therefore expect to find terrain favourable to worms also to be the most congenial to badgers. This turned out to be true in the Badger and Habitat Survey, the results of which pertaining to where setts were sited, we can now turn to.

The percentages of the 1,373 setts in the Survey in various major habitats are shown in Table 10.[324] Setts were commonest along hedgerows which are, of course, readily available over most of the Irish countryside. Typically they were dug where there was a dry ditch beneath the hedge. Hedges, together with sites around trees or bushes (tree lines, woodland and scrub), accounted for 80% of all setts. Such sites are generally similar to those in Britain, but there setts were rather less common under hedges and more often in woodland. But then in Britain there is a greater proportion of land under timber. A sett in woodland or scrub was usually near the edge, so that there was just as ready access to open ground as from beneath hedges. Generally there was a strong preference for cover and the overall picture is one of badgers making their homes either under hedges or near the edges of patches of scrub or woodland, from which they could conveniently sally forth at night to forage for earthworms on adjacent pastures. Nevertheless, it will be plain from Table 10 that, presumably where there was insufficient suitable cover, the animals sometimes had no alternative but to to dig their homes in open grassland. Setts were rare on bog and moorland and were commonest in regions with good quality pasture. Indeed the number of clans per unit area was strongly associated with the proportion of cattle pasture in the surrounding terrain, but negatively correlated with the proportion of pasture grazed by sheep. All of this may be put down to availability of earthworms, which are most abundant on the superior grassland most commonly reserved for cattle, less plentiful on the poorer land used for grazing sheep, and scarcest of all in the acidic soils of moorland. Setts were also rare in young conifer plantations, because these are usually planted on poor quality land, often bog. All of this means that the counties with the highest badger densities were those with the best agricultural grazing land, namely Kilkenny, Limerick, Louth, Meath, Offaly and Westmeath.

Table 10 - Percentages of 1,373 setts in the Badger and Habitat Survey in various habitats.

Habitat	Percentage of total setts
Hedgerow	35
Tree lines	8
Woodland (all kinds)	18
Scrub	19
Grassland	13
Other	7

During the Badger and Habitat Survey setts were also found in cliffs, quarries, open-cast mines, riverbanks, roadside verges, railway embankments and on built-up land. The late C. Douglas "Jimmy" Deane (1913-92), formerly Deputy Director of the Ulster Museum, once found a sett in a room in the lower storey of an abandoned farmhouse in Co. Wicklow and another under a road in Co. Down which eventually caused part of it to collapse under a lorry.[68] Perhaps the oddest situation for a sett is on Coney Island, Co. Sligo. In July 1977 I examined a stranded sperm whale *Physeter macrocephalus* on the beach there. In due course the body was buried and eventually badgers dug a dwelling for themselves amongst its bones.

Badger clans are highly territorial. Fortunately zoologists can map the boundaries of territories using a simple technique which, although devised in Great Britain where zoologists have also actually seen badgers from neighbouring clans fighting on the border between their territories,[186] has also been widely used in Ireland, by Sleeman in Co. Cork,[317] by O'Corry-Crowe in Co. Offaly[249] and by Feore in Northern Ireland.[126]

The edge of a territory usually lies along topographical features such as hedges, fences, roads or at places where there is a sharp division between one type of habitat and another, for example between woodland and pasture. However, on uniform habitat, for instance on grassland, badger paths sometimes delimit the perimeter of a territory. The clan probably marks the border of its property with scent in more than one way, but the most obvious one is with latrines, and it will be recalled that badger faeces are scented with secretions from the anal glands. Although badgers often dig latrines near a sett or elsewhere on their property, they normally site them in greatest numbers beside the topographical features marking the edge of their territory. There they serve the same function as a notice bearing the legend "Keep out. Private property".

To map individual territories in a region, a wildlife scientist must first find all of the main setts. Badger bait, in the form of peanuts and syrup or treacle, is then prepared and coloured plastic granules added to it in quantity. In Ireland the granules have been supplied by Athlone Extrusions Ltd., are 3 mm long by 2 mm in diameter, and are available in a range of colours. Each main sett is assigned a particular colour of granule and bait with granules of this colour is liberally ladelled out beside it over a period of at least some weeks. The badgers gobble up the mixture and in due course the granules appear in the faeces in the latrines. By examining these, it is a simple (if messy) procedure to assign individual latrines to particular main setts and therefore to the clans living in them. Thus the territory of each clan is delineated.

Both the area of a territory and the size of the clan living in it vary with the quality of the land and therefore the availability of earthworms in it. This has been nicely demonstrated by Feore in three districts in Northern Ireland, selected to represent good, medium and poor badger habitat.[126] She located all of the setts in each district and worked out the territories using the baiting technique. 95% of latrines in all three districts were within 5m of some sort of natural boundary, including hedgerows, woodland edges, shorelines and walls. She also live-trapped, anaesthetised, tattooed and released all of the badgers in each area. Her work extended over some 2 years.

One of her areas was at Castleward, Co. Down, a National Trust property on the shores of Strangford Lough, which may be considered optimal badger habitat at sea level, consisting of parkland and improved pasture, interspersed with coniferous and broadleaved woodland. The second area was at Katesbridge, Co. Down, at an altitude of 110-210 m above sea level. This was mainly pasture of reasonable quality together with some poorer grassland but negligible woodland. The third region was in upland at Glenwhirry, Co. Antrim, and was predominantly unimproved grazing, with isolated conifer plantations.

Table 11 - Density of setts, average areas of territories and average numbers of adults in badger clans in three districts in Northern Ireland.

	Castleward	Katesbridge	Glenwhirry
Density of main setts/km^2	2.3	0.8	0.3
Density of all setts/km^2	14.3	4.3	1.5
Average area of a badger territory in ha	50.4	127.4	345.0
Average number of adults in a clan	6.4	2.1	2.6

Feore's results are summarised in Table 11 and reflect the quality of the habitat from the badger's point of view. The best habitat, at Castleward, had by far the most main setts per unit area and therefore the densest concentration of badger clans. Moreover, the clans were bigger than elsewhere. As a corollary, a clan there needed by far the smallest area to furnish all of its requirements. At Glenwhirry, with the worst habitat, the density of main setts per unit area was lowest and the average area for a territory greatest. In fact the density of clans there was even lower than the density of main setts suggests. There were three main setts but only two clans, because one clan was using two main setts over 1 km apart - an unusual occurrence. Working out the territories at Glenwhirry was also difficult as there were fewer latrines along their edges. Because the territories were so big, the clans would have naturally needed more latrines in order to mark the borders as efficiently as at Castleward or Katesbridge. In fact the clans at Glenwhirry had significantly *fewer* latrines in their territories than those in the other two areas. It was probably not worth the animals' while expending much effort in laying full claim to terrain so unfavourable, and where badger density and competition for resources is low. Any interloper would probably have to have ranged widely to survive and would therefore have come across a latrine or bumped into one of the residents anyway. In contrast, in areas with high density populations, competition for resources - mainly earthworms, of course - is bound to be intense and so territories are demarcated and defended vigorously.

That badger density varies with habitat has a bearing on any estimate of badger numbers in a region calculated from numbers of main setts multiplied by average clan size. Earlier in this chapter the estimated total number of adult badgers in Northern Ireland was given as 52,000 based on Smal's figure of 5.9 adults per clan. In the light of her work at Castleward, Katesbridge and Glenwhirry, Feore recalculated the number to allow for variation in habitat and came up with the much lower figure of 38,000. It is possible that, for the same reason, Smal's figure of 200,000 for the Republic may also be overestimated.

Whereas badgers generally confine themselves to their territories, occasional excursions outside are not unknown. Sleeman, who radio-tracked badgers in Co. Cork, recorded two sows who strayed more than 7 km from their home setts.[317] However, the area was highly disturbed and such results are decidedly atypical.

Why are badgers sociable? Why do they defend territories in groups rather than individually as most other mustelids seem to? Earthworms, which play such a great part on shaping the lives of these mammals in other ways, may again provide an answer. Studies in Great Britain[186] show that when badgers feed on earthworms they give a distinct impression of grazing, of a steady, continuous process of seizing and pulling out worms that have partially emerged from their burrows to seek dead leaves or other provender on the surface of the soil. So badgers generally choose places where there are ample worms out at night and where they can take them easily, which dictates short vegetation. On grassland, where the majority of worms are almost certainly caught in Ireland, the grass has to be short for efficient feeding, which indicates land grazed by farm animals. Despite the worm's chaetae, which enable the tail end to grip the sides of its burrow when the head end is being tugged, badgers acquire the knack of extracting the majority of worms in one piece.

Earthworms do not come up everywhere every night. The conditions for greatest emergence occur when it is pitch dark, when there is little wind, when it is warm - a temperature of around 10°C is best - and particularly when it is wet. The grass is usually sufficiently wet for emergence for about two days after the last rain. Worms will still come out if there is a good dew fall, but dew is often extremely localised - one field may have plentiful dew while a neighbouring one has not. A combination of short grass and meteorological factors, but especially dew, means that on a given night in a particular district there will usually be ample food in the form of worms in some places, the badger's "feeding patches", but not in others close at hand, and that the situation will change from night to night.[186]

All of this means that to obtain earthworms with ease on most nights a badger must have a repertoire of places in which to forage, some conducive to the worms being available on some nights, some on others. It follows that not only will the animal need a large territory, but one that is likely to snake out in various directions to include all the necessary feeding places, in other words a territory with a highly irregular and therefore extra long perimeter. It takes more time to mark and patrol the edges of a large territory, and particularly so if the perimeter is disproportionately long. The reader may recall that a major drawback to being a mink is the necessity of maintaining a linear territory. The difficulty for badgers is overcome if individuals join forces. Together they can defend a much bigger joint territory including places where worms are readily obtainable on different nights.[186]

The reproduction of badgers in Ireland, or at least that of the sows, has been examined in some detail. In the late 1980s, as part of a major study of bovine TB, the Department of Agriculture in the Republic systematically removed badgers from an area of some 600 km[2], mainly in eastern Co. Offaly. The results of this experiment are considered later in this chapter. A by-product was a large number of badger carcasses and during 1989-90 Drs R. Whelan and Tom Hayden, at University College Dublin, acquired a sample of 548 females, with dates of death, spread through almost every month of the year. Of these 152 were yearlings - animals born in the previous year - and 396 were in their second year or older.[373]

The reproduction of badgers, like that of mink, stoats and several other mustelids, features delayed implantation. Hayden and Whelan were *inter alia* able to extract the blastocysts (which the reader may recall is the stage of development of the embryo at which delayed implantation intervenes), record any *foetuses* (embryos after implantation and therefore at a more advanced stage of development) and also evidence of *lactation* (production of milk).

The results indicated that there may be at least some weeks difference between one year and another in the timing of the mating and whelping seasons. In 1989 most births were between 22 January and 4 February, whereas in 1990 they occurred mainly between 18 and 28

February. There were, in addition, some early and late births up to 10 days outside these periods. Suckling extended from February to June. Most successful copulations in spring were between 17 and 31 March in 1989 and 8 and 21 April in 1990 so that by the end of April in both years 60-65% of the sows were pregnant with blastocysts. There was little further change in this value until August, when the proportion of sows containing blastocysts began to increase again until by September 80-90% of sows were pregnant. Rather fewer yearlings became pregnant than older animals. We can deduce from this that the badger has two mating seasons in Ireland, or at least two peaks in mating: in March-April and again in August-September. This does not complicate the scheduling of the subsequent whelping season, because all of the blastocysts remain undeveloped in the uterus until December or early January, when a proportion of them attach and in due course develop to term. In fact a surprisingly high proportion of the blastocysts failed to implant and were lost. So, whereas 80-90% of sows were pregnant with blastocysts in September, after implantation only 65-70% still contained young. Moreover, because the average number of blastocysts per pregnant sow was 2.7 and the average number of foetuses per sow that was still pregnant was 2.9 (range 1-4), it may be concluded that in most cases the sow either lost the lot or that every blastocyst implanted successfully. Losses were much commoner among yearlings than in the older adults. It is reasonable to infer that the average litter size at birth was also 2.9 cubs with a range of 1-4.

A little simple mathematics shows that if 65-70% (say 0.65 as a decimal fraction) of female badgers produce 2.9 young each year, then the average number of cubs over all females is around 2.9 x 0.65 = about two. Assuming an equal sex ratio in the population, this means that, for every male and female, two cubs are produced annually. In other words the population doubles. So 50% of badgers must die every year if we are not to be overwhelmed by them. Furthermore, as infant mortality is virtually certain to be higher than that for adults, it seems that an adult badger must have a rather better than even chance of surviving from one year to the next, and that a good proportion of badgers, once adult, must survive for several years.

Fortunately we have some fairly, if not entirely, reliable information on the ages to which Irish badgers live, again from carcasses of animals "removed" in attempts to limit the spread of bovine TB. This work was also carried out at University College Dublin, by Drs A.I. Il-Fituri and Tom Hayden.[164]

BADGER

Table 12 - The ages of 54 adult Irish badgers as determined from incremental lines in the cementum of the teeth. Note that these results were not verified against teeth from badgers of known age and are therefore provisional.

Age in years	Number of badgers
1	17
2	10
3	9
4	6
5	4
6	3
7	2
8	2
9	-
10	1

Mammalian teeth, including our own, are bound to the gum by a substance known as cementum, which forms a thin layer over the roots and is laid down throughout life. Environmental effects result in fluctuations in the rate and nature of the addition of cementum. In particular there are often distinct differences between winter and summer in many species of mammal, which result in a series of incremental lines rather like the growth rings in a tree trunk. These can often be used to indicate the age of the mammal. The lines must be viewed under the microscope on a fine slice of a tooth, which is usually first decalcified in acid, then sectioned and stained. Sadly, the technique is not without snags. It is not valid for all mammals, or even all those which live in latitudes where there are major changes in climate from summer to winter. Even for a given species, it may work with one population and not another. The reasons for this are unclear. To be certain one must have specimens of known age from the population studied, so as to authenticate the reliability of the procedure and, for wild mammals, such specimens are usually difficult to come by. As this technique has been shown to work well with some populations of the badger but not with others, specimens of known age are indispensable in obtaining dependable results. Unfortunately, none were available for the Irish badgers, so the results obtained by Il-Fituri and Hayden from a sample of 54 badgers, shown in Table 12, must remain provisional. It appears that a significant number of the animals, once adult, may survive for 6 or more years. One had apparently reached 10 years of age.

We now come to the knotty problem of badgers and tuberculosis in cattle, and knotty it is, for there are few more emotive issues surrounding Irish wildlife. At one extreme are some farmers who will be satisfied only with the total extermination of the badger, which is impractical to say the least. On the other are animal lovers who dismiss the evidence, because it is largely circumstantial, and look upon the beast as a harmless old gentleman who is unreasoningly, even maliciously, accused and persecuted for a crime of which he is entirely innocent. Nevertheless, both sides would probably agree that the recurrence of tuberculosis in cattle, which is caused by the micro-organism *Mycobaterium bovis* and can be transmitted to humans with potentially fatal results, cannot be ignored. Most reasonable people would also feel unable to discount the damage by bovine TB to the dairy and beef industries.

Tuberculosis in cattle is characterised by tubercules or lesions (damaged areas) in the lungs which are readily visible to the naked eye during a *post mortem*. Usually the disease is detected at an early stage during routine testing, when the animal is not yet infectious to others, and it is slaughtered. If not, the lesions may also spread to other parts of the body. Transmission from an infectious cow (i.e. one that can pass the disease on) normally originates in the lungs. The bacteria pass out in an aerosol - the fine mist of droplets carried in the breath - and this may be inhaled by other animals. In addition, if an infected cow swallows its sputum, bacteria could also be present in its faeces. The disease might then be picked up by any creature which noses around cattle dung.

The only significant wildlife reservoirs of bovine TB in the British Isles appear to be badgers and deer. The disease was first recorded in a badger in Britain in 1971[241] and in Ireland in 1973.[244] In badgers the lungs are probably also the primary site of infection, but the kidneys are usually infected too. Lesions on the skin are not unusual. In advanced cases with open lesions, bacteria pass to the outside from the kidneys via the urine, and also at the opposite end from the lungs, particularly during coughing and sneezing. If sputum is swallowed, the faeces will also be contaminated. It is important to realise that generally only a minority of badgers have bovine TB and it has been speculated that some individuals may be resistant.

Passage of the disease from one badger to another is easy to understand. In the close confines of the sett, the aerosol route must be greatly facilitated. Bites, received either in the settling of dominance within the clan or during territorial disputes, are another mode of badger-badger transfer and probably account for most of the lesions on the skin. Transmission from cattle to badgers is more problematical. Because of routine testing, cattle with TB tend to be identified and slaughtered before they reach an infectious stage; they cannot therefore infect badgers. Moreover there are areas in Great Britain where TB is present in badgers, but not in cows. Infection there must be badger to badger.[241] A cow which slipped through the net and became infectious might pass the disease to a badger via an aerosol if the two came close enough for long enough, but such an association is improbable. Badgers, except perhaps when diseased or ailing, tend to avoid cattle. A badger might also pick up the bacterium when hunting for beetles in the dung of an infectious cow.

It may come as something of a revelation to those who are unaware of it, but there is no direct proof of transmission of tuberculosis from badger to cow whatsoever. All of the evidence is circumstantial, but none the less powerful for that. The badger's champions dismiss the evidence simply because it is circumstantial, but who would doubt that George Joseph Smith was guilty of murder when all four of his brides happened to drown in their baths at a time when he was the only other person at hand? Sir Bernard Spilsbury's demonstration that accidental drowning was impossible in the baths in question was elegant but, many would say, superfluous. Transfer of TB directly from badger to cow is probably something which does not happen often, and to be lucky enough to observe an act suspected of leading to such transmission under natural conditions, and furthermore to check both cow and badger both before and after is asking a great deal.

An experiment in Great Britain in which infected badgers were housed with uninfected cattle demonstrated conclusively that the disease *can* be passed from badger to cow, but this does not reflect normal conditions in the field. Indeed, it took 6 months in such artificial circumstances for the transfer to take place. Whereas badgers tend to avoid cattle, they commonly forage for earthworms on pasture and leave behind urine and also puss from any TB abscesses on the skin. Latrines are also a potential source of infection. However, even this may be an unlikely route, for other work in Britain has shown that most cattle avoid grass contaminated with badger faeces and urine for some weeks, after which the bacteria may have been killed by light, dessication or other micro-organisms. Unfortunately, and perhaps surprisingly, it is not clear just how long the bacilli can survive under such conditions. A minority of cows are not so fastidious and these may be the ones which become infected. A further possible factor is that overstocking may lead to cattle grazing areas infected with badger urine that they would otherwise ignore. Dairy cows kept indoors and fed on concentrates may also be at risk. Dairy nuts are concocted so as to be exceedingly tasty to cows, and even if they are contaminated with badger urine there is sometimes no alternative but to eat them. Badgers apparently also find them delectable and in Britain have sometimes been observed to visit farm buildings to pilfer them.[241] Sleeman has described how a badger visited a milking parlour in Co. Cork, although he did not see what it did there.[321] Cattle might also pick up TB bacteria if they were to investigate a recently dead badger which was infected.

The circumstantial evidence for badger to cow transmission in Britain, although substantial, is outside the scope of this book. There is also compelling circumstantial evidence for Ireland. In the Republic from 1984 the veterinary scientists of ERAD (the Eradication of Animal Disease Board) have "removed" badgers from every county, mainly in areas with high levels of TB in cattle. Of a total of 3,909 *post mortem* examinations of carcasses up to 1989, which included a few from 1980-1984 found dead, 664 (17%) were infected.[82,85] It is therefore indisputable that a significant minority of badgers carry the disease. They are the only other mammals besides cattle to do so in any number over wide tracts of the Irish countryside. So when there is a TB breakdown in a herd, in other words when an individual tests positive for tuberculosis, and the cause cannot be attributed to anything else, such as TB in a neighbouring herd or the buying in of beasts, the badger naturally falls under suspicion. This grows stronger if badgers on the farm are killed and shown to have tuberculosis lesions.

The removal of badgers by ERAD was most intense in seven districts in Cos Cork, Offaly, Galway and Longford. It began in 1984-86 and, when the results were assessed in 1988, there had been a marked fall in the number of reactor cattle over the culling period in all seven.[248] Such results are suggestive but not conclusive. It is well known that the frequency of TB in cattle fluctuates from year to year - a very important point - and it might simply have been falling in these districts at the time. A valid experiment would require that the parameters be settled beforehand, with an experimental area or areas, where badgers are culled, and a control area or areas where they are not; but at the time such an approach would have been unrealistic. The vets had to start with little prior knowledge, and with finite resources, on what was the new problem of bovine TB in badgers in the Republic. They were bound to begin in a practical way, gathering information as they went. Starting an expensive experiment along the lines that I have outlined with limited previous knowledge might not have produced any valid conclusions.

With the basic information gained in the 1980s, ERAD initiated the East Offaly Badger Research Project in 1989 in an area of high incidence of bovine TB in eastern Co. Offaly, but including small parts of the contiguous counties of Meath, Kildare and Laois.[85,91] The vets divided the region into a project area in the centre, a buffer zone 1.5 km wide around this, the two totalling some 600 km[2], and a bigger outer control zone. The project area contained about a thousand herds of cattle; the control zone about 3,000, with on average 50 animals per herd. From 1989 to 1994 a total of 1,735 badgers were systematically removed from the project area and buffer zone. None were taken from the control zone. During the period 1988 to 1994 there was a total of over two million TB tests on cattle in the project and control areas. Whereas the reactor animals per thousand tests dropped from 3.91 in 1988 to 0.79 in 1994 in the project area - a fall of 80% - that in the control zone fell from 3.39 to 2.11 - a fall of only 38%. Bearing in mind the fluctuation in the incidence of the disease, these results are impressive.[84] The difference continued up to 1997, the latest year with results to hand.[83] Work on badgers and bovine TB in the Republic continues with similar removal experiments in progress in the four counties of Cork, Donegal, Kilkenny and Monaghan.[147]

O'Corry-Crowe used the bait marking technique to study badgers living in an area of some 16 km[2] within the project area in east Offaly. He first worked out the territories in this way in 1989 before there had been any culling, and then again in 1990 when about half of the badgers had been removed. The farms in the district were highly fragmented, each consisting of up to eight individual parcels of land; badger territories were on average almost three times larger than the average farm. In practice a badger territory overlapped between six and 14 farms. If, as seems highly probable, badgers were a source of TB to the cattle there, then the land division among the farms was clearly most conducive to the spread of the disease, which infected up to nearly 30%

of the herds during the period of study. Removal of 50% of the badgers, however, did not apparently reduce the potential for badger to cow transmission. Rather, it appeared to enhance it. Presumably because the social structure of the clans in the district was upset at least temporarily, the average area of the territories increased and with it the numbers of farms that the clans visited. This result seems somewhat at odds with the overall findings of the East Offaly Badger Research Project. However, the situation is complex and it may mean that partial removal of badgers from an area may sometimes be worse than no removal at all.[249]

Whilst I have no direct experience in this field, I might perhaps be forgiven for a slight reservation on the results of the East Offaly Badger Project. One of the conditions of this experiment was that badgers were not killed in the control zone. Sources in the Republic, which I believe to be reliable, inform me that farmers illegally kill many badgers, often by snaring, and also stop up their setts. During the Badger and Habitat Survey, Smal found that 15% of all setts had suffered disturbance, mainly by digging or blocking with branches, stones or even slurry, and that over 90% of this was unlicensed, and therefore illegal. It seems highly probable, therefore, that badgers were also killed and their setts destroyed within the control zone in east Offaly. If such low level attempts at control were to aggravate rather than restrict the spread of bovine TB, the difference between the percentage of reactor cattle in the project area and the control zone might have been less if the badgers in the latter had been left alone. Such reservations might also apply to the four county trials in progress mentioned above.

Bovine tuberculosis is less of a problem in Northern Ireland than in the Republic and early research seemed to indicate that transmission from badgers was of minor importance. A recent case study, directed by Dr Owen Denny of the Veterinary Service, has caused some revision of this point of view.[77] In the study 215 case dairy herds, those which had had one or more positive TB reactions in home-bred cattle in the previous 2 years, and 212 without, were selected randomly. Herds in which purchased cattle were found to be reactors were specifically excluded. Data on the Department of Agriculture Animal Health Computer, together with information collected in person by trained investigating officers, were used to examine a range of possible risk factors. Where there was any doubt about a farmer's response, the investigating officer was instructed to verify the details himself by inspection. Statistical analysis highlighted two main associations with TB breakdowns: the presence of badgers on a farm and contiguous neighbours who had had confirmed TB breakdowns. While an association between badgers on a farm and TB in the cattle has been demonstrated, this falls short of establishing that badgers transmit the disease to cattle. Association is not proof of causation. However, this work constitutes strong circumstantial evidence.

If it were accepted that badgers are a significant source of infection for cattle then, apart altogether from considerations of wildlife conservation, extermination of badgers over large areas of Ireland is simply not a practical solution. It would be extraordinarily expensive and would have to be continued more or less indefinitely. As badgers can only be killed under license, and Ireland's international obligations make any change in the relevant legislation governing this unlikely, any culling would have to be undertaken by staff employed by the state and not by farmers. Such staff would have to be paid. An inordinately expensive and apparently largely ineffectual programme of culling in south-west England, the only area of Great Britain where badger related bovine TB is reckoned a major problem, should serve as a warning to Ireland.

One way of solving the problem of TB in badgers might be to vaccinate them, and, among others, Paddy Sleeman is currently researching the feasibility of such an approach. Catching and vaccinating the animals wholesale would, of course, be quite impractical, but a vaccine might be administered in a bait. In order to determine the effectiveness against TB of baiting the animals in the field, it is important to be able to say whether any badger "removed" and inspected for TB has, in fact, eaten such a bait. This requires the inclusion of a suitable chemical marker in the bait which leaves a permanent and detectable trace in the animal's body. Experiments so far have shown not only that the markers work but that most badgers will consume the baits, which are, incidentally, coated in chocolate.

The practicality of vaccinating *all* Irish badgers in this way remains open to question. Whereas it has obvious merits, perhaps the development of an effective vaccine for cattle would be more appropriate and cost-effective. Unfortunately any vaccine for cattle would upset the current TB testing programme: for instance because a cow so vaccinated would develop antibodies to TB and thus test positive for it.

Some people conveniently forget that cattle may catch TB from other cattle, or even from the spreading of slurry, and that improved husbandry will reduce prevalence, particularly the immediate testing of any cattle suspected of infection. Double fences between adjoining farms would probably prevent transmission from one herd to another, but how many double fences does one see in Ireland? Whereas it must be acknowledged that the badger does play some part in transmission of the disease to cows, undoubtedly appropriate cattle-husbandry is as much of an issue as the badger.

Even before the discovery of its infection with bovine TB, the badger had an inexplicably bad press here. When working on foxes in Northern Ireland in the 1960s, I was astonished to discover the degree of such prejudice. More than one country person whose livestock had been depleted by foxes actually expressed gratitude that the visitor had not been a badger! Many of the badger's supposed crimes in the matters of poultry and lambs can be laid at the fox's door. Such ill will and prejudice finds its most horrifying expression in the illegal pastime of badger baiting. A flavour of the modern version is provided in extracts of a report of a successful prosecution for it in Cork City in 1985 in which the judge imposed a total fine of £250.[12] The defendant was not present in court but a statement from him was read out in which he said that he had been involved in it all of his life.

> The sport is the dog runs towards a badger in a set time. If he fails the time, he is lifted. He is then to stay quiet. The slightest whimper and he is lifted. [Then] he is to stay with the badger for six minutes and that's it. If the dog is doing well he is given so many points. If he is being torn apart by the badger he is lifted out. The statement described the special hole dug in the ground for the baiting, which can be opened to release the dog if he is being torn....Four dogs won prizes, having stayed six minutes.

In the 1960s I knew experienced fox-shooters who preferred to take the terriers with which they raised the foxes away from burrows if a badger was even suspected of being in residence, preferring to leave rather than risk a confrontation between their dogs and brock. I particularly recall one large terrier, appallingly mutilated about the head, which was said to have an incorrigible penchant for badgers.

Baiting is not the only indignity the animal has had to suffer in the past, as a report from *The Field* of 1892 explains.[272]

The badger in this part of the Co. Cork is not rare....Farm hands occasionally capture unwary ones, and offer them for sale as pets, or to test the mettle of the national terrier, or to be converted into bacon. A badger's ham is often seen suspended from the rafter of a farmer's kitchen....

In fact badgers were being eaten in Ireland over a thousand years before this date as there are references to salted badger meat, or perhaps fat, in early Irish law texts.[174] In Rutty's *Natural History* of Co. Dublin (1772) there is another reference to the beasts in a culinary setting.[292]

> The Badger, *Gray, Brock* or *Bawson*
> Besides the medicinal use of its Fat, the flesh, when roasted, is good food, like pig's flesh, and makes a good Ham; and the Skin is tanned for Breeches, waistcoats, &c. and is sometimes dressed with the hair of furriers.
> The Hair makes pencils for Painters.

And finally, a quotation from a curious book of 1917, *The Wild Foods of Great Britain,* by L. Cameron.

> The hams, when cured by smoking - over a fire of birchwood for preference - after the manner used in curing bacon, are a decided delicacy, and may then be cooked and eaten either hot or cold. They are commonly treated in Germany, and to a less extent in Ireland.

On at least one occasion in the intermittent war in Ireland between mankind and the badger, the latter has seized the initiative. This account, which is once again by G.H. Pentland, and therefore reliable, was of an incident concerning four ladies out for an afternoon walk in Co. Clare in 1901.[273]

> A large badger ran past them, actually touching the feet of one of them, and entered a hole in the bank. Presently it showed itself, but disappeared when one of the ladies clapped her hands, and the ladies walked on. The Badger emerged again and followed them. A Retriever which was with them attacked it, and a fight ensued, in which the Badger seems to

have got the upper hand, for the dog ran away and took no further part in the proceedings, and the Badger again went to ground. Three of the ladies then went to tell the gamekeeper what had occurred, and the fourth remained to watch the earth. Luckily for her at this juncture she picked up a stick, for the Badger came out again, passed her, and gave chase to her three companions, and failing to catch them, returned and attacked her, in a determined manner. She got behind a tree and dodged it as best she could, striking at its head as it followed her round the tree. Meanwhile the shrieks of her companions had brought assistance. The coachman was first on the scene, followed by the keeper and gardener, but before they arrived, the plucky lady had stunned the Badger, and was no longer in danger.

I hasten to add that this is the *only* reliable instance of a badger attacking humans in Ireland that I have been able to find and was no doubt provoked by the dog. Badgers, unmolested, are harmless.

Chapter 4. Spraint - the Otter's Calling Card

> Observe, that where otters are plenty, they have certain treating places on little islands in the river, or some little point of land lying contiguous to the water where he can come easily unto it out of water, and in those places every otter that comes up or down the river rubs, tumbles and scrapes the ground, and makes his spraints....
> *The Experienc'd Huntsman.* Arthur Stringer. (Published in Belfast in 1714.)

Although more robust than the stoat and mink, the otter still possesses the short legs and elongate body typical of the mustelids. Indeed it evokes an unmistakable impression of streamlining, especially in the water, where this is most beneficial. The animal is all sinuous curves. The ear flaps scarcely break its smooth outline and even the tail tapers gradually from base to tip, quite unlike that of the mink. When wet, the thick fur adds a further element of sleekness, or at least until the animal shakes itself, when the coat takes on a distinctly spiky appearance. The hair of most mammals consists of a short underfur and longer guard hairs. Because the latter extend above the underfur, and overlie it, they give the coat its apparent colour. Although the otter's guard hairs, which are long and stiff and possess an almost metallic lustre, are wetted, this is never completely true of the much denser underfur. This is coated with oil from skin glands, traps a layer of air next to the skin and thus improves insulation against heat loss when the animal is submerged. The otter also has webbed feet.

Otters are surprisingly big animals. I dissected the carcasses of 29 adults from Co. Galway, mostly trapped by locals for their pelts in the early 1970s before the species was legally protected, and which therefore arrived with me minus their skins. It is generally considered that a skinned carcass weighs about 80% of an intact one,[45] so the full body weights would have ranged from 6.9-10.8 kg for males and 4.8-7.4 kg for females. The length of the head and body of the largest male was 749 mm and the tail 481 mm.[100] Much greater weights have been quoted for Irish specimens, including one of 15.9 kg (35 pounds) from the River Shannon.[8] However, record specimens of any animal, unless verified, are always open to suspicion. An otter killed on the

Sixmilewater at Castle Upton, Templepatrick, Co. Antrim, a century ago was said to be 60 inches (1.5 m) in length.[215]

The otter may be found almost anywhere in Ireland where there is water with fish in it, both inland and on the coast. It is predominantly nocturnal and perhaps most often spotted by anglers inland at dusk, when they are waiting patiently for the fish to bite. However, it is also seen fairly often in full daylight, especially, but by no means exclusively, in unfrequented districts.

On the shores of Shetland, where the animals are regularly abroad both day and night, their hunting behaviour has been observed in some detail. Despite the beautiful illustrations in Victorian and Edwardian natural history books of an otter chasing a shoal of fish, the animal prefers to concentrate on prey which it does not have to pursue: slower swimming fish or those which lurk on the bottom. In Shetland otters usually fish close to the shore in water less than 8 m deep. Dives average 23 seconds and rarely last longer than 50 seconds.[187] So there is comparatively little time for pursuit. The otter has therefore usually to work hard to catch its prey under water and its life is no more leisurely than that of any other Irish mammal. Nevertheless, whereas the mink commonly operates both on land and in water, it is in the latter element that the otter forages almost exclusively, and it is more fully adapted to it. The reader may recall from Chapter 1 that the mink is much less well equipped as an angler, diving on average for only 10 seconds and at most for only half a minute.[88]

The ancient pastime of hunting spawned its own vocabulary, most of it now redundant; hunting itself was known as *venery*. One term from the language of venery still extant is *spraint,* meaning otter faeces, and its use can be traced back at least to a manuscript of 1410. Its survival owes much to the significance to the otter of its droppings - for they are not something merely to be disposed of as quickly as possible - and thus to the naturalists who study the animal, as this chapter will make plain. Spraints are vastly more useful in detecting the presence of an otter than any other sign. Whereas tracks sometimes provide an indication of the presence of the beast, the naturalist looking for infallible signs is much more likely to encounter spraint first.

In view of the frequency with which wild otters must produce spraints, I sometimes ponder the popular notion of the animals as charming pets. A dog will generally evacuate its bowels once or twice a day, something which owners accept as a minor inconvenience and necessary evil. Dogs might not be so popular if this were necessary much more often.

When fresh, spraints are slimy and usually black, although those composed of the remains of freshwater crayfish are a reddish brown. As they age, they dry, become friable and fade to a pale grey. They do not last long: one study in Scotland showed that 50% of them had disintegrated within a fortnight.[170] A rise in the water level in a river or lake will wash them away almost immediately. They are exceedingly variable in shape, ranging in size from little more than a smear a few millimetres wide to cylindrical monsters, 10 cm long by 1.5 cm in diameter. The biggest ones that I have seen all contained crayfish. Two features are useful in recognising spraints. First, they usually contain fish bones, which are clearly visible and give the spraints a jagged outline: some examples are illustrated in Fig. 11. Second, the smell is absolutely diagnostic: a pleasant, sweet, musky, fishy odour which is quite unmistakable. Once inhaled, it is never forgotten. Unless the spraint is very fresh, it has to be held within a centimetre or two of the nose in order to sniff it. The fragrance is derived from the animal's scent glands, which open at the anus and into the rectum. Although the droppings of otter and mink may be voided at little distance from one another, it is nearly impossible to confuse the two. Besides having a more or less standard shape (Chapter 1), mink droppings have a foul odour. Sometimes a spraint may be found which consists entirely of a yellow jelly, without any prey remains at all. This is often said to be pure secretion from the anal glands, but it may simply be excessive mucous exuded from the walls of the gut. Mucous has to be produced in some quantity there to lubricate the passage of the faeces which, being mostly composed of fish bones, possess all manner of sharp points and rough edges.

Spraints are deposited at conspicuous places near the water, often on minor topographical features. After a little experience, it is easy to spot regular sprainting places, which are called *seats*. With practise, one almost begins to think like an otter and at times one can predict seats with a degree of accuracy that is almost uncanny. Collecting spraints is also satisfying: a walk along a river bank on a fine day to gather them is to me unquestionably the most agreeable sort of routine fieldwork.

Inland the commonest type of seat is a small, grassy mound beside a watercourse, which may be the direct result of manuring by otters over the years. This also results in the grass on such mounds often being a bright green, which contrasts with the surrounding vegetation. Such seats can often be sighted from a considerable distance. Sometimes an otter will scrape grass and other vegetation together into a heap and leave a dropping on top. The reason for such behaviour is obscure but it may be because the animal wishes to spraint somewhere where there is no regular seat or suitable alternative feature at hand. The uprooted vegetation withers and the patch quickly turns brown.

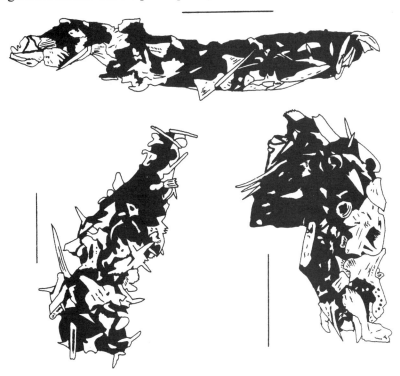

Fig. 11 Three spraints showing the typical irregular outline due to fragments of fish bones. The line scale given with each is 1 cm.

A large rock, either beside a river or projecting from its surface, often forms a seat, especially if there are no others near it. Such rocks are usually flat and it seems that the animal will ignore them unless there is a comfortable area of more or less level surface on which to squat. Logs and tree stumps are similarly used, as are hollows under tree roots eroded by the water. Otters often pay special attention to rocks and ledges under bridges, which are probably the best place for the novice to begin looking for spraint. In Co. Mayo I was once shown spraints beneath a platform supporting a religious statue beside a river. Evidently it served the same purpose as a bridge for an otter.

Otters commonly defecate where a drain or brook enters a river or at the confluence of two watercourses. Up until 1999 at least, there was a well-marked seat at the southern tip of Jordan's Island, which lies in the River Corrib north of Galway City. The river flows around both sides of the island and the seat is situated just above where the two channels rejoin. It can be viewed from the footpath on the Quincentennial Bridge, which spans the river immediately to the south. Although some distance away, the seat is obvious most of the time, for it is usually a bright shade of green.

Where an otter uses an overland route regularly, perhaps between two lakes or cutting across a meander in a river, there is almost invariably a distinct path through the vegetation and this may be marked with spraints at either end and sometimes along its length.

Perhaps the most interesting places where I have gathered spraint were the limestone islets in Lough Leane at Killarney - Crow, Otter and Swallow Islands - each only a few hundred square metres in area. I found droppings on exposed rocks there but most had been left in the fissures and caves which seam the limestone. This involved some crawling within confined spaces and minor hazards in landing from a rowboat, kindly made available, complete with outboard motor and pilot, by the Supervisor of Killarney National Park.[115]

On the coast, seats are often on grassy mounds at the limit of grass cover at the edge of the shore. Big rocks on the upper shore are sometimes favoured, but in my experience it is rare to find spraint on them unless they are isolated, or particularly large and flat, or both. Otters often defecate on piers, particularly on steps leading down into the water. On the other hand a pier with steep sides and no steps is simply not worth inspecting, because the animals cannot scale it. Grassy tidal islets of a few hundred square metres are also a good place to look. I have even found seats on patches of sea pinks, and Pat Smiddy tells me that he has had the same experience.

If there are otters about on the coast, then one of the most likely places to find spraint is around any fresh water, for instance a stream draining into the sea, or a pool above high tide mark. Fresh water is important to coastal-dwelling otters for bathing. Experimental work on captive otters in Scotland showed that unless they were able to rinse the sea water from their fur every day or so, they were reluctant to enter salt water at all. The salt from the sea dries in the pelt and presumably interferes with its thermal insulation. It probably prevents the oil from the skin glands from coating the underfur properly, so that a layer of air can no longer be retained next to the skin.[187]

The home range of an otter, because it lies either along a watercourse, on the edge of a lake or on a stretch of coast, is linear, like that of a mink, and therefore expensive in terms of energy. However, because an otter is much bigger than a mink, it has relatively less surface area. In addition the body is not so thin and the diet, which is is mainly fish, which are rich in oil, is also probably more nutritious.

Otters can defecate in the water perfectly well and often do so,[187] but obviously often take trouble to deposit their droppings at special places on dry land. One explanation is that they might be being used to mark out a territory.

Extensive, long-term observations in Shetland on coastal otters showed that individual otters, or bitch otters at any rate, do not hold separate territories. Several adult females share a territory, but defend it jointly against other groups of females. The arrangement is similar to the badger clan, although otters are not social in the sense of a group all sharing the same underground home. The reason for the joint territory may be that certain places in it are only good to fish under certain conditions, just as certain fields provide an abundance of earthworms to a badger clan on some nights but not others. Perhaps one female otter on her own could not hold a big enough territory to provide suitable feeding places all of the time.[187] While the females in a particular group on Shetland did not live together, and each tended to specialise on a particular length of shore, there was considerable overlap in their home ranges and they appeared to be quite tolerant of each other. Work in mainland Scotland points to a similar social structure on rivers and lakes. Male otters operate over much greater distances. Their ranges overlapped those of several females, which they tolerated, but they seemed to be much less forbearing of each other.[187] The overall pattern, therefore, is of several otters, all probably known to one another, sharing the same length of watercourse or shore.

If spraints were being deposited on land to mark out a joint territory, one would expect to find them mainly at its edges, as with a badger clan, and therefore mostly in concentrations at widely separated points on river bank, lake edge or sea shore. In fact, they are pretty homogeneously spread and so this is unlikely to be their main function.

The primary significance of fresh spraint at a seat is more likely to be a short-term message to the other members of the group that the site has just been fished. This would be of mutual advantage. If one animal eats all or most of the fish that are easy to catch at a particular spot, then fresh spraint will signal this to the next otter, which will not waste time in foraging there. Shortage of suitable fish is likely to be temporary, as other fish move in within a short time. Spraint beside fresh water near the sea is more difficult to explain, but beside a freshwater pool *might* indicate that an otter has washed in it recently and that it was therefore less suitable for bathing for a time.[187]

Many fish living on the sea shore or immediately below low tide mark tend to migrate into deeper water in winter, and since otters operate in water in the shallows close to shore, prey for coastal-dwelling otters is likely to be scarcer in the colder months of the year. So signalling that a particular spot has been fished should therefore become more important. If such reasoning is sound, then there should be more spraints at coastal seats in winter than in summer. This is exactly what has been found in the two published Irish studies in which spraints were gathered from a stretch of coast throughout the year.[182, 239]

Spraint serves two functions for naturalists. It indicates presence of otters and the remains in it yield valuable information on the otter's diet. Sometimes it is also possible to make deductions on the behaviour of the animals. Spraint is easy to find; the animals that produce it are not. It is hardly surprising, therefore, that most of the recent information on Irish otters has been derived from spraint and deals with the distribution and diet of the otter.

The only survey of otter distribution in Ireland that can be regarded as in any way comprehensive was undertaken by Mr and Mrs Peter Chapman from January 1980 to February 1981 for the London-based Vincent Wildlife Trust.[46] With admirable logic, they chose Athlone, which is almost in the centre of Ireland, as a base for their operations. They covered half of Ireland by investigating every other 50 km square of the National Grid. This produced a map rather like a chess board: presence or absence was examined in the "black" squares, but not in the "white". Within each 50 km square they visited as many of the 10 km squares as they could and checked for signs of otters along watercourses, lakes and coast. They simply walked along 600 m of bank or shoreline at each place (on a river this could be both banks for 300 m) and looked for spraints and tracks. Whereas they noted well-worn paths, apparent rolling spots, fish remains and sightings reported by locals, they did not accept them alone as convincing evidence. If there were few potential seats, the Chapmans sometimes extended their searches to 800 m of bank. Very often they started searching at a bridge, which not only provided convenient access to a river, but also a likely place for a seat. Some rivers where they found no signs had recently been in spate and so faeces nearby could have been washed away. Such rivers they usually visited again at a later date.

Of the 2,373 sites that they checked, 91.7% yielded signs of otter. The Chapmans found rather less signs on high ground, although they did come across some at altitudes up to 549 m (1,800 feet - in Co. Cork). The animals were also scarce or absent in polluted waters. In this context it is worth noting that they found evidence of otters at every place where they looked on the west coast of Ireland, but at only half of the places on the east coast. In 1982-83 I collected spraints with a student down the western shores and found that if a coast looked promising, then it was virtually certain that we would find spraint somewhere. As the human population is concentrated in the east, the fewer signs on the shores there could be put down to greater human disturbance. However, this is probably less of a problem for otters than has been previously thought, and the Chapmans still found signs within 4 km and 2.5 km respectively of the centres of Dublin and Cork; Belfast was on a "white" square and was not surveyed. They also discovered spraint at the edge of a car park in the middle of the city of Limerick and in the centre of Galway City. So the relative scarcity of signs on the east coast was most likely due to pollution.

Between January 1990 and November 1991, Miss Ruth Lunnon, of Trinity College Dublin, with the help of conservation rangers and volunteers from The Irish Wildlife Federation and Ulster Wildlife Trust, resurveyed a subsample of 268 of the Chapman's sites, widely distributed throughout the country. Signs were found at 89.6% of these, a figure which does not differ significantly from the Chapmans'. This resurvey too showed a strong negative correlation between the presence of otters and pollution. Nevertheless, Lunnon found signs of otter on the River Liffey in the centre of Dublin.[194] In addition, as recently as 1994, she also recovered spraint at Heuston Bridge and where the River Dodder meets the Liffey at Ringsend.[237]

Besides these two country-wide surveys, there have been some smaller-scale ones, but with much more effort per unit area. From May 1988 to May 1990, Dr Liam O'Sullivan, of University College Cork, surveyed the Munster Blackwater, together with its major tributaries. The main channel of this river extends for 168 km and takes in parts of

Cos Cork, Kerry, Limerick, Tipperary and Waterford. This has been the only attempt at a comprehensive survey for traces of otter on the catchment of a major river in Ireland. O'Sullivan used the same techniques as the Chapmans at a total of 195 places. Of these he found spraints, tracks or both at 97%. The Chapmans had included part of the Blackwater in their work and noted otters at 83% of sites in this, a figure somewhat smaller than their overall national average. O'Sullivan, however, recorded signs at 95% of sites in the area worked by the Chapmans.[263,264] The explanation of the apparent rise is unlikely to have been a genuine increase in the population of otters during the 1990s. It probably merely reflects the relative intensity of O'Sullivan's technique, concentrated as it was on a relatively small area. If this is so, then the Chapmans' overall figure of 91.7% of 10 km grid squares positive for the country as a whole probably *underestimated* their presence.

Pat Smiddy investigated presence of otters in his home territory of east Cork and west Waterford between 1982 and 1992, particularly on the Rivers Bride, Womanagh and a small part of the Munster Blackwater, and also along the coast. He too found widespread and abundant traces of them, even around Cork Harbour to within 2 km of the city centre, sometimes on jettys adjoining heavy industrial installations.[332]

These studies all show that the otter is common in Ireland, although this must be a relative statement. As the beast is aquatic, it can only occur where there is water. It is also a medium-sized carnivore living in a linear range, which means that, even in the best of circumstances, it will be thin on the ground. On Shetland and in mainland Scotland, the only places in the British Isles where scientists have estimated numbers in populations regarded as undepleted, there was an average of only one otter per 15 km of stream and slightly less than one otter per kilometre of coast.[187]

Pollution evidently influences presence locally in Ireland and high levels of pollutants have been measured in a few Irish otters, with potentially sub-lethal effects including impairment of reproduction. Contaminants will also strike at the animals indirectly - if there are no fish, there will be no otters. Arterial drainage adversely affects fish populations in Ireland, which locally may take some years to recover, and in turn the populations of otters.[266] Otters require fish above all things to survive, for it is certain that otters in Ireland feed overwhelmingly on fish.

OTTER

In the summer of 1993, Dr Denis Tangney did a small spraint survey in Connemara National Park, at Letterfrack, Co. Galway. Yet again, this showed the animals to be widespread, both on the coast and in streams and lakes.[344] Readers who have visited the district will recall that the most beautiful feature of this highly scenic area is Kylemore Abbey, a castellated mansion, at present a girls' boarding school, which stands beside Poulacappul Lough and against a backdrop of well wooded cliff infested with rhododendron. The Abbey attracts tourists in droves, particularly in summer. Despite this, Tangney found spraint all around Poulacappul Lough except on the lawn immediately below the Abbey. Evidently the animals were not bothered by human disturbance but drew the line at sprainting on a place which is walked over by scores, if not hundreds of people every day. I found remains of saltwater fish in the droppings that Tangney collected from the sea shore, but they also occurred in those he found on the upper reaches of the streams, over 3 km from the sea. This shows that at least some of the otters were foraging both on the sea shore, where the fish were caught, and along the full lengths of some of the streams, where the faeces were voided.

Before turning to the diet of Irish otters, it is worth considering a further small-scale survey, on Roundstone Bog, Co. Galway, which produced more in the way of information than the mere presence or absence of the animals.[188] This is an area of unspoiled bog in Connemara strewn with numerous lakelets. In 1989 Miss Mary Kyne, a former student of mine and then a school teacher, decided to look for spraint there during her holidays in June and July. She searched beside streams, along lake shores and on paths throughout an area of about 26 km², which contained some 80 lakelets, varying in size from less than 0.5 ha right up to Lough Fadda and Lough Boland, both of which are about 0.5 km². The summer was exceptionally dry, which made for relatively easy walking over the ground, much of which was usually waterlogged.

Altogether she found 98 seats and 81% of these lay either on or close beside paths running between the lakes, often on small mounds or on rocks. The paths were well beaten, but narrow, so that the surrounding ground vegetation overhung them. They were therefore easily distinguished from sheep tracks and she concluded that they had been made by the otters themselves. Indeed, because of their distinctive appearance, a few were recognised as otter runs even if, at the time she inspected them, there were no spraints on them. Some of the seats lay along their length, others at their ends, within a metre of a lakelet.

Besides the spraints on the runs, she found 5% of them beside streams and 14% elsewhere on lake shores.

It will be remembered that the Chapmans considered a check on 600 m of bank or shore to be adequate to detect the presence of otters. If such criteria had been used on Roundstone Bog, the animals might not have been recorded there at all! Once again, this suggests that the Chapmans' figures for presence of otters in Ireland are conservative. Spraints were thin on the ground on Roundstone Bog and therefore otters also. There were few places where the animals could have lain up securely, although this should have been possible on some of the islands in the lakes. However, shortage of such sites is unlikely to have limited numbers. All but three of the 180 spraints collected from the seats contained only eel remains. The otters were therefore virtually exclusively dependent on eels, and there may have been few enough of these. So food may have been the factor which limited numbers.

There was never more than one path between any two lakes and generally it ran in a straight line. It also avoided rocky outcrops and patches of thicker vegetation and tended to keep to the lowest ground. The runs therefore represented the shortest and easiest routes between the lakes. One cannot avoid inferring that the runs were chosen for maximum efficiency in travelling between the lakelets and therefore for exploiting the area most effectively.

Spraints for analysis are best collected in specimen tubes or envelopes and then dried, after which they can be stored almost indefinitely. To analyse a spraint, some zoologists simply break it apart and then try to identify the hard parts. An easier procedure is to place it in a specimen tube, if it is not already in one, cover it with hot water, add a pinch of the effervescent denture cleaner Steradent, and leave it overnight. Next morning all of the hard parts will have separated from the gungy matrix and it merely remains to wash them in a kitchen sieve and put them on a sheet of paper to dry. Everything identifiable is then squeaky clean and bright and can be examined under a low power microscope.

In the course of the past 20 years zoologists have analysed almost 10,000 spraints in Ireland. So we now know a great deal about the diet of the otter here. Several of my students have cut their teeth in mammal research by collecting and analysing spraint. It is cheap, instructive and interesting. The largest single piece of work was undertaken by Mary Kyne, who analysed the spraints which Chris Smal had obtained incidental to amassing mink scats from his study areas around

Mullingar (Chapter 1). She also included some others that she gathered herself from the River Brosna, where it flows out of Lough Ennell. Altogether she analysed 2,349 spraints, mostly collected every month from each locality, from several different years.[189] The results are summarised in Table 13 together with those from some other studies of material taken from beside fresh water in other parts of Ireland, including the catchment of the Munster Blackwater River,[265] Lough Leane and Muckross Lake, Killarney,[115] Lough Feeagh, Co. Mayo,[146] and from the catchment of the Clare River in Co. Galway in 1981-82[207] and again in 1991-92.[32]

A comparison of the diets of otter and mink around Mullingar (Table 1) shows that otters caught far fewer mammals and birds, but almost twice as much fish and frogs as the mink did. The otters ate fewer crayfish and the figure in Table 13 is also probably overestimated relative to mink because, whereas the mink culls off the carapace, the otter apparently does not. So there would presumably be more shell per crayfish in the droppings of the otter. As a large proportion of a crayfish is indigestible, the otters probably take fish where there is a choice, whereas mink, with their poorer diving ability, tend to select crayfish because they are easier to catch.

Because the otter is clearly a much more proficient fisherman than the mink, which forages as much on dry land as in the water and catches more crayfish when they are available, potential competition between these two mammals is minimised. It is apparent from the results in Table 13 that, in districts where there are no crayfish, fish are often even more important to otters than around Mullingar, and frequently make up over two-thirds of the diet. At some places, as exemplified by Roundstone Bog, virtually the whole of the diet is fish. In any competition between the otter and mink, it is the latter which probably suffers. Not only is it less skilled at hunting under water, but it is much smaller and in any hostile encounters is bound to come off worst. Neither is there any question of mink "ganging up" on an otter. It will be clear from Chapter 1 that adult mink do not readily forge alliances with one another. An otter is also likely to be the victor even in the only conceivable sort of encounter where there may be more than one mink: a mother with her kits.

Unfortunately we have only a general idea of the prey species that were available to otters in all of the places where spraints were collected near fresh water in Ireland. We have, for instance, no precise figures on the relative numbers of the various species of fish, as might

Table 13 - Percentage frequency of food categories found in spraints in some investigations of the diet of otters in freshwater in Ireland. + = present.

Location -	Mullingar	Blackwater Catchment	Killarney	Lough Feeagh	Clare Catchment	
Habitat -	Lake & river	River	Lake	Lake	River	
Season -	All year	May-Sep	June	Aug-Jan	Jul-Dec 1981-2	All year 1991-2
Total mammal	-	2.0	-	-	0.4	0.2
Rabbit or hare	-	0.3	-	-	0.1	0.2
Brown rat	-	0.4	-	-	0.3	-
Bank vole	-	0.6	-	-	-	-
Field mouse	-	0.4	-	-	-	-
Unidentified	-	0.3	-	-	-	+
Total bird	1.9	9.8	0.8	1.0	3.4	1.4
Rails	1.1	4.5	-	0.2	1.0	0.1
Ducks/geese	0.2	0.3	-	-	0.6	0.6
Passerines	-	0.7	-	-	-	-
Waders/gulls	-	-	-	-	0.4	-
Pigeons	-	0.4	-	-	-	-
Unidentified bird	0.6	3.9	0.8	0.8	1.4	0.7
Total fish	48.1	68.7	98.1	84.3	45.0	57.8
Perch	28.4	-	3.2	-	11.6	9.3
Pike	4.8	-	-	-	3.6	6.9
Trout family	1.6	25.0	18.6	39.8	5.7	4.9
Carp family	7.4	7.5	-	-	-	17.1
Eel	4.9	26.9	56.5	33.3	7.3	2.0
Stickleback	-	1.6	-	5.1	9.7	9.3
Shad	-	-	19.8	-	-	
Other fish	-	7.7	-	1.6	-	6.0
Unidentified fish	1.0	-	-	4.5	7.1	2.3
Frog	15.0	0.3	1.2	3.7	9.0	9.3
Crayfish	29.1	-	-	-	31.3	23.6
Other invert.	5.5	16.6	-	11.0	11.0	4.2
Grooming fur	0.3	-	-	-	-	+
Unidentified/other	0.3	2.6	-	-	-	3.5
Total spraints	2349	303	171	266	781	1790

have been obtained by electrofishing. Nevertheless, it does appear that the otter is an opportunist, taking whatever is readily available but selecting the more nutritious items when there is a choice. This will be obvious from what follows.

It is apparent from Table 13 that, generally speaking, when an otter preys upon birds, they are likely to be water birds of some sort. Of course the mink concentrates on these too, but it takes more non-aquatic birds, such as passerines (Table 1). Where waterfowl are abundant, the otter will exploit them. On Lough Ennell in the Mullingar district, where it will be recalled there were few crayfish but abundant aquatic birds (Chapter 1), the latter made up 10% of the diet.[189]

Where crayfish are available, Irish otters eat more of them in summer than in winter and *vice versa* for fish. In winter crayfish tend to spend most of their time under boulders or in crevices where they might be less assessable to otters, but this is unlikely to be the whole story. Both kinds of prey are cold blooded, so their body temperatures are directly dependent on that of the surrounding water. Generally speaking the metabolic rates, and therefore the potential activity, of cold blooded animals increase by a factor of two or three with every rise of 10°C in ambient temperature. In winter, therefore, fish and crayfish are, perforce, rather less active and, in the open at any rate, both would be easier to catch. Given the choice, however, an otter prefers a succulent fish to a crustacean, so much of which is indigestible shell. Feeding experiments in Sweden with captive otters have verified that when both preys are on offer, the otter chooses fish.[90]

Although the otter eats more fish overall in winter, it takes much fewer eels, because these are then torpid and may bury themselves in mud, where they are difficult to find. Bear in mind that the time available to an otter to locate and seize prey during a dive is short. In other seasons, where eels are available, they are probably the favourite food, being slow swimmers, and therefore easy to bag, besides being rich in oil and far from bony. Eels are known to be relatively scarce on the Clare River catchment and around Mullingar, which accounts for the comparatively low levels in spraints there. It is no coincidence that these two localities are also the ones where crayfish are available. Eels prey heavily upon juvenile crayfish and so where eels are numerous the crustaceans are scarce or absent. It is noteworthy that on only one tributary of the River Bann, which is famous for its eels, are there any crayfish.

In Britain scientists have devised a method of estimating the lengths of several species of fish from their vertebrae recovered from spraints.[379] For each species of fish, a number of individuals of many different sizes was obtained, their lengths recorded, their vertebrae extracted and the lengths of these also measured. For each species fish lengths were then plotted against vertebra lengths on a graph and equations derived to convert the length of a vertebra to the approximate corresponding length of its owner. All of the analyses of spraints collected near fresh water in Ireland in which eel vertebrae have been measured indicate a maximum length of eel around 70 cm,[32,189,207] which was about the upper limit in a sample of over 5,000 eels from 18 widely scattered localities in Ireland measured by fisheries research staff. [236] So otters may successfully tackle the biggest eels.

Wherever the sizes of salmonids have been estimated from intact vertebrae in spraints collected in Ireland, the maximum length has always been about 25 cm and the average around 15 cm, in other words smallish specimens.[32,189,207,265] Salmonids are fast swimmers and speed increases with length. On the other hand, fragments of much bigger vertebrae turn up in spraints more rarely, demonstrating that, on occasion at least, otters secure much larger fish, probably adult salmon or sea trout. Furthermore, should an otter manage to secure a fish as big as a salmon, it could take a hearty meal from it without eating any vertebrae at all.

Because salmonids are rapid swimmers, the extent to which otters feed on them may depend in part on the availability of alternative prey. On the Clare River catchment, for instance, salmonid figured in the diet at only 5-6% and yet the trout fishing there is considered excellent. On the other hand, on Lough Feeagh, where virtually the only fish available are eels, sticklebacks and salmonids, the last comprised nearly 40% of the diet (Table 13).

Lough Feeagh is an important lake for salmonids and the Salmon Research Trust's laboratory stands on its southern shore. Here the passage of trout, salmon and eels in and out of the lake is monitored by trapping on two steep, fast flowing streams: the Salmon Leap and the Mill Race. Trout breed from October to January and salmon spawn from November to December, so adult salmonids are most numerous in the lough at the end of the year[146] and, since it is winter, eels are then difficult to catch. This corresponded with a peak in the consumption of salmonids by otters, salmonids appearing in over 90% of spraints collected in November and December.

Regrettably, the otters did not always confine their attentions to the lough itself. In 1988 an enterprising individual found that it was much easier to steal the fish held in the Trust's traps and so became a thorough nuisance. Eventually it was caught and taken in a van to be released in Connemara National Park. On the way it managed to escape from its cage, much to the consternation of the driver. Consequently the "release" was unavoidably premature and short of the Park!

Another instance of the importance of salmonids at the end of the year was provided by spraints collected from October to December from the Agivey River, in Co. 'Derry - my first experience of spraint analysis.[122] There were salmonids in 80% of these. For some time I was concerned that there must be something faulty with my identification when spraint after spraint appeared to contain the one thing. Eventually I contacted Mr Ken Vickers, of the Fisheries Division of the then Northern Ireland Ministry of Agriculture, who assured me that he would have expected small salmonids to be very common in the river and especially at the end of the year, when trout would have been breeding.

Sticklebacks are normally a minor item on the otter's menu. Indeed Table 13 exaggerates their importance. Readers may remember from Chapter 1 that percentage frequency is derived from the number of spraints containing a particular prey. But remains of sticklebacks tend to make up only a small part of a spraint, and so percentage frequency overestimates their significance. Thus, for instance, the 9.3% for percentage frequency for the Clare River system in 1991-92 in Table 13 corresponds to a figure of only 2.6% by percentage weight of remains in the spraints. Probably sticklebacks are so small that they are not worth making much effort for and, as they do not shoal much, there are limited opportunities for catching several at once. In some Irish studies, the otters preyed on these fish much less often in summer than in winter.[146,182] As they tend to hide in water weeds and aquatic vegetation grows more profusely in summer, they may be less vulnerable then.

The carp family or "cyprinids" are mostly slow swimmers and are potentially good prey for otters. The figures in Table 13 probably reflect both abundance and ease of capture relative to other fish. It may be recalled from Chapter 1 that it is only possible to distinguish individual species of cyprinid from the teeth on their pharyngeal bones (Fig. 5) and that these usually only turn up occasionally in spraint. In Ireland roach, rudd, dace *Leuciscus leuciscus* and minnow *Phoxinus phoxinus* have all been detected at one time or another. Note that in the Clare River catchment the otters caught no cyprinids whatsoever in

1981-82 but in 1991-92 these fish comprised 17% of the diet, all of those identified being roach. Roach invaded the Clare system in 1982 from Lough Corrib, where they had been introduced, and had spread into all of the tributaries of the Clare by 1991. They were evidently welcomed by the otters.

Minnows have hardly ever turned up in spraints in Ireland. Like sticklebacks, they are tiny fish, and probably seldom worth the otter's while bothering about. However, of 67 spraints I analysed from Poulacappul Lough, in Connemara, in summer, 47(70%) contained minnow, often in large numbers.[344] Their small size meant that the fish were eaten whole and the pharyngeal bones were there in profusion - a startling prospect under the microscope. Minnows often go about in shoals and large shoals may have been abundantly available in Poulacappul Lough. Maybe the otters found it profitable to lunge into these, ingesting several at a mouthful.

In freshwater, other fish are usually unimportant to otters. One interesting exception is provided by the Killarney shad, or goureen, a local form of the twaite shad *Alosa fallax*. Twaite shad normally migrate between fresh water and the sea; the goureen is an exception, living the whole of its life in the lakes of Killarney. Next to the brown trout, it may be the commonest fish in the area.[43] In analysis of spraints from Killarney's lakes (Table 13), the shad was second only in importance to eels. The otters took shad in greatest numbers from the western side of Lough Leane, beneath Tomies Wood, where the goureen is believed to be most abundant.[115]

For many years our first year students routinely dissected frogs, which furnish an excellent training in the techniques of dissection, and which our technicians collected annually from a series of disused cattle troughs in roadside fields around Headford, Co. Galway. Depending on the weather, in Ireland frogs usually emerge from hibernation in February or early March and make their way to minor watercourses, ponds and other standing water (such as that in the cattle troughs), where they congregate in multitudes to breed. The technicians were normally in the field in the course of our department's research at least every week during the winter and, from about the end of January, they checked the troughs for frogs on the way home, descending on them to harvest the animals once there were enough. I was thus in a position to know exactly when the breeding season was at its height. Up to 1984 no one in Ireland had analysed spraints collected during late winter, so I decided to remedy this. Once I knew that the frogs were hard at work reproducing themselves, I quickly visited a total of 13 places in Cos Galway and Mayo to collect spraints and managed to accumulate a total of 119. Of these 78% contained frog bones, and in most cases nothing else.[110] Since then the analyses on the Clare catchment and around Mullingar have yielded similar results, with frog locally the most common prey in late winter or early spring. Frogs in a breeding pond must present something of a bonanza for an otter. There are often a lot of them there and space in which to escape and cover in which to hide may be negligible. During courtship the male frog seizes the female and then holds her tightly for perhaps several weeks before spawning. The couple are thus even less advantageously placed to scarper than in other seasons and, for an otter, it is a matter of "catch one, get one free". Outside the breeding season it is a different story. In summer especially, frogs are usually terrestrial and often live away from water. Then otters rarely collar them.

Some perceptive readers will have realised that, besides the common frog, Ireland can also boast of a species of toad, the natterjack *Bufo calamita*. To be sure, its origin is dubious and it may originally have been introduced by man. It is also restricted to part of the Dingle Peninsula and the north of the Iveragh Peninsula in Co. Kerry, from which some have been recently taken and settled on the Slobs in Co. Wexford.[185] Natterjacks are much more at home in warmer climates and the British Isles are a northern outpost where the animals only survive in the warmest places. These amphibians spend much of their time in burrows and so need a loose substrate in which to dig. In Britain and Ireland they are therefore mainly confined to sand dunes and lowland heath. They spawn much later than frogs, generally in May, and are much rowdier beasts. A common frog cannot raise much more than a gentle croak. Male natterjacks, when assembled in their breeding ponds at night in hundreds to await the females, advertise their presence vigorously in songs so loud that they can be heard several hundred metres away. Irish ornithologists may brag of the spectacular nature of breeding colonies of seabirds. A pond of breeding male natterjacks, all hollering at the tops of their voices for mates, is for me spellbinding, even if it usually requires a torch to see the blighters.

Do otters afford these breeding "Kerry toads" the same attention as they do frogs? For some years in the 1980s my colleague, Dr Kieran McCarthy, regularly surveyed the toads' breeding sites in Co. Kerry and, with his help during an excursion of a couple of days in May 1985, I collected 89 fresh-looking spraints around breeding ponds and on nearby watercourses. Analysis revealed that most of the diet was eels. There was not a single bone from toad or frog. Moreover, we found no spraint beside any of the toads' breeding ponds that did not also contain fish.[113]

Breeding natterjacks must be even more obvious to an otter than frogs. Furthermore, they often choose shallower water in which to breed, so would be even less able to escape. Toads crawl rather than hop, so even this might tell against them. Nevertheless, they are left strictly alone. The answer must lie in the poison glands in their skin. Most species of frog and toad have some such glands, but there are relatively fewer toxic substances in the skin of the common frog than in that of the natterjack. The secretion of the poisons is most effectively triggered by pressure, as that exerted when the toad is seized in the jaws of some would-be predator. Although toad skin produces toxins in small quantities, they are about as effective as strychnine. Dogs and cats which make the mistake of seizing toads in Britain do not die as a result, but are sick enough to regret their actions. The poisons do not render toads invulnerable. Some animals, notably seagulls, are able to rip open the skin of natterjacks and eat the insides only.[24] It seems likely that few otters in Co. Kerry have learned this trick. Most of those which try to eat a natterjack probably only do it once.

The "other invertebrates" in the studies in Table 13 usually included a significant proportion of *Dytiscus* diving beetles. As these beetles are a few centimetres long, they are worth catching. However, some of the other invertebrates, such as tiny snail shells, probably came from the guts of fishes which the otter had eaten.

So much for fresh water, but what do otters living on the Irish coast eat? The first research in Ireland in this direction was by Mr Kevin Murphy, one of my students who now, perhaps appropriately for someone interested in otter food, works as a consultant to fish farmers. I recollect field work with him with some nostalgia, for he was the type of student - I have fortunately met more than one - in whose company it is impossible to feel anything but cheerful. His approach was two pronged. First, we made excursions to all of the counties on the western seaboard and collected above likely-looking shores. In all we visited 14 different stretches of coast and accumulated 394 spraints. It was quite clear from the contents of these that the otters were feeding principally on a wide range of small fish living on the shore or in the shallows of the sea, together with some crustaceans, molluscs and a few frogs. The latter were almost certainly overestimated as a fraction of the annual diet, because almost all of the collecting was done in February and March, when frogs would have been at their breeding sites.[240]

Murphy's second programme was to collect and analyse spraints each month over a full year from seats at selected stretches of shore along some 15 km of the south side of Galway Bay. The shores here are sheltered and rocky, but often partially overlain with sand and other sediment. The surrounding district, the Burren, is a famous area of limestone and all drainage is underground. Freshwater fish were therefore practically unavailable. However, because the River Corrib and other rivers flow into Galway Bay in its innermost part in the east, the seawater, as one moves towards the east along Murphy's study shores, becomes gradually less salty.[239]

Murphy's overall results from Galway Bay are shown in Table 14. Evidently the diet was broadly comparable to that on the rest of the west coast: mainly a great variety of small fish commonly found on the shore or below low tide mark. It should be emphasised that some of the families of fish were represented by more than one species in the spraints and that the "other and unidentified fish" included even further kinds. Some of the fish species eaten are very small or thin or bony, or sometimes all three, including pipefish, gobies, sticklebacks and butterfish *Pholis gunnellus*. Together such unappetising fare accounted for 22% of the diet. The three-spined stickleback *Gasterosteus aculeatus,* may, incidentally, be found both in saltwater, particularly in estuaries, as well as in fresh. In fact nearly all of the remains of three-spined sticklebacks occurred in spraints at the extreme eastern end of Murphy's study shores, where the water was least salty. The sea stickleback *Spinachia spinachia* is a strictly marine species. Conger eels *Conger conger* grow to sizes quite unmanageable by an otter and, judging by the bones recovered, the specimens taken were very small. The figures for crustaceans and molluscs probably overemphasise their importance, because both yield comparatively more hard parts than do fish. The crustaceans eaten were by no means confined to crabs, of which there were several species, but *inter alia* included shrimps, sea slaters and sand hoppers.

The most important prey fish were rocklings and eels, followed by wrasses, gobies and sea scorpions which, despite the name, are fish.

Most adult eels spend the greater part of their lives in fresh water, before migrating to the Sargasso Sea to spawn. The young travel in from the Atlantic and, being highly sensitive to fresh water, find their way up

Table 14 - Percentage frequency of food categories found in spraints collected over one year on the north-east coast of Inishmore, Aran Islands, and on the south shore of Galway Bay. + = present.

	Inishmore	Galway Bay
Eel	7.2	14.1
Conger eel	0.7	4.8
Pipefish family (Sygnathidae)	+	1.0
Three-spined stickleback	1.5	1.0
Sea stickleback	+	2.4
Rockling family (Gadidae)	27.0	17.6
Wrasse family(Labridae)	25.4	9.8
Goby family (Gobiidae)	1.7	9.7
Blenny family (Blenniidae)	4.2	2.2
Butterfish	1.0	8.3
Sea Scorpion family (Cottidae)	6.6	7.1
Flatfish families	1.9	1.9
Other and unidentified fish	2.2	1.5
Crustacean	6.4	13.8
Mollusc	9.7	4.3
Purple sea urchin	3.6	-
Bird	0.1	0.3
Frog	-	0.2
Other	0.9	-
Total spraints	1510	1026

rivers, where they grow to maturity. There is an important eel fishery on the River Corrib. Juvenile eels undergo a period of acclimatisation in coastal waters before moving into freshwater and a proportion settle for longer on the sea shore. The latter tend to be small, and the great majority of eels in the spraints from Galway Bay did not exceed 20 cm in length, as estimated from their vertebrae; almost all were less than 30 cm long. The amount of eel in the spraints increased along the shore from west to east, which corresponds to the decrease in the saltiness of the sea and doubtless more eels settle on the shores with lower salinity.

Just as in fresh water, the otters ate a lot less eel in winter. In summer 60-80% of the spraints contained eel, whereas in January this had dropped to around only 20%. In summer they were therefore by far the most important prey fish. Rocklings, wrasses and sea scorpions, which together made up over 29% of the diet, followed a reciprocal pattern, and were eaten most often in winter and least in summer. This was not because they were scarcer in summer, for, as mentioned already, some of the fish on the shore and in the shallows move into deeper water in winter. (Wrasses do not do this and may indeed be easier to catch in winter, as detailed below.) It appears, therefore, that in south Galway Bay, the availability of eel may have a profound affect on the extent to which otters catch other fish throughout the year.

Probably because of the restricted time that they can spend under water, the otters of Galway Bay rarely caught what zoologists know as *pelagic* fish: those which normally live on the surface or at least off the bottom of the sea away from the shore, and which are usually relatively fast swimmers. Admittedly seven spraints contained pollack and six sprat *Sprattus sprattus,* but in five cases these occurred together in the same spraint. This material was gathered at a time when it was said that there had been shoals of pollack close inshore in Galway Bay. Although pollack are probably usually too quick for otters, if there were enough of them around, the otter might have a better chance of catching the odd one. Pollack sometimes feed on sprat and the former may simply have been pursuing the latter into the Bay at the time. At any rate, the simplest explanation of both of these pelagic species appearing together in droppings is that the pollack had eaten the sprat and had in turn been dined upon by an otter.

Miss Sharon Kingston, another of my students, analysed spraints, again gathered every month for a year, from the north-east coast of Inishmore, the largest of the Aran Islands, which lie in a line running north-west to south-east across the mouth of Galway Bay.[182] Geologically the Arans are an extension of the Burren and Inishmore lies only 34 km from Murphy's study area. There is no freshwater drainage at all on the north-east coast of Inishmore, even underground. The shores studied were predominantly rocky and are mainly exposed. Although Kingston did all of the analysis, most of the collecting was by Dr Michael O'Connell, a schoolteacher on the island and also a part-time marine biologist; he earned his doctorate for research in the school holidays and in his spare time. As most of this was on the island's fish, we had a good idea of the relative abundances of the species available to the otters. We were therefore eventually able to deduce from the results that the otters were to some extent opportunists. However, although the ate more of the most abundant bottom and shore-dwelling fish, they largely ignored the pelagic ones, some of which were very common. As explained below, they also probably tended to favour the more nutritious non-pelagic species.

The results of the analyses are also displayed in Table 14, and although the diet was again of fish of the shore and shallow sea, the proportions of the various kinds differed. For one thing, eel was much less important. Eels are scarcer on Inishmore because there is little freshwater outflow and none at all on the north-east coast. Although the otters ate less eel in winter, even in the summer months eels never appeared in more than about a third of the droppings. The most important prey were rocklings and wrasse, both of which were caught more often than in south Galway Bay (together 52% as compared to 27%). Both are larger than most of the other fish consumed, which means that they might be underestimated by percentage frequency. In fact by percentage weight they amounted to 70% of the diet. The otters also ate much fewer fish that are small, thin or bony (pipefish, gobies, sticklebacks and butterfish) than in south Galway Bay (4% as compared to 22%).

The results of other studies of the diet of the otter on the Atlantic coast of Europe show rocklings to be the most consistent major food item. A combination of characteristics renders them particularly suitable prey: they are slow-swimming, easy to catch, palatable and large relative to the other fish on the shore and in the water immediately below it, and therefore provide a greater return for catch effort. Wrasses are also relatively large and slow swimmers but are warm water fish and are near the northern limit of their geographical range in Scotland. They are therefore of little significance to otters there, as has been shown by spraint analyses. On the other hand, in Portugal they have been shown to be important prey.[182]

Inishmore, with its predominantly rocky exposed coast, provides better conditions for both rocklings and wrasses than south Galway Bay, because rocklings prefer exposed shores and wrasses rocky bottoms and on both counts Inishmore qualifies as the better habitat. So two particularly suitable groups of prey fish are commoner. O'Connell concluded that two species of wrasse were "very common" and two rockling species "common" and another "frequent".[182] So conditions are more congenial for the otters of Inishmore than on the coast of nearby

Galway Bay. This explains why they eat relatively few small, thin or bony fish, which represent a poor return on catch effort, the most efficient foraging strategy obviously being to select the largest and most nutritious items available.

The otters evidently secured more wrasse in winter and this may be because they are essentially fish of warmer climates. They do not migrate offshore during the colder part of the year but become torpid and hide amongst rocks. Enterprising otters which could find them would therefore be rewarded for their efforts.

In both Kingston's and Murphy's studies, the otters ate fewer blennies and butterfish in winter. Many butterfish have by then probably moved to deeper water. The commonest of the blennies, the shanny, which is abundant on Inishmore, breeds from April to August in Co. Galway[182] and the male guards the eggs, so male shannys may be particularly vulnerable to marauding otters in summer.

The purple sea urchin *Paracentrotus lividus* is a widespread and attractive feature of Inishmore's shores. These animals burrow readily into the soft limestone and groups of them, each in its own pocket in the rock, are a spectacular feature of rock pools. On the mainland in Galway Bay they are exploited commercially as they are esteemed as a delicacy on the continent. It seems from Table 14 that the otters may sometimes feel the same way about them. The urchin's sexual organs, which are what gourmets eat, reach maximum size in December and the market for the animals is predominantly in December-January. The otters know when they are at their best too, for although they eat them throughout the year, consumption rises to a peak in November, December and January, when Kingston recovered bits of their shells from some 20% of spraints.

Chapter 5. The Poor Pine Marten - an Arboreal Weasel

The poor pine marten I continually fret over.
R.M. Barrington. Quoted in "The pine marten." C.B. Moffat.
The Irish Naturalists' Journal 1927.

The inclusion of the word "pine" in the name of the pine marten, the rarest of all Ireland's mammals, is misleading, for the species is by no means especially associated with pine trees. The Irish name is "cat crainn" or "tree cat", which has led to confusion in the older literature between the marten, the domestic pussy and the wild cat, if such ever existed in Ireland. For the same reason the exact meaning of the word "cat" in Irish place names is open to dispute. The natives of Co. Clare still commonly refer to the animal as the "marten cat". Perplexingly, in parts of the west of Ireland in the nineteenth century the beast was only known as "madra crainn" or "tree dog"![154,377]

Cat crainn is correct in regard to its commonest habitat, to the overall size of the animal, and to the length of its legs relative to the body, for they are proportionately longer than those of most mustelids. Seen fleetingly and from any distance - usually loping across a road or along the edge of a wood or forest clearing - the dimensions and agility of the animal certainly convey the impression of a cat. This has been brought home to me on a number of occasions by people who, having encountered a marten for the first time, and unaware of what it was, described it as "like a cat". However, at close quarters the animal is scarcely feline. The head is pointed and the ears rounded, both features far from cat-like. Moreover, the tail is as bushy as a fox's. The body is also to the usual mustelid specification, being much more elongate than a cat's. Again, while the feet have stout claws, which are useful in ascending trees, they cannot be retracted into sheaths, as can those of a cat.

In Ireland adult male martens weigh in at 1.5-2.0 kg and females at 1.1-1.5 kg. The length of head and body is some 40-60 cm and the tail a further 20 cm or so.[151]

The crowning glory of the marten is its fur, which is long, dense, soft, warm, chocolate brown and gorgeous. No one, least of all a hardened academic zoologist like myself, could dispute that the marten is a beautiful animal. The very length of the fur undoubtedly makes the beast look decidedly bigger than it is. The hair is darkest on the legs; there is a cream, or pale yellow or orange patch on the throat; and the fur within the ears is also cream. In Ireland this is unique amongst wild mammals and therefore an unfailing guide to recognising the beast on a brief sighting. I particularly recall a summer night in 1985 in the wild, wooded country in the east of Co. Clare, when I saw three martens, two in the evening and the third in the small hours on the road as I was driving home. The last was running away from the car but stopped and turned to reveal the diagnostic creamy insides of its ears.

As will be seen later, in Ireland the marten's pelt was nearly its undoing, for in former times this was greatly esteemed. Some indication of its value may be had from the close relationship of the animal with the sable, with which it actually hybridises where the geographical ranges of the two species overlap in Russia. From ancient times sable was of all furs the most valued. The Russian royal house derived a substantial revenue from the sale of the pelts, having first selected the choicest ones for its own adornment. Originally the sable was found from Scandinavia through to eastern Siberia, but was exterminated in Europe and over much of its range in Asia by fur trappers. Several hundred thousand skins were traded annually during the eighteenth century in the city of Irbit in western Siberia, a figure which had dropped to 25,000 by 1910-13. Subsequently, the Soviet government instituted a programme of protection and reintroduction.[246] The pelt of the nearly-related pine marten, if not quite so prized, is at least a close also-ran.

Most of the research on pine martens in Ireland in the twentieth century was done in the 1970s and 80s by Mr Paddy O'Sullivan, who was originally a forester but moved into Dúchas. In the early 1970s it was thought that marten numbers were declining and he was handed the task of finding out something about them.

He surveyed their distribution in the 26 counties of the Republic. In this he operated conservatively, recording presence only where he saw an animal or a dropping.[262] Fortunately martens tend to deposit scats in conspicuous places, along paths, on forest roads, on large stones, on logs and on the tops of walls; and a marten scat is easy to recognise. It is perhaps 8-10 cm long, cylindrical, with a constant diameter of around 1 cm along almost all of its length, but often pointed at the ends, and has a distinctive pleasant sweet smell, rather like violets. Indeed the marten itself is a delightfully perfumed beast, the old English name of sweet mart being used to distinguish it from the polecat, also known as the foul mart, and with some justification.

O'Sullivan also studied a population at Dromore Forest, an area of mixed woodland near Ennis in Co. Clare, which includes two lakes and a river. He did this both by direct observation and by noting signs of the animal's presence and activity. He investigated the diet at Dromore by analysing scats,[364] and numbers and home ranges by a trap-mark-recapture study, similar to Chris Smal's on the mink. He marked all of the animals he trapped with ear tags.[260] The information thus derived was supplemented with that from a female which was radio-tagged and tracked both day and night for over 3 weeks. In addition O'Sullivan ran a captive breeding programme at Dromore with a view to reintroducing the animals into localities from which they had disappeared. Part of his research has still to be published, but he kindly filled in some of the points for me in broad terms.

Table 15 - Types of habitat in which pine martens were recorded (by direct sightings or scats) at 97 localities in the survey of the 26 counties of the Republic 1978-80. Figures are percentages of total localities.

	Percentage of total localities
Deciduous woodland	16.5
Deciduous woodland underplanted with conifers	46.4
Mature coniferous woodland	6.2
Thicket stage coniferous woodland	15.5
Scrub	5.2
Scrub on limestone pavement	10.3

The animals in Dromore Forest had distinct habitat preferences. They liked areas with well developed ground vegetation and avoided open spaces and places where ground vegetation was sparse. This O'Sullivan put down to suitable prey - about which he knew a great deal from his scat analyses - being commonest in ground cover. In his survey of the Republic, he recorded martens at 97 localities. The habitat types at these are summarised in Table 15. Although his data are somewhat biased, because he tended to look in places where, from his experience at Dromore, he expected to find the animals if they were around, the figures in the table emphasise that the beasts lived mainly in woodland. They were scarcer in mature conifer, which has relatively little ground cover, and instead tended to chose the early thicket-stage of afforestation. He intensively surveyed three areas of state forest which harboured martens and also contained plantations of sitka spruce and lodgepole pine varying in age from 5-20 years. In these it was the thicket stage that the animals favoured.

On the limestone pavement of the Burren district in Co. Clare he also recorded the animals in hazel scrub. Much of the Burren is bare limestone hills, with low vegetation confined to the many cracks and crevices in the rock. However there are also larger areas of soil of glacial origin, which are frequently colonised by thickets of small hazel trees. It is widely known that martens are common in the Burren. When one is spotted there, this is usually on open pavement, presumably because it would be nigh impossible to *see* one in the scrub. However, there are still enough reports to justify concluding that martens in the Burren spend significant time in the open. Moreover in the 1950s and 1960s there were still pine martens on the Inishowen Peninsula in Co. Donegal, which is almost entirely open and windswept. Elsewhere in their geographical range outside Ireland they sometimes live in the open on high ground, moorland and along coastlines.[151]

The martens at Dromore were generally solitary and had distinct territories of some 13-15 ha, the boundaries of which they marked with scats, which accounts for the latter being deposited in such prominent places. The territories were rather smaller than those investigated elsewhere in Europe, which suggests that the habitat in Dromore Forest was particularly favourable.

Although martens are excellent climbers and those at Dromore often made dens in trees or scaled them after food, they did not travel through the treetops but on the ground on a series of regular paths.[364]

It may come as a surprise to readers that, in addition to O'Sullivan's work, there is a significant body of reliable information on the biology of the pine marten in Ireland from the seventeenth century. Arthur Stringer, huntsman to Lord Conway on his great estate at Portmore on the eastern edge of Lough Neagh, was a man well before his time. Exasperated with contemporary treatises on hunting, which were manifestly unreliable and often plain hearsay, he wrote his own based exclusively on his personal experience, entitled, appropriately enough, *The experienc'd huntsman*. This he published in Belfast in 1714.[341] Besides instructions on hunting, he also wrote reliable accounts of the "nature" of the various beasts of the chase. Indeed so painstaking were some of his observations that it was not until the mid twentieth century that they were verified.

Stringer had much to say on what he called "the martern". He too considered that the best habitat was in woodland with abundant, thick ground cover.

> A rank wood is their province....The place where they seek prey is most commonly in the rankest coverts of wood, either standing or lying in hedge groves and old hedge roots, or in rank coverts of thorns and brambles....The marten always runs the foulest and worst ground where lying wood is rankest, as also briars, thorns, and old hedges....

It is also apparent from Stringer's observations that, whenever he observed the animals on the move, this was, as at Dromore, on the woodland floor. He normally raised a marten from a tree but, once disturbed, the animal's first concern was to reach the ground, sometimes by jumping from a great height, and then to make off. However, this is qualified by the statement "....but when the hounds come near him, the tree is the last refuge." While this does not preclude movement from tree to tree, it appears that this is not the common mode for travelling any distance.

Stringer considered the marten to be primarily nocturnal, usually spending the day aloft, sometimes in a hollow tree, or in a hole in a trunk, or lying in the fork where one of the larger branches grew out from the bole, or taking its ease in an old bird's nest. There are other recorded instances of bird's nests as resting places. The great Belfast naturalist, William Thompson (1805-52), mentioned martens occupying magpie's *Pica pica* nests at Malone House and Belvoir Park, both now in the suburbs of Belfast, and at Tollymore Park, near Newcastle, Co. Down.[350]

Stringer particularly noted that martens which spent the day in the tree tops chose trees well girt with ivy "so that crows and birds cannot see them and there they lie round like a cat" - a splendid description of a marten asleep. This suggests that he believed that any individual rash enough to expose itself in the canopy during the day was likely to be mobbed by birds. Perhaps he had witnessed something of the kind himself. At any rate, mobbing certainly occurs, one such incident being witnessed in 1897 in Co. Waterford.[132]

PINE MARTEN

....I was walking through the beautiful woods of Curraghmore....when I heard a regular uproar of birds....Walking in the direction as quietly as possible, I expected to see a Fox carrying off a young bird. Among the branches of some low oaks was a large party of Blackbirds [*Turdus merula*]; one of them, a fine cock with bright orange bill, being greatly excited, scolding away at the top of his voice, and with outspread wings facing a point from which he expected trouble for himself and his family; and there among the leaves, lying close along a branch, was a Marten, crouching low as if he was going to spring....

During the winter Stringer noted that every few days a marten would often shift from one tree den to another. A female with young, on the other hand, was constant to the same den. The tree in which this was situated was easy to spot, as there were not only copious droppings on the ground immediately beneath it, but "reliques of every thing they prey on as feathers, pieces of fowl, rats, mice...." He stated that there were generally three or four young in a litter, which compares with 1-5 for captive animals in Britain and Germany.[151]

Stringer gives no indication of the whelping season and no relevant data from O'Sullivan's work have been published. Fortunately, there are a few snippets of information on breeding in Ireland from the nineteenth century. Three young, with their eyes not yet opened were found in a den on 18 April.[356] A female caught on 31 March was pregnant[33] and another in captivity gave birth on 31 March.[51] Finally, Dr Caroline Shiel told me of a lactating female she found as a road casualty on 26 April. These suggest that the young are born in March or April, which are the generally accepted months of whelping outside Ireland.[151] Observations on one Irish litter showed that the young are white at birth but become brown within about 4 weeks.[51]

O'Sullivan analysed a total of 831 scats that he had collected at Dromore from 1973 to 1976. The results are given in Table 16. Some of the food categories are rather more precise than in the tables for the other mustelids in previous chapters. This was because O'Sullivan knew the area and the prey available intimately, and was therefore able to narrow down potential food items: for example ducks rather than ducks and geese, and waders rather than waders and gulls, because there were no geese and gulls in the forest.

Undoubtedly martens have extraordinarily catholic tastes and a

Table 16 - Percentage frequency of food categories in 831 marten scats collected at Dromore Forest, Co. Clare, 1973-76. + = present.

	Percentage frequency		Percentage frequency
Total mammal	13.3	**Total invertebrates**	47.9
Hare	5.3	Ground beetles (Carabidae)	16.3
Field mouse	5.7	Bees, wasps*, wax, honey	6.0
Red squirrel	0.5	Earwigs (Dermaptera)	0.2
Pygmy shrew	0.5	Woodlice (Isoptera)	0.2
Stoat	+	Snails (Mollusca)	0.6
Unidentified mammal	1.2	Earthworms (Oligochaeta)	24.6
Total bird	20.1	**Total fruit**	16.9
Rails (Ralliformes)	0.9	Rowan berry	2.5
Ducks (Anatidae)	0.5	Blackberry	2.5
Passerines (Passeriformes)	10.5	Wild strawberry	1.2
Waders (Scolopacidae)	+	Wild raspberry	0.4
Game birds (Phasianidae)	+	Rose hip	0.4
Pigeons (Columbiformes)	1.1	Wild cherry	1.0
Hawks (Falconiformes)	+	Hazel nut	3.8
Unidentified bird	5.1	Crab apple	1.2
Egg shell	1.9	Hawthorn berry	+
Frog	1.2	Beech nut	+
Lizard	0.6	Ivy berry	3.7

*(Hymenoptera)

menu of greater diversity than any other Irish mammal, barring the domestic rodents and ourselves. Whilst the great range of comestibles strongly suggests opportunism, the proportions of four overall categories, namely mammals, birds and their eggs, invertebrates and fruit, remained much the same from one year to another, suggesting some sort of system in foraging.

As there were no rabbits in the district at the time, O'Sullivan had no difficulty in distinguishing Irish hare *Lepus timidus hibernicus* in the droppings, presumably the remains of young. The nimbleness of the field mouse does not entirely save it from the marten, and it appeared in the scats with much the same frequency as in the guts of Sleeman's stoats (Chapter 2). Unlike the stoat, however, the marten may find shrews distasteful, for they turned up only seven times. Likewise, only seven scats yielded remains of red squirrel *Sciurus vulgaris*. This contradicts the many statements in the older literature that squirrels are a staple prey. Stoat appeared in one scat only and the stoat in question may have been found dead. Certainly the martens fed on carrion, for they partook of a dead calf which had been dumped at Dromore by a farmer.

The birds eaten were mostly passerines, which might be expected, for small birds of this type - finches, tits and the like - would predominate in woodland with heavy ground cover. The martens caught many more birds in January and February than in the other months and this may reflect birds enfeebled or killed by harsh winter conditions. On the other hand, most of the avian remains which O'Sullivan could not identify were from May and June. These may have been from fledglings, which would have been numerous in these months. It is difficult to identify feathers from fledglings in scats any further than just "feathers". As martens are at home in trees, it is inevitable that they will purloin eggs and nestlings.

Although there were both a river and lakes at Dromore, the animals evidently relied little on these for prey, eating few waterfowl and frogs and no fish whatsoever. Neither was there any indication of increased predation on frogs in late winter or early spring, when, as we have seen in Chapter 4, frogs are at their breeding sites and most vulnerable.

Earthworms were detected in the scats from their chaetae, which survive digestion, just as they do through the gut of the badger (Chapter 3). A proportion of the chaetae might have arrived at second hand as it were, for at least some of the birds the martens devoured could have contained earthworms. However, judging by the much higher frequency of chaetae than birds in the droppings, one can only conclude that earthworms are probably the most important single component in a remarkably varied diet. In this aspect of its menu the marten comes close to rival the badger. Ground beetles - the big black, semi iridescent, crunchy-looking insects which often scurry out when one turns over stones - are clearly also acceptable morsels.

Bees, wasps, wax and honey appeared in the scats only from June to October; outside this period few nests of bumble bees and wasps would have been available. I was once on hand when O'Sullivan was analysing scats and recollect the sweet smell which emanated from those containing bees and wax. Maybe a little of the honey had actually survived passage through the gut. There are two references to the raiding of bee hives in Ireland from the nineteenth century, but both are hearsay. One was published by a naturalist who did not give a locality lest someone should try to trap the animals.[152] The other, near Ennis, Co. Clare, was reported by the Commandant of the Royal Irish Constabulary in Dublin.[19]

> I was fishing in the neighbourhood, and the local sergeant of police told me of some Martens having robbed two bee-hives, and, traps having been set, two were caught in one night, and the sergeant made the skins into a cape for one of his children.

The martens at Dromore evidently had a sweet tooth and this extended to fruit. As might be surmised, the importance of the latter varied seasonally and peaked in autumn and early winter. Individual fruits also appeared in the scats only in particular months and fairly predictably: rowan berry (August-December), blackberry (August-November), wild strawberry, raspberry and cherry (predominantly July), hazel nut (September-November), crab apple (November-January) hawthorn berry (October-November) and ivy berry (February to June). In 1974, when a freak storm in April blew down many trees, ivy berries were abundantly available at ground level and the martens fed on almost nothing else for a month.

Stringer particularly referred to "berries" as food, but the fondness for fruit is mentioned again and again in the subsequent literature, including the following: from Co. Wicklow in 1891,[19]

> They are most readily killed when the bilberries are ripe, for they are very fond of the berries.

from Co. Clare in 1892,[19]

> In the woods in the neighbourhood there are a great many cherry trees which bear a good deal of fruit, and the owner of the property was lamenting the fact that the Marten-cats eat them all, and was puzzled to know how they managed to get at the fruit at the ends of the long pendulous boughs.

from Donegal in 1894 - a comment on individuals which had been trapped some years before,[270]

> They used to steal plums on the garden wall, and be caught on the top of it, but I cannot tell the dates or numbers.

from Pontoon, Co. Mayo, in 1956 concerning another frugivorous marten, this time in a garden,[258]

> He was actually sitting under a gooseberry bush, eating the gooseberries, and the bush was not more than three yards from the window we were watching from. After having sampled several he went round the house to....a tin with honeycomb in it which was floating on water. He had a smell of it and then a taste...as he was determined to take it he submerged his head completely and got hold of it and went off in triumph.

and finally an extract from a letter dated 1968 to Douglas Deane and now in the archives of the Ulster Museum, concerning a further invasion of a garden at Pontoon.

> I can also confirm their taking our strawberries as last year which were disappearing although netted and we could not think what was taking them until a pine marten was found in the strawberry bed under the netting. It obviously knew its ground and made a hasty retreat when discovered.

I have myself analysed two small groups of marten scats from Co. Clare. One consisted mainly of feathers of passerine birds and ivy berries but also included field mice, a lizard, beetles and snails.[116] The other comprised rabbit or hare, field mouse, brown rat, passerine birds, frog, beetles, earthworms, ivy berries, and sloe fruits.[104] Both sets of results differ little from O'Sullivan's, apart from the rat, which appeared in only one scat. Stringer remarked that

> ....the greatest part of their food is birds, rats, mice, snails and berries, though they will not stop to kill hens and ducks.

His inclusion of rats is perhaps significant, for in his time the common rat was the black species *Rattus rattus* (Chapter 12), which is a great climber and, if arboreal in Ireland at the time, may have come the marten's way more often than the brown does today.

Occasionally the pine marten attacks poultry, sometimes polishing off more birds than it actually eats. The phenomenon of "surplus killing" by mammalian carnivores is discussed in Chapter 1. The more serious charge of preying upon sheep appears repeatedly in the Irish literature, but it seems unlikely that such a small animal as a marten would be much good at it. However, as carrion is evidently acceptable fare (and also refuse, as will be recounted presently), it is perhaps not impossible that a marten might occasionally scavenge around lambing fields and be thus falsely accused. Like other mammals which attack their prey in the throat, such as the mink, martens are sometimes said to suck blood. Thus one, killed after having been alleged to have disposed of 17 pheasants, was said to be "gorged with blood".[19] The matter of supposed blood-sucking by mammals has already been dealt with in Chapter 1.

For a wild member of the weasel family, the marten sometimes shows remarkably little aggression to man if cornered. When O'Sullivan caught martens for breeding stock, they could all be handled without gloves within 2 or 3 days; none ever bit him. He was even able to manage some of the animals which he had trapped for the first time without gloves. Furthermore, this otherwise elusive mammal not infrequently visits or dens in buildings, even inhabited ones, and on occasion exhibits an astonishing lack of shyness of man. Mr Éamon de Buitléar, Ireland's famous wildlife cinemaphotographer, found a female and her young in a deserted cottage in the Burren, where the chimney pot furnished a convenient entrance and look-out post. There is a delightful photograph of a cub perched on it in one of his books.[76]

One winter I was called by the caretakers to an imposing summer residence at Ballynahinch, Co. Galway. They were worried about a marten which had been using the basement. When I arrived it transpired that their chief concern was about a window frame that the animal had thoroughly chewed. The window was open, so it took me some time to ascertain that the damage had been done when it had been shut, with their visitor, which had been using it to come and go, inside. The poor creature had been imprisoned and had simply been trying to escape. At around the same time, at Derryclare Lake, only a few kilometres distant, a female marten raised a litter of cubs above a suspended ceiling at a fish hatchery. In this instance the owners were more considerate and left the door ajar so that she could enter and leave as she pleased.[120]

Another visit to a building - this time of a particularly pertinacious intruder - comes from Portlaw, Co. Waterford, in 1936.[144]

> About 9 o'clock the same evening the parlour-maid came to me to say that there was a strange animal in her pantry, and on going to investigate I found the Marten with its head in a refuse bucket. It did not move until I clapped my hands....At 11.30 the same night my daughter heard the claws of an animal on the oilcloth in the passage outside her bedroom, and, thinking it was one of our dogs, opened the door to let it in, when, to her astonishment, she found it was the Marten....Two evenings later the servants were having their supper when one heard something moving in the scullery, but thinking it was one of our cats they took no notice until they had finished....On going to the scullery the cook found the Marten with its head in the refuse bucket there and as she entered it jumped out of the open window but returned as soon as she left the scullery....

Michael Viney described a close encounter that one of his correspondents had with a marten in Co. Clare. The man and his brother heard their Jack Russell terrier barking furiously at something under the deep freeze in their utility room. They dismissed the dog and eventually a pine marten emerged. Some time later their mother went to use the washing machine and found droppings in the drum! The marten, or at least a marten, was a frequent visitor to their dust bin.[359] I recollect being told by the hotel staff at Pontoon in 1969 that pine martens often raided the bins there and I have correspondence on other cases of scavenging of a similar kind.

Douglas Deane recounted how he inspected a marten den in the attic of a house on the shores of Lough Conn, Co. Mayo, where the animal had made itself comfortable by pulling the stuffing out of a mattress. The makeshift bed was surrounded with the remains of woodpigeons. While the couple who owned the house were watching television, there was sudden pandemonium as the marten rushed in followed by their spaniel, raced round the room, up the curtains and across the backs of their chairs. Eventually the beast was secured and released unharmed.[69]

Mr John Higgins, conservation ranger in the Clonbur-Cong district of Cos Galway and Mayo, tells me that he is called in to eject martens from attics and outbuildings two or three times a year. Where the intruders are females with cubs, he normally waits until the latter are full grown. He has observed that the animals commonly strip away fibre glass insulation in roof spaces to make nests for themselves. Personally I would not fancy fibre glass as bedding.

The hunting of martens in Ireland for their fur goes back at least to Mediaeval times. Gerald de Barri of Wales, perhaps better known as Giraldus Cambrensis, mentions it in his *Topographica Hibernica,* published a few years after 1185. The following is a translation from the Latin.

> Martens are very plentiful in the woods, in hunting which the day is prolonged through the night by means of fires. For night coming on, a fire is lighted under the tree in which the hunted animal has taken refuge from the dogs, and being kept burning all night, the marten eyeing its brightness from the boughs above, without quitting its post, either is so fascinated by it, or, rather, so much afraid of it, that when morning comes the hunters find him on the same spot.

It is doubtful whether Giraldus witnessed this himself and, as he was evidently both superstitious and gullible - the *Topographica* includes miracles and mentions some mammals in Ireland for which there is no other evidence whatsoever - the above account may be fiction.

The *Libel of English Policie,* a poem written around 1430, particularly mentions the animal's skin as an item of trade.

> I caste [venture] to speke of Ireland but a lytelle,
> Commoditees yit I woll entitelle,
> Hydes and fish, salmon hake and herynge,
> Irish woolen, lynyn cloth, faldynge [coarse woolen cloth],
> And marternus [martens] gode, bene here [are their]
>   merchandyse,

At about the same time Nicholas Arthur of Limerick is mentioned as trading in horses, falcons and skins of otters, martens, squirrels and other furs.[21]

In her scholarly account of Anglo-Irish trade in the sixteenth century based on English customs accounts and post-books, Dr Kathleen Longfield shows that the value of exports of hides and skins from

Ireland at the time to have been second only to that of fish.[193] Pride of place among pelts was given to that of the marten. In the early 1500s a skin was valued at about 1/- (1 shilling = 5p), in comparison to that of the otter at only 5d (5 old pence = 2.1p). By the close of the century marten skins were changing hands at 3/4 (16.7p) each. Although this was partly due to inflation, it was also possibly because of a scarcity of the animals. Of course the skins would have retailed for greater sums abroad, not only to allow some profit to the merchants, but also because of an import tax in England, as is illustrated by the following entry for Bristol in 1504-5. The word "dicker", meaning ten, used as a measure of skins, dates from Roman times.

> Boat called the Magdalen of Waterford, of which William Pembroke is Master, came from Ireland 25th day of June and has in the same....William Artgeppoll native, 100 calveskins, value £1 5s 10d. sub. 1/3....William Stempe native, for 5½ dickers of salt skins, value £3 13s 4d., sub. 3/8. 4 dickers of wild animals value 10/-, sub 6d. 100 rabbit skins, value 5/-, sub. 3d. 7 marten skins, value 7/-, sub. 4½ d, 35 wolf skins, value 4/4, sub 2½ d....

There is an interesting reference to martens in manuscript material of 1520 concerning the Earl of Ormonde illegally taxing the natives for the upkeep of his establishment, which was evidently run on a magnificent scale.[7]

> ....not only for his horsemen, kerne and galloglas [retainers and men at arms], but also for masons, carpenters, [and] taillours, being in his own works, and also for his sundry hunts, that is to say, 24 persons with a 60 greyhounds and hounds for deer hunting, another number of men and dogs for to hunt the hare, and a third number to hunt the martin, all at the charges of the King's subjects, meat, drink and money; the whole charges wherof surmounteth 2,000 marks by year.

Unfortunately the exports of skins from Ireland in the seventeenth century remain to be researched. Dr Robert Hunter tells me that the relevant documents are in the Public Record Office in Chancery Lane in London. They consist mainly of numerous vellum port books, many of which are damaged, from various sea towns in Britain, with four or more from each decade. Extraction of representative data on skins therefore presents a sizeable task. The *Calendar of the State Papers (Ireland)* lists the exports for 1665, but there are no marten skins.[210] With the colonisation of Ireland from Great Britain during the period, and the consequent destruction of the Irish forests, conditions for pine martens cannot have improved. In 1600 about an eighth of Ireland was clad with ancient woodland. Some 200 years later the area under timber had been reduced to a fiftieth, the bulk of the clearance having taken place by 1700.[205] Although martens may live in the open, their primary habitat is woodland, and individuals surviving its clearance in the 1600s would have been at the mercy of those who coveted their fur. An indication of the situation may be had from a letter written in Dublin in 1638 by Lord Deputy Strafford to the Archbishop Laud, then Primate of England.[153]

> Before Christmas your Lordship shall have all the Marten skins I could get either for love or money since my coming forth of England, yet not to the number I intended. The truth is that as the woods decay, so do the Hawks and Martens of this kingdom. But in some woods I have, my purpose is by all means I can to set up a breed of Martens; a good one of these is as much worth as a good wether [castrated male sheep], yet neither eats so much or costs so much attendance; but then the Pheasants must look well to themselves, for they tell me these vermin [i.e. martens] will hunt and kill them notably.

In his book *De Regno Hiberniae Sanctorum Insula Commentarius*, published in 1632, Peter Lombard particularly remarks on the value set upon the fur of the marten in Ireland.

Arthur Stringer published his book in 1714, which he states was based on 35 years experience, and which therefore pertains to the late 1600s and early 1700s. He was clearly still able to find a sufficiency of martens in north-east Ireland at this time. Stringer recommended hunting the beasts at night with a small party of men and a few dogs which have been trained to bay only when they have scented a marten. When the quarry was thus located, it was shot if on a branch or, if in a hole, either encouraged to bolt or extracted. The book gives detailed instructions on how to follow the animal and how to proceed in a variety of situations. Because the sport was undertaken entirely on foot on rough terrain, Stringer considered it suitable only for young men that are "able to take great pains".

Records of exports from Ireland in the eighteenth century are held in the Public Record Office in Kew in London and data may be readily extracted from them. I collated those on the pelts of wild mammals from 1697 to 1819.[109] Over this period skins of, among others, a total of over 10,000 deer, 49,000 foxes and 42,000 otters were exported, but apparently not one from a pine marten. It can only be concluded that the animal was so scarce that any skins which came on the market were snapped up locally. Indeed there is evidence of a trade in imported marten skins in Rutty's *An Essay toward a Natural History of the County of Dublin,* which was published in 1772.

> *Martes. Marter Johnstoni.* The Martern or Marteron....
>
> It is found at *Luttrel's town.* It destroys rabbits and poultry, and is almost as mischievious as the fox.
>
> The Skins come from different parts of the world and make a great part of the Furrier's trade. Tippets and Muffs are made of them. The *Sable* is a species of Martin, the hunting of which makes great part of the revenue of the Czar of *Muscovy....*

Note that, despite the lack of exports, the animal was still said to be found in Co. Dublin.

Not only were martens hunted for their fur. Like several other carnivorous mammals and birds they were classed as "vermin" to be destroyed out of hand. Indeed in 1787 an Act of George III *inter alia* offered a bounty of 5s (25p) for each one killed. This legislation was repealed under William IV.

With the coming of the nineteenth century and the rise in game shooting, persecution of martens in Ireland almost certainly intensified and they were trapped wholesale, both deliberately and accidentally. In the late eighteenth century in Britain and Ireland, driven game and large game bags were considered unsporting, but this attitude was to be changed dramatically in the 1800s by a number of factors. The development of faster-burning gunpowder resulted in shotguns with shorter barrels, which were easier to manage. The introduction of the percussion cap in 1818 greatly reduced the proportion of misfires and led eventually to breech-loading in the 1850s. By 1874 manufacturers had developed choking whereby a gun barrel tapered somewhat towards its mouth and thus discharged a closer pattern of shot. Hammerless ejectors appeared in 1875, which expelled spent cartridges automatically and made for quicker reloading. All of this encouraged the wider ownership of shotguns. Alongside developments in gun technology, the rise of the middle classes meant many more persons with money to spend, and how better to spend it than to copy one's aristocratic betters in the slaughter of game birds? The parallel development of the railways meant easier and quicker access to shoots.

The demand for game birds resulted in widespread intensive stocking, equally intensive keepering and the attempted extermination on the shoot of every carnivorous mammal - except usually for foxes which were preserved for hunting - every member of the crow family and every other kind of bird with a hooked beak. Again and again in the Irish literature of the period martens are noted in snares or traps and again and again it is remarked that they are easy to catch. Traps were normally of the gin type which often cause horrific damage to limbs held in them, even if the victim does not try to bite off its leg in its frantic efforts to escape. More than once such maimed specimens were considered to be rare enough to be sent to Dublin Zoo, but not, of course, of sufficient importance for the trapping to cease. However, there are at least two instances, at Tomgraney, Co. Clare,[154] and Curraghmore, Co. Waterford,[287] of the owners of estates protecting martens. Unfortunately, the following extracts are more typical.

In the neighbourhood of Killarney, more especially near Lough Carragh, still fairly numerous. In 1856 no less than ten were trapped on the Lansdowne estate. In April 1877, Lord Kenmare's keeper at Killarney spoke of trapping Martens there as no uncommon thing, and mentioned 7s 6d as the usual price he got for the skins....a correspondent....wrote:-
"At a shooting I rented some years ago, in the wilds of Co. Kerry, Martens were very common in the dense rocky mountain woods, and we used to trap numbers of them....I found the best plan....was to tether a young live rabbit to about 18 inches of string fastened to a peg, and place four strong steel traps around it, lightly buried under the grass, and secured by chains to other pegs...."[154]

I often see lamentations over the destruction of wild animals, and the Marten Cat mentioned as one that is almost extinct. About a fortnight since, the keeper on a distant mountain [in Co. Sligo] reported to me that he had found the feathers of several Grouse [*Lagopus scoticus*], destroyed by vermin. I gave him half-a-dozen traps, and desired him to bring me the animals or their skins. I received today....the eleventh Marten Cat, some of them being of immense size....it does not appear to me that there is any fear of the animal becoming extinct....[376]

Of an individual trapped at Bryansford, Co. Down, in 1891 and stuffed, there is the melancholy observation that it had probably been trapped before "as one foot was gone"[270] and there was a similar incident in Co. Wexford in 1937.[137]

William Thompson, who died in 1852, was of the opinion that the pine marten was present in every county in Ireland.[350] I have tried to piece together its distribution from after his death until O'Sullivan's survey in 1978-80[262] and have considered the period 1870-1975. Because the animal is so rare and striking, its appearances were often noted and many have been published, usually as short notes in such periodicals as the *Irish Naturalist,* the *Irish Naturalists' Journal,* the *Zoologist* and *The Field.* I have combed the literature for such records and have also written to museums both in Britain and Ireland for information on their holdings of Irish martens. Besides skins, skulls and stuffed specimens, the Natural History Museum in Dublin also holds manuscript records. The late Douglas Deane's files in the Ulster Museum have also been useful. He had evidently intended to write an article on the animal but did not get round to it. There were even a few snippets from Dublin Zoo on animals which had been sent there. All of this data I treated cautiously. Where the approximate year when a marten was noticed was not clear - and many people mentioned the animal as having been about in a particular district at an indefinite time in the past - or where a statement was vague hearsay, I rejected the information. To avoid possible duplication, I also counted two or more animals from the same year in a particular district as a single record. I have gathered the records - 216 in all - into three groupings, each representing approximately periods of 35 years: 1870-1905, 1906-40 and 1940-75. The results given by county are shown in Fig. 12.

Just because there were no records for a county during a particular period does not necessarily mean that there were no martens there. For instance, they were present in Donegal between 1870 and 1905 and 1941 and 1975, so it is unlikely that there were none there in the intervening years. Nevertheless, it seems probable that the absence from a county of any record from 1905 to 1975 means that the animal had become extinct there by or in the the twentieth century. The factors influencing the numbers of records in each of the periods are complex and any attempt to disentangle them is speculative and largely futile.

Over the whole 106 years the animals were recorded in a total of 30 counties, the exceptions being Cavan and Carlow, for which there is a doubtful record. So Thompson's surmise regarding the marten's presence in all 32 in the mid-nineteenth century was probably correct. Whatever the causes, thereafter the trend was certainly downward. Although interest in the natural world in Ireland increased in the 1960s and 70s, the number of records, and of counties with martens, still fell from the second to the third period. Moreover, five counties yielded records for the first period only, suggesting that the marten was no longer present in them by the middle of the twentieth century.

O'Sullivan used as the unit of his survey[262] (undertaken from 1978 to 1980) the 10 km squares of the Irish National Grid. From studies of the literature, records from the Biological Records Centre and information from taxidermists, gun clubs and the media, he compiled a list of squares with previous reports, or even strong rumours of the presence or former presence of martens. He then looked in all potential habitat in each square for droppings. When a square was found to be positive, he stopped searching and moved on, but he first checked all the contiguous squares. Altogether he investigated 428 squares. Although O'Sullivan's technique was conservative, he has had by far the most experience of the animal of anyone in Ireland and he was employed by the state, which managed most of the potential habitat in the Republic. So it is absolutely certain that any of the squares he recorded as positive held the beasts at the time. In addition it is just possible that a few of the animals might have been around in a small minority of the squares he designated as negative.

Fig. 12 (Opposite) Numbers of records of the pine marten in Ireland by county for the periods 1870-1905, 1906-40 and 1941-75. Counties with no records are blacked out. An Antrim, Ar Armagh, Do Donegal, Dn Down, Fe Fermanagh, Ld 'Derry, Mn Monaghan, Ty Tyrone, Du Dublin, Kd Kildare, Kl Kilkenny, La Laois, Lf Longford, Lo Louth, Me Meath, Of Offaly, We Westmeath, Wx Wexford, Wk Wicklow, Cl Clare, Co Cork, Ke Kerry, Li Limerick, Ga Galway, Le Leitrim, Ma Mayo, Ro Roscommon, Si Sligo, Ti Tipperary, Wa Waterford.

141

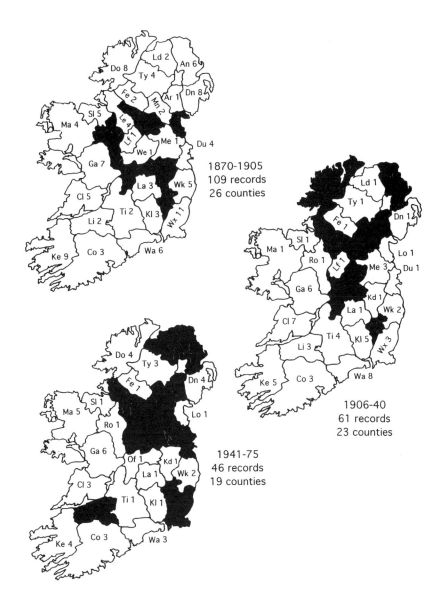

1870-1905
109 records
26 counties

1906-40
61 records
23 counties

1941-75
46 records
19 counties

Only 97 (23%) squares proved positive. These are shown in Fig. 13 and revealed an appalling contraction in the range of the animal during the previous few decades. It was only common in the western counties of Galway, Mayo, part of Sligo and particularly in Co. Clare. It had disappeared from Co. Donegal, where, from Deane's files, it is clear that it was still present in the 1960s. It was gone too from Wicklow, where one was seen by a naturalist of impeccable credentials in 1961.[235] O'Sullivan recorded the animal in only one grid square apiece in Cos Cork and Kerry. Yet Deane's files include three instances from around Killarney in the late 1950s and early 60s and it seems to have been locally common in Kerry around 1920, with instances near Miltown, Beaufort and Caragh Lake.[287] Martens had been trapped at Gougane Barra in Co. Cork in 1973 and at Killarney in 1973 and 1979. O'Sullivan scoured the relevant grid squares but could find no traces of any martens there. Apart from the west of Ireland, there were outlying populations on the Slieve Bloom Mountains in the centre of Ireland, in the Boyne Valley in Cos Louth and Meath, and at Curraghmore in Co. Waterford. It is worth adding that most of the records from the previous hundred years from Waterford had also been from this district, where, it may be remembered, for at least part of the time the animal was protected.

What had caused the drastic contraction in the range since the Second World War? O'Sullivan came up with a convincing argument. The increased demand for meat and wool in the 1950s led to government grants and low interest loans to improve farming on uplands. To protect their sheep from foxes, dogs and other predators, the farmers habitually laid baits liberally laced with strychnine, which was copiously available for the purpose. Little or no precautions were taken and the poison was usually put on fresh meat and often placed at the edges of woodland, which farmers often regard as a haven for foxes. As much of the upland was commonage - in other words it had several owners simultaneously - the same areas were often poisoned repeatedly by different farmers. We know that the marten eats a wider range of food than practically any other Irish mammal and that it will readily feed on refuse and carrion. At Dromore O'Sullivan found that it often hunted along walls and at the edge of woodland. So the poor pine marten was, of course, in grievous peril from such indiscriminate poisoning.

During his survey, O'Sullivan talked to families in Co. Clare who had, for up to four generations, trapped and snared the animals for their pelts, operations having continued right up to 1976, when the Wildlife Act protected the species. Despite this, the marten population in Clare

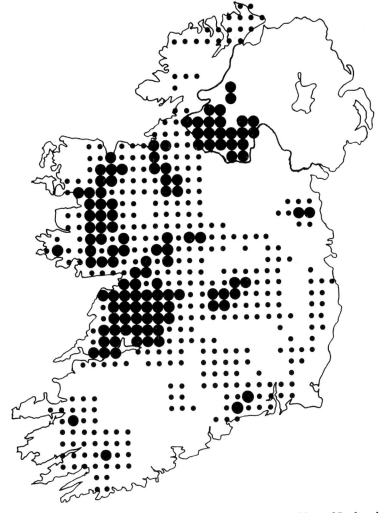

Fig. 13 The distribution of pine martens in the Republic of Ireland in 10 km squares of the National Grid as determined in 1978-80. Large symbols indicate positive squares, small symbols squares which were checked but proved negative. Also shown are approximate positive squares in Northern Ireland in 1993.

was apparently unaffected and O'Sullivan, who was stationed in Clare throughout the 1970s, considered that martens were common over most of it. Although there is much suitable habitat in Clare, the key factors may have been that sheep farming was relatively unimportant and that commonages are rare. Consequently the frequency with which strychnine was laid was likely to have been low. While O'Sullivan's theory cannot actually be proven, it is convincing, and no other more credible explanation has been put forward.[262]

The saga of the poor pine marten may yet have a happy ending. The availability of strychnine in the Republic was severely restricted by a series of government regulations in 1982-84 and virtually prohibited in 1991. O'Sullivan's captive breeding programme had run up to 1986. By this time there were indications that populations in the wild were beginning to recover. So the programme was terminated and the captives gradually released into the wild at Killarney, where happily their descendents thrive today. There is evidence that the marten is now expanding its range in the Republic. Since 1980 I have had reliable records from several districts, and even a few counties, in which O'Sullivan considered it to be missing. In 1981 an experienced naturalist told me of one poisoned at Castleconnell, Co. Limerick, which had been stuffed and displayed in a pub. However, Castleconnell is close to the county border with Clare, so this is hardly remarkable, although it is on the opposite bank of the River Shannon. I also had reports from foresters in 1986 of the animals being established 3 km further east at Newport, Co. Tipperary. In 2000 a naturalist informed me that there are now martens in the district around Donegal Town (Co. Donegal) and, in the same year, two were found dead at an undisclosed location in Co. Wicklow. My informant supplied a photograph of one of these and there is no question as to the animal's identity. I have had other reports from the Glenmalure and Roundwood areas. Pat Smiddy has noted five road casualties in east Cork and west Waterford since 1991, probably the result of an expansion of the stock at Curraghmore.

A resurvey, either by appropriately qualified civil servants, or by others funded by the government, is overdue.

In 1993 the Ulster Wildlife Trust and the Northern Ireland Forestry Service collated information of presence of martens in state forests around Northern Ireland. The 10 km grid squares which were unequivocally positive are shown in Fig. 13.[13] Such records are confined to Co. Fermanagh and the southern part of Co. Tyrone, where the animal is said to be doing well. There is also an isolated population at Baronscourt, further north in Tyrone, where numbers are believed to be falling. Such a distribution again seems to point to a diminution in the range from the previous few decades, for there were several sightings in Co. Down in the 1940s and 50s. In the well wooded district around Finnebrogue, near Downpatrick, one was trapped in 1947[340] and another spotted crossing the road in 1951.[62] Yet another was caught in a trap at Tollymore Forest Park in 1958[283] where, it will be recalled, Thompson, more than a century before, mentioned that martens were present. Since then there has also been a perfectly reliable record from Killyleagh in 1978, where one was knocked down on the road near the castle.[50] Killyleagh is but 8 km from Finnebrogue. Whether strychnine, which was readily available for poisoning "vermin" in the 1950s and 60s, has played any part in a probable decline of the animal in Northern Ireland is open to argument. At any rate, in 1972 the Welfare of Animals Act effectively ended its availability.

Chapter 6. Irish Bats

...bats are worth a little watching, and have secrets in their economy which must be found out by watching, and which the most skilful scrutiny of the largest series of dead specimens will not suffice to disclose. It is true, no doubt, that the time available to most of us for this kind of observation is rather limited; but to those who realise how attractive, not to say exciting, a form of pursuit it may become, opportunities will, every now and then, suggest themselves.

"The duration of flight among bats." C.B. Moffat. *The Irish Naturalist* 1905.

Several kinds of mammal have taken to the air, but whereas some rodents and marsupials and a few others can glide, bats are the only mammals which have achieved powered flight, a talent which has had a profound effect on almost every aspect of their lives. In particular flight dictates a rigid specification and, although there are nearly a thousand species of bat, the amount of anatomical variation is relatively small. This often makes it difficult to tell them apart. Even children can identify most of Ireland's mammals, at least when they see them stuffed and static in a museum, because they are so diverse. But few children, or even adults, could distinguish most of our native bats in the hand. In fact there are nine species of bat in Ireland which are listed in Table 17.

Bats' wings give a spurious impression of size and whereas the biggest bats of all, the fruit bats commonly called flying foxes, may have a wingspan of 1.7 m, their maximum recorded weight is only about 1.5 kg, equivalent to an adult rabbit. Close up most people are surprised at just how small a bat appears. Even then, because the body is clad in dense fur, it still looks bigger than it is. Most bats are mouse sized or smaller and Kitti's hog-nosed bat *Craseonycteris thonglongai,* found only in Thailand, is the smallest of all mammals and tips the scales at only 2 g. Readers will recollect from the Introduction to the Weasel Family that small size means relatively high surface area and that disproportionate amounts of energy are required to maintain body temperature. The wings of bats also increase their surface area, but at rest they are either folded up or are closed around the animal, when they act rather as an overcoat. Besides, during flight the heat generated incidental to the energy expended, is more than enough to maintain body temperature.

Table 17 - Wingspans (in millimetres) and body weights (in grams) of adults of the bat species found in Ireland. In extreme instances dimensions may be greater than those below. Measurements marked with an asterisk are exclusively Irish. Otherwise they include, or are entirely from, specimens from Great Britain or, for Nathusius' pipistrelle, mainland Europe.

Species	Wingspan	Body weight
Vesper bats (Family Vespertilionidae)		
Common pipistrelle *Pipistrellus pipistrellus*	200-234	5.0-7.4
Soprano pipistrelle *Pipistrellus pygmaeus*	199-232	4.3-6.8
Nathusius' pipistrelle *Pipistrellus nathusii*	220-250	*7.7-9.5
Long-eared bat *Plecotus auritus*	255-265	*6.5-8.4
Natterer's bat *Myotis nattereri*	250-300	7.0-12.0
Daubenton's bat *Myotis daubentonii*	220-245	8.5-11.0
Whiskered bat *Myotis mystacinus*	210-240	4.0-8.0
Leisler's bat *Nyctalus leisleri*	290-320	*14.0-20.1
Nose-leaf bats (Family Rhinolophidae)		
Lesser horseshoe bat *Rhinolophus hipposideros*	225-250	4.0-9.4

Worldwide, the many species of bat have adapted to a variety of foods, including fruit, nectar, flesh (in the form of frogs, lizards, birds, rodents and other bats), fish and blood, but the majority feed on insects, or related animals such as spiders and centipedes, and it is to this majority that all of the Irish bats belong. Incidentally, although only three species of bat, the vampires, all of which are confined to Central and South America, feed on blood (which they lap rather than suck), and two of these prefer to attack birds, this does not prevent some feeble-minded radio and television persons from trying to turn interviews with bat specialists into showcases for Count Dracula. The same persons also still feel compelled to ask whether bats entangle themselves in women's hair, when almost everyone who is interested will already know that the answer is a resounding "No".

The following general description of the lifestyle of bats applies to those in Ireland, although it draws upon information from Britain and further afield to furnish a more comprehensive picture.

Although bats' wings may look like leather in some illustrations or on dessicated museum specimens, and the Irish for bat - sciathán leathair - translates literally as " leatherwing", in life the wings are soft and flexible. They are dark coloured, almost hairless membranes composed of two thicknesses of skin with a layer of connective tissue sandwiched between containing nerves, blood vessels and a little muscle. The wing is spread over the greatly elongated fingers of the hand, which act rather like the ribs of an umbrella, and along the length of the arm. The thumb is free, very short and bears a curved claw, as do all the hind toes. The wing extends down to the legs and also between them as the *interfemoral membrane.* There is a gristly or bony projection sticking out from the heel, called the *calcar,* which runs along the edge of the interfemoral membrane and braces it. Thus all of the arms and legs are within the wing except for the thumb and feet. (Fig. 14)

Generally bats can manoeuvre more precisely in flight and execute much tighter turns than birds, abilities especially useful when pursuing insects. Two properties of the bat's wing contribute to such finesse. First, a bird has effectively only one finger, which, with the arm, lies along the leading edge of the wing, and the flight feathers stick out behind. In flight the bird has relatively little direct control over changes in the shape of the feathers, which, like hair, are effectively dead structures. In contrast, the bat has four fingers running through its wing, which is all living, flexible tissue, and so the animal can alter the shape of the wing at will. Second, air can pass between feathers, but this is impossible through the membrane of a bat's wing. By the way, any small perforations in the wing membrane soon heal over.

The power for flapping the wings comes from strong muscles in the upper chest and back. With so much muscle in the region of the shoulders, it is no surprise that they are robust, with big shoulder blades and stout collar bones anchored on the first ribs, which are particularly strong, as is the upper part of the breastbone. Otherwise, the skeleton of a bat is remarkable in being both sturdy and lightweight, the latter vital to an animal which flies. Along with the enlarged chest comes a large heart. The heart of a mouse-sized bat is perhaps three times as big as that of a laboratory mouse. This is necessary to supply blood rapidly to the flight muscles, and with it oxygen and fuel. As a means of progression, flight is exceedingly energy demanding.

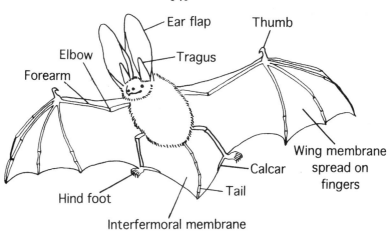

Fig. 14 Some important features of bat anatomy.

A great deal of fuel is burned during flight and consequently much heat is produced. However this can be rapidly dissipated from the relatively enormous surface of the wings. When the body temperature reaches a critical level, at about 40-42°C, valves in the blood vessels in the wings dilate and the wings are flooded with warm blood which is cooled by the air passing over their surface.

In bats the *femur,* the upper bone of the leg, is twisted through 180° along its length, with the net effect that the knee bends in the opposite direction to that of other mammals - a bat therefore genuflects back to front. As this occurs the interfemoral membrane, with the assistance of the tail and calcars, is thus brought forward to form a pouch. This can be used both as a kind of air brake and to catch insects. When a bat is hunting flying insects it may catch them in its mouth or in the pouch; it then bends forward and seizes the prey in its jaws. It may even bounce an insect off the wing and into the pouch.

The claws help a bat to move about, often on vertical surfaces, when it is not flying. The thumb claw is also commonly used in grooming and the hind claws enable the animal to hang securely, upside down, when it is at rest. Then, with the weight of the body transferred to its feet, a tendon in each toe automatically pulls down the claw so that both feet are furnished with a row of hooks from which the bat is safely suspended, even when it is asleep. A bat may also sometimes hang for a time by one foot and use the other to comb its fur. Good grooming is essential for a flying mammal and, besides painstakingly combing out its hair, a bat will also lick over its wing membranes.

Most male mammals have at least some of their external genitalia decently hidden. Although the scrotum may or may not be slung outside the body, the penis is normally withdrawn into a fold of skin. The exceptions, with both scrotum and penis on display, are the primates (obvious on ourselves and monkeys in the zoo) and bats. Sexing bats is a doddle.

The ears of bats are big and their hearing acute. They can also hear extremely high pitched sounds, far above those that we can, such sounds being described as *ultrasonic*. In all Irish bats, except for the lesser horseshoe, there is a very obvious structure within the ear, called the *tragus,* which looks like a smaller ear. The tragus of the lesser horseshoe merely forms a flange at the base of the ear. In fact most mammalian ears have a tragus, including are own, although it is usually relatively small. Look at someone's ear and you will see it as a small fleshy point sticking up and obscuring part of the bottom of the ear hole. The tragus in bats is believed to improve perception of the direction from which sounds come.

Although their eyes are relatively small, bats can also see fairly well, the old simile "blind as a bat" being untrue. The power of scent is also reasonably good, or at least much better than our own. Mothers, for instance, have no problem in recognising their offspring by smell in the midst of a host of others.

The teeth of babies are hooked spines which are present at birth or shortly thereafter. These enable the young to grip the mother's teats and in early life the baby remains attached to the mother in this way except when she is out foraging, when the youngster stops at home. If necessary, however, the baby can also be transported in like manner. Female bats have only two nipples, on the upper chest. Such positioning is important if offspring are to be carried in flight, as the load is then suspended from the mother's centre of gravity, an important consideration in any flying machine if the trim is not to be upset.

The adult teeth are markedly different. The canines are long and pointed, like those of a dog or cat, and are used to puncture the hard outer covering, the *exoskeleton,* of an insect and so gain a grip on it. Typically the teeth in the cheeks behind the canines each have a sharp, W-shaped, vertical flange which, working alongside that on the tooth opposite, is well adapted to chopping up insects.

Bats are pretty strictly nocturnal and are only occasionally abroad in the day. Pat Smiddy watched one flying over the Owenacurra River in Co. Cork at 6.20 pm in May 2000. He was able to catch it when it landed on bankside vegetation and it proved to be a whiskered bat. Whereas I once saw a bat at Oughterard, Co. Galway, on a bright sunny afternoon in June, most of my daytime sightings have been in winter, particularly in December around the old quadrangle building at NUI Galway, where I was employed for over 30 years. Indeed most daytime flying appears to be in warm periods in winter. As a number of bats were brought in to me from around the university during my sojourn there, and all proved to be pipistrelles, it seems likely that most of the day-flying individuals were too; more of pipistrelles flying in winter below.

In late summer bats build up body fat and then hibernate, dropping their body temperature to 1-2°C above that of the surrounding air and burning fuel at perhaps less than 1% of that when not hibernating. Energy consumption is thus reduced to a minimum in a season when there are relatively few insects available. But energy is still required to keep the metabolism ticking over and consequently by spring body weight may have fallen by 20-30%. During the winter bats will arouse from time to time and may forage on nights when it is warm enough for there to be sufficient insects around. For some, and possibly all individual bats, topping up fat in this way is essential in order to survive the winter. Young of the year, in particular, may have had insufficient experience in hunting by autumn to lay down enough fat and some undoubtedly perish in the winter as a result. Unfortunately, arousal itself requires burning extra fuel in order to warm up. Moreover, arousal does not necessarily occur at night. If a bat wakens during the day and needs to feed, it may be important to hunt immediately as waiting until dusk unfed in an aroused state would involve further fuel costs. In addition, if its fuel reserves are low it may be beneficial to hunt during the day when there will be more insects on the wing. All of this shows that, for bats in winter, energy balance is often at a critical point. Disturbance of winter roosts by human visitors, especially if repeated, causing unnecessary arousal in inclement weather and sapping vital fat reserves, may be fatal.

Sadly, bats abroad in daylight are in peril from predation by birds. Calculations based on various observations on birds chasing and killing bats flying during the day in Britain, together with figures on mortality obtained during the experimental release of bats in full daylight in Australia, suggest that a bat flying in daytime is at least one thousand times more likely to be killed by a bird than one flying at night.[336] Such a risk underlies the dire circumstances that force a bat to hunt in broad daylight: the absolute necessity of replenishing fat.

Even at night bats may be caught by owls. Owls regurgitate the indigestible parts of their prey - the bones, fur and feathers - as pellets, analysis of which provides an excellent record of the animals eaten. Skulls from several species of bat have turned up in the pellets of Irish owls from time to time. Mr Frank King, the doyen of Kerry ornithologists, observed a barn owl *Tyto alba* catch a bat in the full glare of a floodlight in a neighbour's garden.[180] The owl had abruptly changed direction in flight to catch a passing moth at the same instant as the bat had homed in on it, almost colliding with the owl, which instantly grabbed it in its left foot. The owl flew off into the darkness with the bat clasped in its talons.

During the day bats roost in buildings, crevices, caves and in hollow trees. Similar sites are used for hibernation, then called *hibernacula*. It is essential that the temperature in hibernacula should neither fall below freezing nor rise above the body temperature of the bat in hibernation, otherwise its temperature would rise too and with it the burning of precious fat. There should also be little in the way of draughts, which might chill the body. Bats will arouse and move if a roosting place becomes unsatisfactory, for instance if ambient temperature falls too low. Thus at the beginning of the winter some bats may install themselves towards the mouth of a cave and during hard weather move further in. Roof spaces rarely make satisfactory hibernacula. There is nothing between them and the cold air outside but slates or corrugated iron.

Most surveys of bat roosts in Ireland have been of those in caves and buildings, because they are easier to find and householders will often report the presence of the animals. We are largely ignorant of the extent to which trees are used. Unfortunately, a systematic survey for bats in crevices and crannies in trees would necessitate the use of an endoscope, a probe which allows the user to look into cavities. Endoscopes are now used routinely in medical diagnosis and in industry, during engineering inspections and in security searches. At least one heavy duty, industrial type would be essential for a survey of bats in trees in Ireland but the cost - perhaps £10,000 - has so far proved prohibitive.

In spring bats leave their hibernacula and usually form small temporary groups, sometimes of mixed sex and often in roosts which are used only in spring. Activity depends on available insects and therefore on the weather. If it turns cold the animals may return to hibernation.

Gradually the females from the smaller groups merge to form a nursery colony, consisting of pregnant bats, adult females which are not pregnant, young females of the previous year (which generally have not mated) and maybe a few immature males. The colony often settles in a single roost for most of the summer and the same roost may be used by the same colony year after year. However, some colonies may move about from one roost to another, carrying any youngsters born in summer with them.

A warm nursery roost is an advantage and it is therefore often, but by no means always, in a roof space. Slate, being dark grey, warms up rapidly in sunshine and the heat is radiated within. Anyone who has entered a roof space on a sunny summer day will appreciate just how warm it can be. Frequently the bats roost beneath the ridge beam, where the warmest air gathers and heat is radiated directly from the slates on either side. There is also mutual benefit from body heat when large numbers of individuals cluster together. This is probably particularly the case when a nursery roost is not in a roof space but in confined and largely insulated quarters, such as a cavity in a tree or in the stonework of a bridge.

All species of Irish bat may mate in late summer or autumn but ovulation does not occur then. Instead the sperm remains in the female reproductive tract through the winter, a plug forming in the vagina. Ovulation is in spring, rapidly followed by fertilisation. However, there are certainly also successful matings during the winter through to the following spring of females which have not been inseminated in autumn.

Ovulation depends on ambient temperature: a late spring means extended hibernation and delayed ovulation. Furthermore, the length of the pregnancy is affected by prevailing temperature: when it is cool in early summer the young develop more slowly. Temperature influences rate of development of the embryos both indirectly, because low temperatures result in fewer insects and therefore less food for the mother, and directly, because if it is cool the adults become torpid during the day and this slows the development of the embryo. Such daytime torpor saves on energy and is intermediate between the fully active state and hibernation.

Early births are advantageous as they leave more time for the young to gain hunting experience and to build up fat before hibernation. Warm weather in April and May may therefore be critical to the survival of young in the following winter, even though such young are yet unborn and may not even have been conceived! The importance of temperature in ensuring early births accounts for a nursery roost being selected for warmth. In an ideal nursery roost there is a minimum of daily torpor and, because the bats in it are fully awake, they are often heard squeaking during the day. This, if the colony happens to have lodged in a house, does not always endear them to any humans of a nervous disposition also dwelling there.

During the summer, male bats form separate, smaller colonies and some individuals may roost alone. This may not merely be because it is outside the mating season and the males have no interest in the society of the females or their offspring: they would, indeed, be of little assistance to them. The mating season and the winter lie ahead, both of which make calls on energy reserves, and the most advantageous course for males in early summer is to feed and to conserve energy by daily torpor. The latter would be impractical in a warm nursery roost. By the same token, adult females that are not pregnant would seem to put themselves at a disadvantage by clustering together in a warm nursery roost with those that are. They would surely be better off roosting somewhere where they could enter daily torpor and thus save energy. The explanation may be what zoologists call *kin selection*. It seems that females may often rejoin a particular nursery colony year after year and that there is therefore a strong likelihood that a female in a nursery roost will have both mother and sisters there as company. Even if she is not pregnant, therefore, a female, by clustering along with her mother and sisters and helping to keep up the temperature of the colony, is likely to improve the chances of them producing young which survive to breed. Thus, indirectly, by way of her relatives, she can pass on some of her genes to future generations. Blood is thicker than water.

A nursery colony tends to increase in size as early summer progresses as more and more females join it, numbers often stabilising shortly before the babies are born, which is generally in June or July. However, cold weather in late pregnancy can induce abortions and, if the summer is especially bad and insects are scarce, mothers may even abandon their babies. While pipistrelles have been known to produce twins outside Ireland, for Irish species generally, the rule is considered to be a singleton annually. The babies are licked over immediately after birth, both mother and offspring uttering high-pitched chirps the while.

Such early familiarisation with each other's calls, along with scent, ensures mutual recognition later.

Young bats are born with little hair and with their eyes closed. But they develop rapidly and after about 3 weeks are almost as big as the adults and can fly, although nursing continues for a further week or two. Thereafter the mothers and other older females gradually leave and afterwards the juveniles themselves until, usually sometime between the latter part of August and early October, the nursery roost is completely deserted.

As outlined above, the mating season is mainly in late summer or autumn and during it bats roost in small groups. Mating in all European bats is promiscuous. However, the precise social arrangements for it are imperfectly known for most species. It appears that males are often territorial and will defend the females both in the roost and probably its vicinity against their fellows. In some species the males are known to emit special calls, sometimes in flight, usually referred to as mating songs, which advertise their availability and attract females for copulation.

Most of the Irish species do not breed until their second year of life. (Thus a bat born in June 2000 would not mate until autumn 2001 at the earliest.) While this is invariably true of males, female pipistrelles may be mated and give birth during their first year of life. This means that a female pipistrelle can produce young when a year old, but that for females of the other species 2 years elapse before the first young are born. This has not been confirmed for Leisler's bat.

As autumn progresses and it grows colder, bats become torpid for longer periods and eventually enter hibernation. There is some evidence that adults do this earlier than juveniles which, inexperienced as they are, take longer to accumulate sufficient reserves of fat.

Many people erroneously regard bats as winged rodents, and therefore pests, and up until recently bats have had a bad press, mainly through superstition, old wife's tales, ignorance, certain works of fiction and because they are nocturnal. It is not the bats' fault if people have an irrational fear of them. Bats are not a health hazard in Ireland. They do not damage the fabric of houses, although urine from large colonies in certain situations may result in a smell which some people might find offensive. Bats never gnaw electric cables, their teeth being unsuited to this and quite unlike those of rodents. Neither do they bring nesting material into buildings; they do not build nests at all. Their droppings dry and may eventually disintegrate. Unless a bat is handled roughly, it will usually not bite and even then the teeth do not easily break the skin. Far from being pests, bats consume large numbers of insects - sometimes thousands in a night if these are tiny - and might even help to control some insect populations, although this has not actually been proven.

One extremely unrodentlike characteristic is their extremely low rate of reproduction. The females of most Irish species, as explained above, do not produce young until 2 years old, and nearly always a maximum of one annually thereafter. Furthermore, mortality among young is not unknown. Babies may die if they fall onto the floor of the roost or if their mothers cannot produce enough milk. Collisions while learning to fly are sometimes fatal, and insufficient fat deposits result in deaths during the winter. So the build up of numbers is slow. If a colony of bats is exterminated in a district, it may take many years for its numbers to be replaced. In short, recruitment is low and bat populations are exceedingly vulnerable. On the other hand, bats live for a long time. Studies in Britain on the species also found in Ireland show that average life spans of 4-6 years are not unusual and that some individuals, if they outlive the perils of youth, die in their teens or even in their twenties. Bats are the most striking exception to the general rule that small mammals tend to have short lives.

Because they are at such risk, all species of bat are protected throughout Ireland. If bats decide to share your house in Northern Ireland and you find this intolerable, get in touch with the Conservation Branch of the Department of the Environment, or the Northern Ireland Bat Group at the Ulster Museum. In the Republic contact your local conservation ranger or, in Dublin, the Dublin Bat Group. All are licensed to handle bats and most, perhaps all, will be overjoyed to make the acquaintance of your colony. As there are now bat groups in Galway and Cork, they may also be able to help.

Besides unauthorised ejection from their homes and illegal deliberate extermination, bats face other hazards at the hands of man. Those hibernating in caves may be disturbed....or worse, by human intruders. (Some 30 years ago one university student, now respected in the field of Irish natural history, told me that, as a boy, he and his companions amused themselves by setting fire to bats in a cave.) Some treatments of roof timbers are lethal to bats for years. Buildings are altered during renovation or conversion and suitable roosts destroyed. This has been a particular problem in recent years in the Republic with the economic boom. In contrast, the roofs of derelict properties eventually fall in with an identical effect. Felling of old trees, often considered unsightly or dangerous, also removes suitable roosting places. Commercial conifer plantations, apart possibly from their edges, are usually poor habitat for most species. The widespread use of pesticides reduces availability of insects, as does habitat destruction: the removal of hedgerows and trees and the canalisation of rivers all decrease numbers of suitable feeding places. The change from haymaking to silage means that many insects that fly and breed over plants in bloom are reduced in number.

Bats navigate in the dark and catch insects by *echolocation,* which is a much refined version of the echo-sounding used in ships to determine the distance to the sea bottom. Echolocation by bats was first demonstrated in the early 1940s by an American biologist, Professor Donald Griffin. Although bats make squeaks that are fully audible to humans, these are used in communication between individuals. Almost all of the sound used in echolocation is *ultrasonic* - too high pitched to be heard by humans - and is emitted as short pulses through the mouth in nearly all vesper bats and through the nose in noseleaf bats (Table 17). The pulses pass out, bounce off objects, and return as echoes to the bat, allowing it to build up a picture of its surroundings and to locate insects. So the bat does not "see" continuously with sound, but rather in a series of snapshots, each corresponding to a returning echo from a sound pulse. You, dear reader, may be puzzled as to how this can be. How can bats detect tiny insects from echoes when we cannot detect anything much smaller than the side of a house in this way? And why such high pitched sounds? To answer these questions a brief digression is necessary into the physics of sound waves.

A wave, whether on the surface of the sea or a sound wave, is a vibration passing through a medium. This is perhaps easiest to demonstrate with a skipping rope. If one end of a rope is tied to something and the other end grasped in the hand and moved up and down, a series of waves passes along its length. The faster the end is oscillated up and down, which of course requires more effort, the shorter the waves become. This demonstrates that the greater the frequency of a wave (the rate of moving the rope's end up and down), the shorter the wavelength and that higher frequencies require more energy. All of this applies equally to sound waves. Now, suppose that you were to place a brick in the path of the rope and repeated the process. You would find that with short waves (corresponding to higher frequencies) the wave would be interrupted by the brick, but not with long waves. Small objects will therefore interrupt waves of short wavelength but not of long.

Sound waves of high frequency, therefore, which produce high notes, have short wavelengths and take more energy to produce. Frequency is measured in vibrations per second: one vibration per second is one hertz; one thousand vibrations per second is one kilohertz (1 kHz). Middle C on the piano is 0.26 kHz (260 vibrations/second), top D on the piccolo 4.75 kHz. Children can hear up to 20 kHz but the upper limit falls with age. Bats make virtually all of their echolocation sounds above 20 kHz which, by definition, are ultrasonic. The highest pitched sounds produced by an Irish bat are by the lesser horseshoe, at about 113 kHz.

Low frequency sounds carry much further than high ones, which are rapidly absorbed in the air. This is why you can hear a man shouting much further away than a woman. Ultrasounds have therefore to be very loud in order to carry any distance. Pulses leaving the nose or mouth of a bat are about as loud as the unnerving sound of a jet engine at close quarters on the tarmac, or several times the legally permitted level of loudness at a discotheque. So loud are they that they would not only prevent any echo being heard, but might actually deafen the bat if there were not some mechanism to prevent this. In all mammalian ears, including our own, three tiny bones in the middle ear transfer the vibrations caused by sound striking the eardrum across to the inner ear, where the sounds are analysed and the resultant information sent to the brain along nerves. The bones act as levers which amplify the vibrations. When a bat emits an echolocation pulse two special muscles attached to these bones contract and dampen the sound. It is in the interval between one pulse and another that the bat listens for the echo.

The great energy invested in producing high frequency sounds of such intensity is repaid by their short wavelength, which means that very small things will interrupt the waves (like the brick does the short waves with the skipping rope) and cause echoes to return to the bat. At a frequency of 20 kHz the wavelength is 17 mm, at 113 kHz 3mm - lengths in the same size range as insects. However, insects smaller than 3mm can also be caught. The situation is similar to the brick and the skipping rope: the waves continue to be interrupted by the brick even when they are somewhat longer than it. In fact the sounds produced by bats are complex and can provide information not only on the size, but on the shape of an insect and the movement of its wings.

A second advantage to bats of ultrasonic sounds is that they are more readily focussed into a narrow beam. The reasons for this are complex and outside the scope of this book.

The pulses of sound produced by vesper bats last only a few thousandths of a second. Within this short time, the pulse starts high, usually about 70-80 khz and then sweeps down to around 30 kHz. The upper and lower frequencies vary with species. So the sound produced, if we could hear it, would be rather like that of a trombone when the slide is pushed out, in music a *glissando*. Such a pulse is said to be *frequency modulated*. In fact at any point the sound pulse is not a pure note: although most of the sound is concentrated at a particular frequency, some will be both above and below it. So sometimes, if the bat is in the hand, a tiny part of the sound may be sufficiently low-pitched as to be audible to us as a click.

A bat detector converts the echolocation sounds made by bats into ones which we can hear. Basic models start at under £100 and on these the sound pulses come out more or less as clicks. This is because the cheapest detectors can only be tuned to one narrow band of frequency, so you hear only part of the frequency modulated pulse. But by tuning up and down you can find out how high the pulses start and how low they go down. With this and other features of the sound perceived through the detector and with experience, it is often possible to distinguish individual species of bats. But it is not always a simple matter and requires experience, particularly as a given species may vary its sound with a particular situation. More advanced models can also pick up sound over a wide range of frequencies at the same time, so that the whole of every pulse is heard, making identification simpler. The most advanced of all, which are now readily available - at a price - record the sound in fine detail on a microchip. This can be analysed in the laboratory by computer later and the species of bat usually determined with considerable accuracy. In fact bat sounds are rather more complex than described here and such complexity is most effectively analysed and interpreted in this way. But even a simple detector is well worth having and opens a window into the amazing world of bats.

You are standing on a bridge over a slow-flowing stream on a balmy summer night, when all is silent save for the faint swish and gurgle of the water and the gentle rustle of leaves on the nearby trees. Then you switch on the detector and the evening is flooded with sound, the sound of the echolocation pulses of bats as they fly up and down beside and over the water, hunting insects. With the sound as a guide - and the detector is often a great help in locating precisely where the bats are flying - you will then probably be able to pick out individuals in the gloom. Use of a torch in such a situation is dubious: it will affect your night vision and you may not be able to keep it trained on a bat. On the other hand, if you are standing just above the level of the water, and the bat is flying nearby close to its surface, a torch is invaluable. Water, stagnant or slow-flowing, with trees and bushes alongside, is one of the best places to listen for bats - all Irish species of bat may feed around water - but deciduous trees away from water, streetlights, hedges, slurry pits and anywhere there are likely to be night-flying insects are worth checking. A bat detector is as useful to a bat enthusiast as binoculars to an ornithologist.

When a bat is cruising in search of prey, it economises on pulses, because they use up energy. It may then perhaps produce less than 10 per second. Once an insect is located this rises to around 20-40 and, in the closing stages of pursuit and capture when more and more snapshots of what the insect is doing are more and more rapidly required, to around 100-200 per second and the length of the pulses shortens. Because the pulses are then so short and are closer and closer together, they sound as a brief buzz on the bat detector. When you hear such a "feeding buzz", you can be pretty certain that an insect has been caught. On the other hand the pulses are softest when the bat is closing with an insect, because they do not have far to go, and loudest when a bat is searching for prey, as the sound has then to travel farthest. The whole process from locating the insect to capture may take only a few seconds. A given species of bat may also vary the length and sometimes other characteristics of its pulses depending on the conditions in which it is hunting.

Until well into the 1980s the heyday for study of bats in Ireland had unquestionably been the latter part the nineteenth century. John R. Kinahan (1828-63), a lecturer in zoology and botany and secretary of the Dublin Natural History Society, was the first to specialise on bats. He was followed by others including Nathaniel Alcock (1871-1913), a medical doctor, and H. Lyster Jameson (1875-1922), a zoologist employed variously in Ireland and abroad. Jameson appears again in Chapter 12 for his work on house mice. Those readers who have visited the magnificent show caves at Marble Arch, Co. Fermanagh, may recall being told that, in 1895, in company with the French speleologist, E.A. Martel, he was the first to investigate the system of caverns and underground passageways there.

By the end of the nineteenth century it had been established that there were only seven species of bat in Ireland, a figure that remained unchanged until the closing years of the twentieth century. These early workers were assiduous in collecting and identifying specimens and garnering any snippets of information on the habits of individual species. Moreover, they published everything that they could. They were constantly on the look out for material. Jameson even sent out flyers begging for bats, dead or alive, and wrote an article in the *Irish Naturalist,* then a thriving periodical with issues every month, expressly on how specimens might be secured and dispatched to him.[166]

> "....specimens in such cases can best be obtained through farm servants, &c, who may know the part of the building in which to search for them....Workmen engaged in pulling down or repairing buildings frequently meet with colonies of Bats....When Bats are known to inhabit such a place as a hole in a wall, tobacco smoke puffed in will generally succeed in bringing them out....In felling old trees, Bats may be found in holes and crevices and under loose bark....Bats frequently fly in at open windows. I captured four specimens this way last summer and autumn in my room at Killencoole. When Bats are on the wing at dusk a well-aimed charge of No. 8 (snipe) shot will sometimes secure a specimen....Specimens when procured, if alive, should be sent by rail, labelled perishable; if dead they can be sent by parcel post, but it is preferable to send by rail....It is much better in most cases to put the specimens as they are captured into spirits....

I have, by the way, myself seen the tobacco smoke technique used, in the 1970s at Curragh Chase, Co. Limerick. Something that was squeaking and scrabbling in a cavity in the bark of a mature conifer was persuaded to emerge by the gentle blowing of cigarette smoke into the cavity. It turned out to be a couple of long-eared bats.

Charles B. Moffat (1859-1945), who qualified for the law but quickly forsook it for journalism, was regarded by the great Robert Lloyd Praeger as "the most accomplished naturalist Ireland has produced". He was often able to identify individual bat species on the wing, away from the roost - no mean accomplishment without a detector. He continued research on Irish bats from the late 1890s up until the early 1920s and, unlike the others, sometimes stayed up all night to make his observations. Those on the behaviour of Leisler's bat, published in 1900,[218] were not to be superseded for almost a century (Chapter 8).

From about 1920 until 1980 is largely a void in the study of Ireland's bats. At any rate, almost all that appeared in print - apart from some review articles in the *Irish Naturalists' Journal* by Moffat - were occasional distribution records for the bats themselves and for their fleas.

At a meeting in Bonn in 1981, scientists representing 16 European nations expressed concern over the rapid decline in bat populations throughout most of Europe; they called upon governments to give immediate protection to bat roosts. This, together with the catastrophic fall in numbers of the lesser horseshoe bat on the continent, a species which was believed to be still common in Ireland, prompted Dúchas to undertake a National Bat Survey in the Republic, in which bat roosts were located, species in them identified and numbers counted. It should perhaps be pointed out that such counts were undertaken at any time throughout the year, and it will be apparent from remarks earlier in this chapter that numbers of bats in roosts vary seasonally, being largest around June and July in nursery roosts.

The Survey was under the direction of Paddy O'Sullivan, who was a seasoned mammalogist, having studied pine martens in Co. Clare in the 1970s (Chapter 5) and who had also been monitoring bat roosts at the same time. The work was carried out mainly by conservation rangers, who were briefed on where to find bats and how to identify them, and not a little by O'Sullivan himself. The rangers also liaised with pest control agencies and members of the public and responded to those complaining of domestic invasions by bats. Publicity was generated through lectures, articles and via radio and television. For success, a project of this kind requires someone with both a genuine ability to organise and virtually unlimited enthusiasm. Fortunately O'Sullivan had these in full measure and his fervour rubbed off on others. The Survey ran from 1985-88 and located around 1,400 roosts. Unfortunately it was not actually published until 1994.[261] In the same year as the Survey commenced, the Northern Ireland Bat Group was founded. Its distribution records for bat species in Northern Ireland were generously made available to O'Sullivan and were incorporated in the publication, which thus covered the whole of the island of Ireland. While the Survey is invaluable, it is very much in the way of a baseline. An updated version, for the Republic at least - more up to date information is now available for Northern Ireland as explained below - incorporating the additional data that have been gathered since, is already desirable.

In 1983 Dr Kate McAney began her PhD research on the lesser horseshoe bat in Co. Clare, working out of NUI Galway (Chapter 7). She graduated in 1987. The shift from postgraduate research to full-time employment in wildlife in Ireland is not easy; suitable jobs are scarce. Nevertheless, after a few years she was appointed Irish Field Officer for The Vincent Wildlife Trust, a charitable wildlife organisation based in London, a post in which she continues to this day. The greater part of her present duties involve conservation and the raising of public awareness of bats. She brings to her work complete dedication, considerable organisational skill and undoubtedly a little charisma.

These developments in the 1980s have ushered in a much wider interest in Ireland's bats and have been followed by the setting up of bat groups in Dublin, Galway and Cork. Dúchas, McAney in her present capacity and the Bat Groups are at present primarily concerned with the location and documentation of roosts and with conservation. Subsequent pure research on bats was continued at NUI Galway, by Dr Caroline Shiel (Chapter 8), mainly on Leisler's bat, and by two postgraduate students at Queen's University Belfast: Dr John Russ and Mr James O'Neill. During her research at Galway, McAney taught herself how to analyse the droppings of lesser horseshoe bats for diagnostic fragments of insects, and thus work out on what the bats had been feeding. The technique is dealt with more fully in Chapter 8. It was further developed for other bats by Shiel, both as undergraduate and postgraduate, and also by other undergraduates under my supervision: Claire Sullivan, Dan Flavin and Sinead Biggane.[307] Consequently we have a good idea of the diet of most of the Irish bat species; all of the descriptions of the diets below stem from this research. Summaries of the results are given below and in the following two chapters. Where a percentage is quoted it represents percentage frequency in droppings. (An explanation of percentage frequency is given in Chapter 1.) I must stress here that although the first person plural is used of such research, it was my students and not I who did virtually all of the hard work.

In the year 2000, the Northern Ireland Bat Group published up to date distribution maps for all species of bat in Northern Ireland[3] and, as I write, John Russ has a booklet of further relevant information in press, particularly on the habitats which the different species seem to prefer.[285] These he determined by monitoring the presence of species with an advanced bat detector in various habitats in a random selection of 1 km squares spread throughout all six counties. All of the comments below on the present abundance or scarcity of the various species in Ireland are based on the information in the National Bat Survey and Russ's/The Northern Ireland Bat Group's Surveys.[3,261, 285]

The remainder of this chapter deals with individual species of Irish bat except for the lesser horseshoe and Leisler's bat, about both of which so much is now known in Ireland as to justify a full chapter each.

There are three species of pipistrelle bat in Ireland (Table 17). Pipistrelles are often described as a small bats, but it will be clear from the table that several other Irish bats are not much bigger. The clearest distinguishing characteristic of pipistrelles is that the tragus, although longer than broad, is blunt and rounded at the end. In all other species in Ireland the tragus, if longer than broad, is distinctly narrowed at the tip. A further diagnostic feature of the pipistrelle is a lobe of skin, the *post-calcarial lobe,* which sticks out behind the calcar on each side (Fig. 15).

Today it is universally agreed that the common pipistrelle is by far the most abundant species of Irish bat,[3,261,285] something which Moffat, Alcock, Jameson and Kinahan would not have disputed, and in the Irish literature there are many references to it, simply as "the" pipistrelle, or even as "the common bat". Not only is it widespread on the mainland, it has also been recorded on Cape Clear and Sherkin Islands, Co. Cork, and on Great Blasket and Tearaght Islands, Co. Kerry. One was also picked up dead on the South Arklow lightship.[231,243]

Alas! There is not one common pipistrelle but two. In the 1990s in Britain, with the increasing use of bat detectors, it became apparent that some pipistrelles produced echolocation pulses which were higher pitched than others, and that the two types lived in separate colonies. This led to closer examination of the types and the discovery of small anatomical differences. Finally investigation of their DNA (deoxyribonucleic acid - the material on which genetic information is

Fig. 15 (Opposite) Some features used in the identification of Irish bats: the shape of the tragus in the ears of six bat species; the hairy interfemoral membrane of Natterer's bat; the post-calcarial lobe of the pipistrelle; the attachment of the wing to the ankle/toe of Daubenton's and whiskered bats; the nose-leaf of the lesser horseshoe bat. Not to scale.

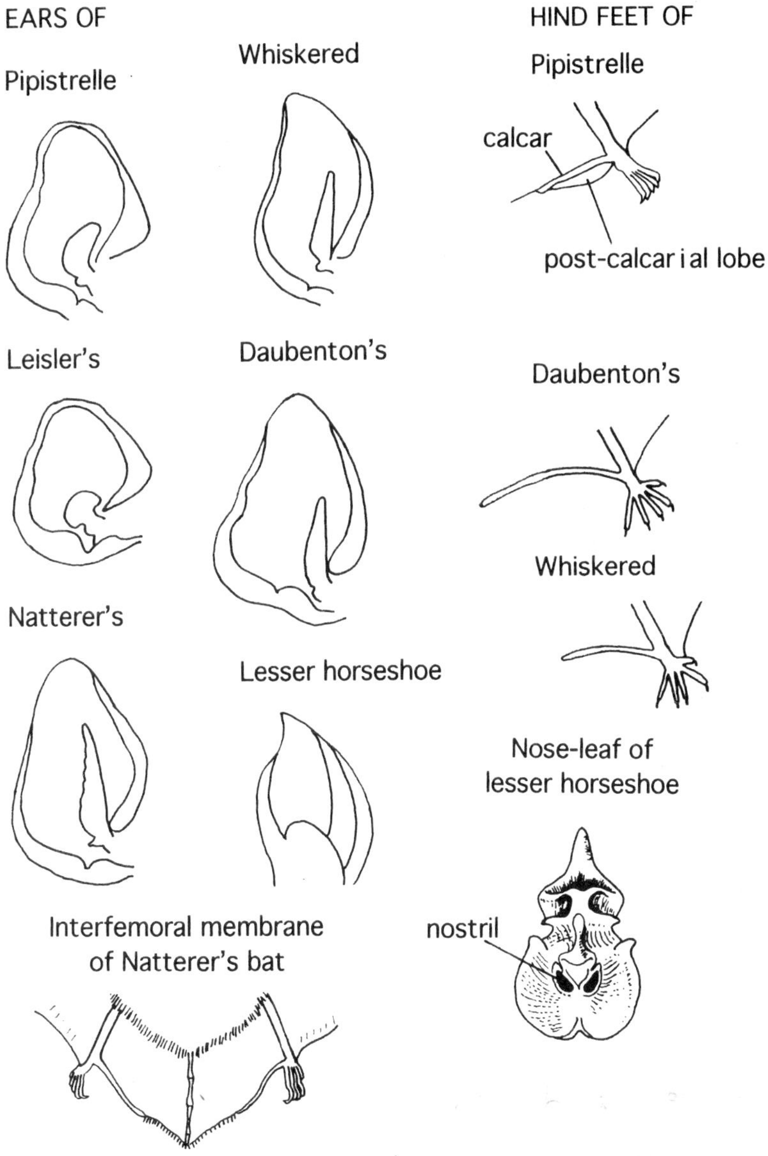

EARS OF

Pipistrelle

Whiskered

HIND FEET OF

Pipistrelle

calcar

post-calcar i al lobe

Leisler's

Daubenton's

Daubenton's

Whiskered

Natterer's

Lesser horseshoe

Nose-leaf of
lesser horseshoe

Interfemoral membrane
of Natterer's bat

nostril

carried) showed that the two were distinct species.[151] That with the lower pitched pulses is still called the "common" pipistrelle, and the higher pitched species, appropriately, the "soprano" pipistrelle. The distinguishing characteristics are otherwise slight. The "common" tends to be slightly larger, has somewhat darker fur and has a darker coloured face, giving it a masked appearance. The "soprano" often has exposed pink flesh on its face and is smellier. Using an advanced bat detector, John Russ first confirmed the presence of both common and soprano in Northern Ireland and it is now known that the two species are widespread there, with the soprano somewhat less numerous.[285] There is no reason for believing that the status of both is not similar throughout the rest of Ireland. Unfortunately, of course, references in the Irish literature to "the pipistrelle" could refer to either. This cannot be helped, but should be borne in mind in the paragraphs below which, for the sake of convenience, assume only one species.

Not only are pipistrelles the commonest bats, generally they also form the largest colonies, sometimes of hundreds. The National Bat Survey found some colonies of around 600 individuals and over 1,000 were counted as they left in the evening from a roost in Co. Antrim on 6 July 1996.[3] The record, however, is from a house near Belfast International Airport, where, in the summer of 2000, Mr Mark Smyth, of the Northern Ireland Bat Group, counted 1,500 as they emerged before he gave up. Pipistrelles have a strict preference for confined spaces, such as behind window sashes, under tiles and weatherboard, behind facia and soffit and within the shallow cavities of flat roofs. A favourite place is the narrow roof space in the flat or gently sloping roof of the single-story kitchen extension that is so often built onto the back of older Irish houses.

The pipistrelle is fond of flitting, of moving into a roost for a month or so and then moving on again. For several years a crevice in the building in which I worked in Galway was thus occupied by up to 48 pipistrelles in late April and early May in this way. The cavity was then deserted, less than 20 individuals returning for the latter part of August and early September. The abrupt arrival of an army of bats often alarms a householder. Some ladies cannot abide the sounds of bats above their heads in the kitchen. And because the colonies are often large, are packed closely together, are vocal, make scratching sounds as they crawl along the ceiling boards, and, it must be admitted, sometimes have a tendency to smell, this species causes more problems with the public than any other.

The pipistrelle is also the bat most likely to be spotted abroad in winter. Thompson mentioned seeing them thus several times,[350] and once observed one pursued by a pair of pied wagtails *Motacilla alba,* an example of the problems bats face from birds during the day. Thompson actually noted the air temperatures when he saw them on 2 days in January 1834, both at about 11°C, and therefore when it was relatively warm.

On a total of 114 evenings, from 1 November 1901 to 28 February 1902, Moffat[219,227] noted the temperatures on which he saw pipistrelles flying near his home at Ballyhyland, Co. Wexford. He concluded that they would certainly fly if the temperature was above 6°C, the probability declining sharply if it were any cooler, so that below 4°C he never saw any on the wing. Numbers of flying insects are closely correlated with air temperature. Recent research on pipistrelles in Cambridgeshire verifies that, in calm air, 6°C is the critical temperature for pipistrelles to hunt in winter, which says much for Moffat's lone observations so long ago.[16]

He noted that this species generally emerges between 12-20 minutes after sunset and that it often hunts up and down along regular beats,[227] observations that have since been repeatedly confirmed.[151]

He also concluded that midges and "gnats" formed an important part of the prey and that the capture of a large prey item resulted in the bat disappearing from its beat for a couple of minutes, which he attributed to its having to perch somewhere in order to consume its prize.

We collected and analysed droppings each month from June to September (total analysed 160) from a window sill under the entrance to a pipistrelle roost, a crack beside a window frame, in a mansion near Bunclody, Co. Wexford.[342] The main items in the diet were caddis flies (Trichoptera 16%) and midges (Chironomidae and Ceratopogonidae 30%), the latter attestifying to the accuracy of Moffat's conclusions. Midges belong to a major grouping of insects, the Diptera or two-winged flies. Most flying insects have four wings. Only the front pair of wings function as such in the Diptera, the hind ones being reduced to club-shaped structures, which are organs of balance. The Diptera include not only midges but several other midge-sized insects, also the daddy-longlegs or craneflies (Tipulidae), the familiar domestic flies and many others. Besides midges, the pipistrelles consumed a further 36% of other Diptera of a variety of families. The remainder of the diet (18%) was a mixture of insects. The best place to catch midges and particularly caddis flies would be near water, as both have aquatic larvae. In Britain vegetation near water is known to be a favourite feeding place for pipistrelles, although hunting may occur in a variety of habitats.[151] The menu as a whole represents night-flying insects. Caddis flies and midges fly at night and the other Diptera caught were predominantly the nocturnal kinds (Nematocera). This suggests that pipistrelles catch their prey while it is on the wing, rather than gleaning insects and similar prey while they are at rest on vegetation or on the ground, as some bats do. While the remains of spiders (Araneida), which do not, of course, fly, were recorded at 2%, it was entirely possible that the bats actually found them in the roost and, as bats void faeces in flight, any collected at the roost would be biased towards anything that happened to be caught there.

Bats, like other wild mammals, suffer from a range of parasites (Fig. 16): in Ireland fleas, mites and ticks have often been recovered from their outsides, and sometimes flukes - smaller versions of the common liver fluke *Fasciola hepatica* of livestock - from their intestines.[112] They also carry nasty looking, spider-like, long-legged insects, about 3 mm long, known as bat flies (Nycteribiidae). These are actually Diptera which, in the course of evolution, have lost their wings altogether. These insects scurry over the body of the bat, the tip of each leg being equipped with a couple of stout claws, which enable them to hang on to the hair. Like fleas, they live by sucking blood. Fleas and bat flies, however, do not have it all their own way, for we have recovered remains of both from bat droppings, evidence surely of successful grooming by the host.

In Ireland the pipistrelle and Leisler's bat have a further tormentor, the bat-bug *Cimex pipistrelli*. This is about 6 mm long and similar in appearance to the infamous human bedbug *Cimex lectularius*. The latter inhabits bedrooms - usually in quantity - lurking in crevices, often in the skirting board, in plaster, or under wallpaper during the day, emerging to visit the sleeper at night for a blood meal, before returning to its lair. The bat-bug operates in much the same way on pipistrelles, except, as you might expect, that it feeds during the day. In Britain the bat-bug is common but, presumably because it lives most of its time away from its hosts, is not often found. In Ireland there are only four records of the species, two from pipistrelles and two from Leisler's bats,[242] but then people do not often search for bat bugs in crevices in roosts. The word "bug", incidentally, which usually indicates a bacterium to a medical man and any sort of creepy-crawly to the general public, has a very precise meaning to the zoologist: insects of the order Hemiptera, all of which have mouthparts adapted to sucking.

Bat bug

Bat fly

Bat tick

Bat flea

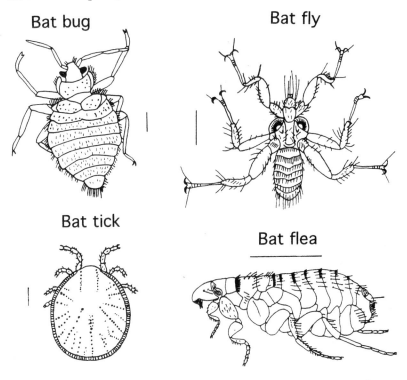

Fig. 16 Some parasites of Irish bats. The line scale given with each drawing is 1 mm.

As can be seen from Table 17, there is a third kind of pipistrelle in Ireland, Nathusius' pipistrelle. Unlike the soprano, Nathusius' has been recognised as a distinct species since early in the nineteenth century. Nathusius' is a bigger bat than the common pipistrelle (Table 17), the fur is shaggier, the hairs on the back have distinctive grey tips and there are a number of other smaller diagnostic features. As a whole these differences are insufficient to alert a bat worker to the fact that the bat is something out of the ordinary on a casual inspection, if there is no reason for suspecting that there are any Nathusius' around. Up until the late 1980s there had been few records of Nathusius' in the British Isles as a whole,[284] and these were thought to be migrants.

One night in August 1996 John Russ, using an advanced bat detector, recorded a bat flying along a line of deciduous trees at Moneymore, Co. 'Derry. On analysis of its echolocation sounds he identified it as a Nathusius' pipistrelle. Subsequently a live specimen was captured in September 1996 in Windsor Park, Belfast, passed to the Ulster Museum, where it was identified, with some perspicacity, by Mrs Angela Ross and Mrs Lynne Rendle, both members of the Northern Ireland Bat Group. In 1997 John Russ and John O'Neill located a nursery colony of around 150 in farm buildings in parkland on the outskirts of Antrim town, which had been in residence there from at least the beginning of May.[286] Using a bat detector, they showed that there were no common or soprano pipistrelles whatsoever in this colony. Since then Russ has found two other colonies in Co. Antrim[3,284] and discovered males roosting singly within 50 m of these. He has, with the aid of his bat detector, also followed the species as it foraged over riparian vegetation, parkland and woodland.[151]

It has now been established, both from bat detector records and from individuals "in the hand" that Nathusius' pipistrelle occurs over a wide area in north east Ireland, in Cos Antrim, Armagh, 'Derry, Down, and Tyrone.[284] Whereas there are also two records - from a bat detector only - in the Republic, from Cos Dublin and Laois,[284] it would be good to have more, with some actual specimens. It is, of course, entirely possible that the species occurs over a much wider area of Ireland than has been determined to date, but we must not jump to any such conclusion without further evidence.

How long Nathusius' pipistrelle has been in Ireland is unknown. As will be clear from earlier in this chapter, there had been very little in the way of study of bats during the 60 years preceding the early 1980s, and so Nathusius' could have been overlooked, especially as it is similar in appearance to the common pipistrelle. On the other hand, there is evidence that Nathusius' has recently expanded its western range in Europe into Great Britain. Thus, in Scotland, although pipistrelles have been intensively studied at the University of Aberdeen for many years, it is only recently that Nathusius' has been recorded in the district.[286] Any evidence of a gradual increase in numbers in the British Isles in the future will support the most likely explanation for it having only been detected in Ireland in the late 1990s - that it is a recent arrival here.

Nathusius' pipistrelles are known to undertake seasonal migrations in continental Europe, moving south west in autumn to a less severe winter climate, and making the return journey in the following spring.

That it has only been recorded in Ireland from May to October[284] seems to hint that our populations may also be migratory. Nevertheless, most records of bats in Ireland tend to be from April to October, because the animals are usually in hibernation for the rest of the year. Moreover, a southern migration from Ireland would have to be to Spain or Portugal, where the species is rare. Besides, winters in Ireland are amongst the mildest in northern Europe.

The long-eared bat is easily identified. As the name suggests the ears are exceedingly long, not much shorter than the entire length of the head and body. When the animal is asleep, the ears are folded back along its flanks, tucked underneath the wings, and then look rather like ram's horns. This may be an adaptation to conserving heat, as erected they add appreciably to the bat's surface area. However, when they are stowed away, the tragus continues to protrude and its appearance can give rise to confusion. One man telephoning me about a long-eared bat he had found described it has having four ears! This species has a placid disposition and will not attempt to bite unless roughly handled. Incidentally, most bat workers refer to the long-eared as the brown long-eared, to distinguish it from a close relative, the grey long-eared bat *Plecotus austriacus,* which occurs in Britain but not in Ireland.

Summer colonies of long-eared bats are unusual in including both male and female adults.[220,252]

Next to the common pipistrelle, and outside the districts where the lesser horseshoe occurs, the long-eared is probably our most abundant bat, although in recent years the Northern Ireland Bat Group has found more roosts of Leisler's bat (108) than of the long-eared (77).[3] Its primary habitat in Britain and in Ireland, where this has been investigated, seems to be broad leaved and mixed woodland but it is also commonly recorded around isolated trees in parks and gardens.[285] It is therefore surprising that it has also been recorded on several largely treeless offshore islands including Rathlin, Co. Antrim, Aranmore, Co. Donegal, Inishvicillaun and Inishabro, Co. Kerry, Achill, Co. Mayo, at the Tuskar lighthouse, Co. Wexford,[243] and recently on Clare Island, Co. Mayo. Although plentiful, the numbers of individuals in roosts are rather lower than for most of our other bats, being observed to be usually "under 50" in the National Bat Survey[261] and "around 20" in Northern Ireland.[3] The preferred roosting accommodation is commodious, such as in attics or under the roofs of churches, so it is not want of space in suitable roosts that limits the numbers there. At the turn of the century Jameson mentioned colonies in churches at Charlestown, Louth and Killencoole, Co. Louth, Castlemacadam, Co. Wicklow, Cappagh, Co. Waterford, and in Kilmore Cathedral, Co. Cavan.[167] These bats may hang up in the open, sometimes on beams, particularly the ridge beam, and are then easily spotted.

Long-eared bats have short, broad, rounded wings of relatively large area, which are best suited to slow flight, to making sharp turns and to hovering. Nevertheless, the animals do not always fly slowly and they are capable of a top speed comparable to some of our other species. However, with the long-eared's wing specification, speed is uneconomic in terms of energy consumption and the enormous ears result in additional drag during rapid flight. The animals progress at such uneconomic speeds only when commuting to their foraging places, thus reducing the risk of predation. On arrival they hunt mainly at a slower, more economical pace.[343]

It will be evident from the above that it is impractical for long-eared bats to fly far to feed. Work in Britain confirms that they forage mostly well within 1.5 km of the roost.[151] Consequently this restricts the overall area over which the whole colony may operate, which in turn may limit the size of the colony - if it were too big there might be insufficient feeding places in the communal home range for everyone.[89] This could explain why even nursery colonies of long-eareds are smaller than those of most of the other common Irish bats, such as the pipistrelle, Leisler's and, where it occurs, the lesser horseshoe.

Long-eared bats emerge relatively late, Alcock and Moffat[2] concluding from their accumulated data that few came out earlier than half an hour after sunset. The average time of emergence in Europe is 44 minutes following sunset.[343] When a bat leaves a roost, the darker the conditions, the better its chances of avoiding predation by birds. On the other hand, the earlier it comes out, the higher the temperature and therefore the more flying insects are available. However, as we shall see, long-eared bats are by no means dependent on flying insects and can therefore afford to set out late. Alcock and Moffat also considered that these bats retired to the roost early in the morning, whilst the light was still dim.

Moffat's observations on the foraging behaviour of the species can hardly be bettered and are all the more impressive for having been undertaken without benefit of any tracking device or night-viewing equipment.[229]

Instead of coursing up and down an open beat, it steals like a little winged burglar in among the twigs of some well-foliaged tree, and spends the greater part of its hunting hours gliding among the branches and preying on the moths and other insects that can be taken at rest upon the leaves. The best way to get a good view of the Long-eared bat's activities is to take one's stand right underneath one of its favourite trees....and watch patiently for the little creature to appear among the twigs overhead. With perhaps half-a-dozen of them circling about us among the leaves we soon cease to wonder that their stealthy movements have so commonly passed unnoticed.

He especially remarked on the ability to hover: "poising, at times, like a humming bird".[3]

Long-eared bats have been followed in the field in Britain and Europe, *inter alia* by attaching reflective markers to them and by radio-tracking, and it is clear that, although they can catch insects in flight, they obtain much of their prey by gleaning it from leaves, and are therefore usually greatly dependent on trees.[343] Remember, however, that they occur on several Irish islands where there are few trees; perhaps there they make do with bushes. Slow flight, manoeuvrability and an ability to hover are, of course, ideal for such foraging. Studies in Britain show that deciduous woodland is important for these bats but that non-native conifer plantations are rarely utilised, except along their edges.[151]

Long-eareds are a rarity amongst vesper bats in producing their echolocation sounds through the nose. These are weaker than those of other Irish species and cannot be picked up on a bat detector at distances over 5 m.[343] Some detectors do not pick them up at all. But since prey items are often on leaves, it is unnecessary for the bat to be able to locate them at any distance. In any event, when the animal is flying amongst a clutter of leaves, the echolocation pulses will generally come up against something within a short distance anyway. Weak pulses are of less use when the bat is catching insects on the wing, but this may be somewhat compensated for by the bat's hearing, which is particularly acute.

Surprisingly, the long-eared sometimes switches off its echolocation when gleaning insects, but it is not then hunting by sight. Although its eyes are relatively large for a bat, they are of limited use in the dark. Superior vision depends on the *absolute* size of the eyes and not their size relative to their owner. This is because the visual elements on the photosensitive layer at the back of the eye, the *retina,* are much the same size in all mammals, from shrews to elephants, and generally the bigger the animal, the bigger the eye and retina in it. Although the long-eared's eyes are relatively large for a bat, in absolute terms they are small. A series of laboratory experiments at low light intensities in Britain showed that long-eared bats can catch moths which are not flying without using echolocation or vision, provided that the moths are moving. Stationary moths, even those close at hand, were ignored. Long-eared bats therefore locate insects on leaves by actually listening for their movements. Hence the enormous ears.[343]

Many species of bat will spend periods during the night at rest in roosts other than those used during the day. Quite often a bat will bring a large item of prey into such a night roost, hang up and take its time over its meal, biting off and discarding appendages of no food value, such as wings. Such rejectamenta are of value to the bat worker as an indication of diet. Moffat remarked upon the night roosts of the long-eared bat, referring to them as "bat-bivouacs".[220]

> The places thus selected for bat-bivouacs have sometimes been set down as "dining-halls", and the phrase at first sight seems well chosen, for the ground beneath where the bats have clustered is often so thickly strewn as to be almost carpeted with the wings of different kinds of moths which the bats have brought in and eaten.

He was particularly favoured by some long-eareds which, for at least each of 7 years, from the end of March until November, night-roosted in a passage beside his bedroom at Ballyhyland. He was thus enabled to conduct his research by intermittently rising from his bed and taking the few steps to the passage. Only those who have worked on bats throughout the night outdoors can fully appreciate the luxury of such a situation. He studied these bats from 1915 to 1919. Among other things he was able to record dates of copulation, which he found to occur in the autumn, from as early as 16 October to as late as 29 November, but more often in the spring, from as early as 5 April to as late as 28 May.[220]

Not only did the bats night-roost beside his bedroom, they also hunted in his conservatory.

LONG-EARED BAT

The specimen which was identified at midnight was in the habit of flying in and out of a conservatory, at frequent intervals, in quest of moths. By watching until it was seen to fly in at 12 o'clock, then shutting the door and lighting a candle in the conservatory, the fact of its being *Plecotus auritus* was clearly ascertained.

We collected and analysed droppings every month from May to September (total analysed 200) from a long-eared roost at Kilmaley, Co. Clare.[306] In view of what Moffat has written, it was no surprise that moths (Lepidoptera) alone comprised 27% of the diet. There was also 11% of caddis flies and an assortment of other prey. Remains of midges were logged at only 0.2%, in marked contrast to their importance to pipistrelles, but then the hunting techniques of the long-eared bat are ill-adapted for catching small insects in flight.

From its hunting methods, one could reasonably conclude that many of these moths were taken while resting on foliage. In addition, a striking feature of the diet was that the bats were eating in some quantity three other groups of prey, all of which were also likely to have been gleaned. Together they comprised 42% of the diet.

(1) Prey which cannot fly at all: centipedes (Chilopoda 6%), spiders (2%) and harvestmen (Opiliones 7%). Harvestmen, also known as harvest spiders, are the spherical-bodied, wingless creatures with eight very long, thin legs.

(2) Prey which do not fly much: nearly all earwigs (Dermaptera 4%).

(3) Insects which do not usually fly at night (mainly Diptera 20%), most of which were represented by a single species, the yellow dung-fly *Scathophaga stercoraria,* the familiar buzzing denizen of cow pats. In Denmark, extensive studies of these insects[149] have shown that, whereas they are indeed diurnal, a minority will remain on cow pats for an hour to an hour and a half after sunset, before flying off to roost. So it is quite possible that our long-eared bats may actually have caught some of them in flight, rather than lifted all of them from vegetation. Presumably because these flies are still on the wing in the dusk and early night, we have recorded them in the diets of all of the bat species in Ireland which we have investigated. They even turned up in pipistrelle droppings, although they comprised only 2% of the diet.

The centipedes in our long-eared droppings were more likely to have been gleaned from the ground than from trees, and here again Moffat's observations may be relevant.[3]

When one of these bats leaves a tree, if its object is merely to pass to another quite near at hand, it darts through the air with a swift arrowy flight; but when a longer expedition is contemplated....The bat plunges headlong to within an inch or two of the ground, and then skims away in a jerking fashion....over the surface of the field.

Unfortunately it is not possible to identify individual species of moths in bat droppings. However, Moffat determined those that had been brought into the night roost beside his bedroom from the wings lying on the floor.[220] These included the peach blossom *Thyatira batis*, buff-tip *Phalera bucephala* (Family Thyatiridae), orange underwing *Archieais parthenias* (Family Geometridae), copper underwing *Amphipyra pyramidea*, light arches *Apamea lithoxylaea*, burnished brass *Diachrysia chrysitis*, shark moth *Cucullia umbratica* (which the bats seemed to catch only in the small hours), dark arches *Apamea monoglypha*, herald *Scoliopteryx libatrix*, angle shades *Phlogophora meticulosa*, silver Y *Autographa gamma* and beautiful golden Y *Autographa pulchrina* (Family Noctuidae). Wings of the last two sometimes outnumbered all of the others, but only when the rhododendrons around the house were in full flower. As the blossoms were much frequented by both of these insects during and after dusk, it seemed likely that the bats were hunting there and may have caught the moths in the act of sipping nectar. Moffat also recovered the wings of three moth species of the Family Arctiidae, some of which are distasteful to bats.[343] These included, from time to time, the tiger moth *Arctia caja* and the buff ermine *Spilosoma luteum*. He found the wings of the closely related white ermine *Spilosoma lubricipeda* only once, even though it was decidedly commoner in the district than the buff. In his experience it was rarely eaten by any bats because of its unpleasant taste. The single occurrence in the passage in his house was during a night of torrential rain, when there must have been little in the way of moths available and the bats were not in a position to turn down any morsel of food, nice or nasty.

He also recovered numerous wings of daddy-longlegs from the passage floor; such prey comprised 7% of the remains in the droppings that we analysed.

Daubenton's, Natterer's and whiskered bats are all closely related and similar looking, although the *modus operandi* of each differs. The ears of all three, whilst large, are nowhere near as big as those of the long-eared and the tragus is slender and tapered or pointed at the tip. Natterer's bat has a fringe of bristles running along the edge of the interfemoral membrane and this alone suffices to distinguish it from any other Irish bat. Separating Daubenton's from whiskered is slightly trickier. The tragus of Daubenton's is convexly rounded along its hind edge; the wing is joined to the leg at the ankle; and the calcar extends about three-quarters of the way along the edge of the interfemoral membrane. The hind edge of the tragus of the whiskered is straight, or even somewhat concave; the wing is attached to the leg on the outermost toe; and the calcar extends about half-way along the edge of the interfemoral membrane (Fig. 15).

Daubenton's bat has also been called the water bat, and with good reason. Whilst other Irish bats may hunt close to water, they also often favour other habitats. Daubenton's bat seems to forage mostly above or in the vicinity of water. It commonly flies close to the surface and can sometimes be followed with the aid of a torch as it skims back and forth, especially if the observer is standing at about water level. Moffat, of course, was well aware of this bat's liking for water.[230]

> The distinguishing trait of this bat is its love for constantly flitting over the surface of water, so low as - on most of the narrow streams that it chiefly frequents - to render it nearly invisible to anyone who is not specially looking for it....On one little stream in County Wicklow, which was found to be a haunt of this bat....it can be watched on any fine evening from the end of March until nearly the end of October....This little stream (which is, in fact, a mill-race) runs for a considerable distance parallel to a small river, the distance between the two being not more than 15 yards; but the bats confine themselves in the strictest way to the narrow mill-race, and never seem to show themselves for a moment on the river, over which both Leisler's Bat and the Pipistrelle may often be seen hunting. This marked preference for one stream over another so very near it....has been persisted in for more than 30 years.

At night it is highly probable that the mill would not have been working and the sluice to the mill wheel therefore closed to conserve water. There would thus have been no flow in the millrace whatsoever.

If this is correct, then the Daubenton's bats preferred to hunt close to the surface of the still water of the millrace rather than over that in the river alongside it, where both Leisler's and pipistrelle chose to fly. The preference for still water is explained in an ingenious series of stereo flash photographs of hunting Daubenton's bats in Britain, in which the animals were snapped both catching insects in flight above the water and also gaffing them from the surface with their feet.[172] Locating and picking insects off the surface would be greatly facilitated by its being smooth. Incidentally, despite this habit, there is no evidence that these bats glean insects from vegetation or from the ground.

Most roosts of Daubenton's in the British Isles have been found near water.[151] Moffat considered Daubenton's to be a "somewhat scarce species" in Ireland,[231] but apparently based this on the different bats that had been sent to him, which are likely to have been biased against any species which does not often inhabit buildings. The most recent data[261,285] show Daubenton's to be common. The most frequent roosting places have been found to be in the stonework of bridges over water. While other species may also pass the day in such retreats, Daubenton's does this much more often. Thus, in a survey of bats in 366 bridges in eastern Co. Cork and west Waterford, Pat Smiddy considered 186 (51%) suitable for bats.[331] Of these 51% showed no signs of occupation, 22% showed signs of recent tenancy (mainly droppings) and 27% actually contained bats. Of these, 75% housed Daubenton's bat. He also found pipistrelle, long-eared, Natterer's and whiskered bats in bridges, but much less commonly. Identifying bats in bridge roosts requires persistence, as it is rarely possible to examine the animals in the hand without extracting and possibly injuring them. Smiddy used a stepladder, a torch and patience.

Further information on bats in bridges comes from a survey of 165 bridges in Cos Sligo and Leitrim from late April to mid November 1998 by Caroline Shiel.[301] Of these she considered 98 suitable for bats and, of them, 67% housed the animals at one time or another over the period of her study. Of 251 bats she was able to identify, 72% were Daubenton's and 26% Natterer's, the remaining 2% being whiskered, long eared and pipistrelle. She also tried to relate occupation of bridges to various habitat factors in the vicinity but came up with only one that was statistically significant, although this was interesting in view of the paragraphs above: Daubenton's bats were somewhat more likely to be found in bridges over rivers which were slow-flowing - and therefore with a smooth surface - or with pools near the bridge.

Most of the Daubenton's roosts in the National Bat Survey were of 10 or less individuals, probably because cavities in bridges seldom afford enough space to admit more of the animals. Both Smiddy and Shiel found that the numbers there were generally very much lower than 10. Nevertheless, much larger colonies of Daubenton's bats are not unknown if there is space for them. Smiddy monitored a nursery roost of over a hundred in an underground passage, possibly formerly a wine cellar, in the bank of the River Blackwater at Glencairn Abbey, Co. Waterford. A few years ago over a hundred were counted emerging from a roost in a bridge near Kilkenny. In the summer of 2000 the aforementioned Mark Smyth counted 280 in the ruins of Shane's Castle on the northern shore of Lough Neagh. There are also several mentions in the older Irish literature of large colonies, but unfortunately without actual numbers being recorded. In Britain colonies of several hundred have been noted.[151]

The greatest threat to Daubenton's bats is the repair of bridges by pressure grouting liquid cement into the crevices.[202] Whereas this is the cheapest way of renovating the stonework - although by no means the most pleasing to the eye or in keeping with ancient masonry - it entombs any bats in residence and may deprive others which roost in the bridge from time to time of accommodation. In Northern Ireland, the Department of the Environment has a most enlightened attitude to this problem. When a local authority supplies a list of bridges due for maintenance, and spraying of cement proves necessary, the Department informs the Bat Conservation Officer at the Ulster Museum, who inspects each bridge and locates any crevices used by bats (usually no more than two or three). The entrances are then plugged, the bridge sprayed and the plugs removed before the cement dries. Crevices may also be left in suitable unoccupied bridges for possible future use, or other steps taken to ensure that the bridge continues to provide acceptable quarters for bats. Unfortunately, the procedure by contractors in the Republic is mostly to ignore the animals. Furthermore, whereas it is an offence for local authorities in Northern Ireland to deliberately block crevices in bridges used by bats, in the Republic the Wildlife Act (1976), while protecting all species of bat, specifically excludes killing them in the course of building or engineering construction. It is said that this deficiency in the legislation is to be remedied before long.

Where a bridge has been widened by building on a new section parallel to the original, there is sometimes a deep crack between the two and this is often a popular residence for bats. On the other hand, when a stone bridge is destroyed and replaced by a modern one of steel and concrete, there is nowhere for them to lodge. In which case it is advisable to provide alternative quarters, perhaps in the form of bat boxes. The latter are specially manufactured to provide artificial housing and are most usually placed in areas where there is a shortage of natural roosting places.

Because Daubenton's bat so often roosts in bridges, the droppings usually fall into the water. Obtaining samples of them is therefore often the foremost difficulty in determining the diet. Fortunately, sometimes a colony in a bridge resides in a peripheral arch, which may run dry during the summer, although heavy rain and the consequent spate will wash any droppings away. We have obtained and analysed material from at least some of the months from April to September from eight different roosts: five from the catchment of the River Blackwater in Counties Cork and Waterford (and all located by Pat Smiddy), and one each in Counties Galway, Kilkenny and Tyrone (total droppings analysed 1,252).[131,342] The food of all the colonies was fairly similar. As with the pipistrelle, which often feeds over water, adult midges (14%) and caddis flies (10%) were important, as were other adult Diptera (30%), two-thirds of which were night-flying types.[131]

There was strong evidence that much of the prey had actually been snatched from the surface of the water. Such items could be divided into three categories.

(1) Animals which live mainly on the surface of the water. These included the larvae and pupae of midges (10%), which cannot fly, are entirely aquatic and may teem on the surface. Also in this group were small numbers of water crickets (Veliidae), predators which skate about on the surface film, and whirligig beetles (Gyrinidae), also carnivorous, which often swim round and round on the surface of slow-moving water.

(2) Aquatic animals which come to the surface from time to time: these comprised the larvae and pupae of caddis flies (6%), plus occasional water beetles (Haplipidae, Dytiscidae, Hydrophilidae) other than the whirligigs.

(3) Animals which are not aquatic and had presumably fallen into the water. Amongst these are spiders (4%) which, if they had not tumbled in were presumably water spiders, although the bats could also have caught some spiders in their roosts. Harvestmen and centipedes were occasionally taken too. Earwigs (1%), which do not fly much, also

qualify for this category, as do at least the majority of the diurnal Diptera (6%), and wasp-waisted insects (Hymenoptera), which fly almost exclusively in daylight, remains of which were recovered sparingly in the faeces.

Together these three groups accounted for over a quarter of the food, demonstrating the importance to Daubenton's bat of gathering prey by surface trawling. The story does not end here, for dead night-flying insects also fall onto the water and living ones strike it whilst on the wing and are unable to take off again. Is there any way that we might get some idea of how important such items might be to the bats? An indication comes from the pre-war work of Dr Winifred Frost (1902-79) on the food of trout in the River Liffey, which she investigated by analysing the contents of some 600 trout stomachs.[141] On the lowland part of the river, two-thirds of the food eaten by these fish in summer as a whole comprised items taken from the surface of the water. Moreover a quarter of the individual items of the food in the stomachs were adult insects that had fallen into the water, including many night-flying types. This shows that such prey are plentiful and it would be extremely surprising if the Daubenton's bats, which we already know snatch other items from the surface, ignored them. So we can conclude that in Ireland Daubenton's bats find considerably in excess of a quarter of their prey by trawling it from the surface of the water.

At some sites in April and May the bats fed much more on midge larvae and pupae than later in the summer. While such prey were probably readily available from April to September, they are tiny and represent a relatively poor return on the effort involved in catching them. Their abundance in the droppings in spring and early summer is probably an example of what ecologists call *optimal foraging*. In April-May insects are comparatively scarce and the bat is probably obliged to take anything it can. In high summer, when insects are abundant and a choice is feasible, it pays to pick larger prey, such as adult caddis flies, which are a better return on catching effort.

Although Natterer's bat is widespread, it is scarce. The Northern Ireland Bat group has discovered a mere eight nursery roosts (plus a few singletons).[3] There were only 44 recorded roosts in the National Bat Survey, 20 of them of singletons, and in a mere seven roosts did numbers exceed 50, although in two there were more than a hundred individuals. A specimen has recently been found on Clare Island, Co. Mayo.

This bat often hides in cracks or crannies, where it is difficult to find. But it also uses open spaces, such as attics and church roofs, apparently preferring older buildings.

Connemara National Park, at Letterfrack in Co. Galway, has been blessed with two large colonies.[120] In the early 1990s there were 40 in the roof of the Monastery Hostel there, besides several individuals in the roof of the Park office. In 1992 a nursery colony of 280 lived in the Gothic Church, 400 m east of Kylemore Abbey. In 1993, during restoration work, numbers there fell to 15, but with the completion of repairs the colony recovered to over 100 in 1995. Over 200 bats were observed leaving on one summer evening in 1998. However, at this stage it was known that other species had also taken up residence in the church as well. At Tully, 5 km north-west of Letterfrack, there is a further nursery colony of some 150. The number of Natterer's bats in the district is probably not unconnected with the abundance of deciduous woodland. A strip of oak-dominated trees, some 4 km long, clothes the precipitous slopes behind Kylemore Abbey. Ellis Wood, although it covers only about 1 ha, lies close to the Park office. Studies in Britain indicate that broadleaved woodland is preferred habitat, although these bats can also operate over more open terrain.[151]

The wings of Natterer's bat are broad and of relatively large area, features which, one may recall from the long-eared, are adapted to slow flight and manoeuvrability. Like the long-eared, Natterer's is a gleaner, capable of seizing prey both from leaves and from the ground. Observations in Switzerland on mown meadows suggest that it may sweep up prey from the grass using its interfemoral membrane and perhaps grasp it in its hind feet.[151] The bristles along the edge of the membrane may therefore enhance its sense of touch. This bat can also land on the ground, run around like a mouse and then fly off with consummate ease.[151] The echolocation pulses, like those of the long-eared, are weak, and for the same reason: it is unnecessary to detect prey at any great distance and emitting less intense sounds economises on energy. Russ's work on habitat in Northern Ireland shows that Natterer's bat is the only bat species in Ireland, apart from Leisler's, to hunt regularly over grassland.[285] However, the two species use contrasting methods of foraging for, whereas Natterer's is predominantly a gleaner, Leisler's bat forages exclusively on insects on the wing (Chapter 8).

We analysed monthly samples of droppings of Natterer's bat collected from a derelict church at Ardagh, Co. Limerick, over one summer (total analysed 200). The results reflected the gleaning habit of this bat, a minimum of two-thirds of the diet being attributable to prey caught in this way, including spiders (12%), harvestmen (5%), wasp-waisted insects (11%), frog-hoppers (Cercopidae 3%) and day-flying Diptera (a whacking 34%). Among the last were yellow dung-flies (14%), flies of the bluebottle family (Calliphoridae 12%), and snipe-flies (Rhagionidae 7%). Natterer's was the only bat we found to take snipe-flies, which are predators that usually sit on leaves waiting to dart out and attack passing insects. The bats took no midges whatsoever or, indeed, any other similar small-sized insects. Compared to the long-eared bat, Natterer's caught very few moths (5% as against 27%), no doubt because the long-eared locates them by listening for their movements, whereas Natterer's is ill-equipped for this, having ears of a size that is nothing out of the ordinary amongst bats, and presumably uses echolocation alone. Perhaps a moth sitting on a leaf with its wings folded "looks" too much like a leaf itself to the bats' echolocation. As with all of the other Irish bats that we have investigated, caddis flies (13%) formed a significant component of the diet.[306]

The scarcest of our bats, excepting perhaps Nathusius' pipistrelle, is the whiskered. Its fur is shaggy and there are several bristles on the lips, chin and forehead, which give it its name. Its presence in Ireland was first established at Feakle, Co. Clare, by Kinahan in 1852, in a remarkably serendipitous incident, considering that he was, at the time, one of the very few people in the island with any interest in bats.[177]

> ....the only Irish specimen known....was obtained....under rather peculiar circumstances, being brought to me by a domestic cat I was in the habit of taking much notice of....At the time I mistook it for the common bat [the pipistrelle], and it was not till some months afterwards....that, on identifying the specimen, I perceived its value.

Cats, of course, are inclined to kill anything small and furry or feathered which they encounter. There are only a few additional recorded cases of cats killing bats in Ireland, but there are instances in Britain of them actually specialising in them. One cat, sitting on a tree branch overhanging a canal, killed over 70 Daubenton's bats within a few weeks.[337]

No further specimens of the whiskered bat turned up until the 1890s, when Jameson began his researches; he was able to come up with a further six records.[167] Other Victorian and Edwardian naturalists added a few more, thinly scattered over the country, but from 1915 until the National Bat Survey began in 1985, there was but a single further addition, by Moffat in 1922 at his home in Ballyhyland.[231]

Whiskered bats are difficult to find as, like pipistrelles, they prefer confined spaces. O'Sullivan[261] remarked that great care is needed in searching for them as they are usually found in roofs, between rafters and felt or in narrow slits where timbers meet, where they are not easy to see. Most of the 34 roosts found during the National Bat Survey were of less than five individuals, although there was one of more than 50. These were also the only bats during the Survey found frequently roosting with other species, including long-eared, pipistrelle, Natterer's and lesser horseshoe. As numbers of whiskered bats are small in such associations, they tend to be overlooked. The Northern Ireland Bat Group has located only eight nursery roosts, plus three singletons.[3]

The above accurately summarises most of what is known of the whiskered bat in Ireland, for little other work has been done on it and we were even unable to obtain droppings for analysis. Indeed, in view of its tendency to share roosts with other bats, collection of faeces unmixed with those of other species may prove challenging. Analysis of droppings in Europe indicates that it too may hunt mainly by gleaning.[358]

There have been rumours in Ireland recently of yet other species of bat being picked up on bat detectors. Until specimens are produced, they must remain dubious.

Chapter 7. The Lesser Horseshoe Bat

While now the bright-haired sun
Sits in yon western tent, whose cloudy skirts,
With brede etheral wove.
O'erhang his wavy bed:
Now air is hushed, save where the weak-eyed bat,
With short shrill shriek flits by on leathern wing.
Ode to Evening. William Collins.

William King (1809-86), Professor of Mineralogy, Geology and Natural History at the then Queen's College Galway, was sitting in the dining room of his house in the neighbourhood of Galway one evening in the middle of June 1858,[181] when one of his

> ....family called out that a bat was flying about the room. No sooner was "Flitty" announced, than the usual preparations were made for its capture, every one present furnishing himself or herself with either a cap, hat, or handkerchief for the purpose. After eluding our efforts for some time, it was at last secured, and forthwith put under a glass for exhibition....I was at once struck with the remarkable appendages surrounding its nose, and saw that it was not one of the usual bats which had been captured in the house on previous occasions. This led me to consult some works I had by me at the time, when I made out that "Flitty" was of the Horse-shoe genus; but I could find no record of any specimens of the kind having occurred to any one else in the country.

The animal was a lesser horseshoe bat - the first recorded instance in Ireland. There had been a dubious assertion of a horseshoe bat in Co. Westmeath in 1845,[177] very much in the nature of an anecdote by a "friend-of-a-friend". No specimen had been produced and Co. Westmeath is well outside the known geographical range of the lesser horseshoe. In Victorian times, certainty usually depended on a specimen, preferably shot and stuffed or, as the couplet so pithily has it: "What's hit's history. What's missed's mystery".

As King observed, horseshoe bats differ from the vesper bats, to which all the other Irish species belong, in the possession of a peculiar structure surrounding the nostrils which looks like a sort of funnel, called the *nose-leaf* (Fig. 15). The lower part of this is horseshoe-shaped, giving these bats their collective name, while the upper is pointed. Horseshoe bats produce their echolocation pulses through the nose and the nose-leaf acts as a kind of megaphone, which can be swivelled about, narrowly beaming the emission in a desired direction. The nose-leaf may also serve to shield the ears from the intense outgoing sound.

The echolocation pulses of horseshoe bats differ from those of vesper bats in being more high pitched; most of the sound in the pulses of the lesser horseshoe is around 110-113 kHz, higher-pitched than that of other Irish bats. Remember that high frequency equates with short wavelength and the ability to detect smaller items. The pulse from a horseshoe bat is also often longer than that of a vesper bat - about a twentieth of a second in the lesser horseshoe - and has a distinct warbling quality when heard through a bat detector. Such a long pulse is also of constant frequency over almost all of its length, but is frequency modulated at both ends: there is a brief rise in frequency up to 113 kHz at the beginning and an equally short fall in frequency at the end. As objects are approached, the pulse shortens, as in vesper bats, and the pulse rate goes up as more and more detailed information is required faster and faster, producing a feeding buzz. The shortening effectively cuts off more and more of the beginning of the pulse until all that is left is the frequency modulated part at the end.

Dr Kate McAney, who did her PhD research on lesser horseshoe bats in Co. Clare in the 1980s and has subsequently been much involved in their conservation in Ireland, has often used a bat detector to listen to them. Whereas she commonly heard the typical long pulses as the bats emerged from or re-entered a roost, she picked them up only occasionally at any distance from the roost, or indeed even when bats were flying around near it.[198] This may be partly due to the animals using shortened pulses but may have resulted from the highly directional nature of the sound focussed by the nose-leaf, so that parts of the longer pulses may be lost when the detector is not in the direct line of fire. Moreover, because the sound is so high-pitched, it does not carry far.

Unlike vesper bats, horseshoes never cram themselves into crevices and crannies. They always hang freely suspended from a horizontal surface by their hind claws. This is facilitated by the wings and interfemoral membrane being attached to the legs well clear of the feet. At rest, the animal wraps its wings around itself like a cloak and they often conceal the body completely. Indeed they act as a kind of overcoat, trapping a layer of warm air round the body. They are also a macintosh, for water dripping from the roof of the roost cannot wet the fur and simply runs off the wing membranes. Although these bats hang freely, a group often clusters closely together. In Ireland in summer the lesser horseshoe roosts mainly in roof spaces, or sometimes under the roofs of old or derelict buildings, and in winter in caves, mine shafts, tunnels, cellars, ice houses and other underground passages.[261]

When alighting, a horseshoe bat turns a partial forward-somersault in the air, at the same time closing its wings, and in a flash is suspended. Finding a suitable place to grip and latching onto it in an instant is a signal indication of the precision of this beast's echolocation.

The entrance to a roost of vesper bats is often small. The animals fly up to it, alight and crawl or squeeze through. A horseshoe bat needs to fly directly through the roost entrance to its perch within, so the opening must be wide enough to do this, although it need not be quite as wide as the full wingspan, for the bat can draw its wing tips closer to its body when entering.

In March of 1859, the year following Professor King's discovery, Frederick Foot (1830-67), Assistant to the Irish Geological Survey, who was working in Co. Clare, discovered lesser horseshoes, most of them males, hibernating in a cave at Quin and in Ballyallia Cave, near Ennis.[135,177] On his first visit to Ballyallia he saw "four in the innermost, and two in the middle compartment....all in nooks". The inner section of the cave was then "tolerably dry" but a week later it was "dripping with water, there having been a good deal of rain in the interval" and there were no bats there, whereas he found five in the middle one - an example of how bats will move position in hibernacula if conditions become unfavourable. In January 1862 he found lesser horseshoes in two other caves in the county.[136] Foot concluded that "there is every reason to believe that the lesser horse-shoe is *the* bat of this part of the country." While such a deduction was unjustified - because this bat roosts openly it is the one most likely to be seen hibernating - there is no doubt that even today, as we shall see, it is probably second only in abundance to the pipistrelle in parts of Clare.

Foot forwarded some of his animals to Kinahan, who, overjoyed, included a detailed description of them in the first review paper on Irish bats, which he read to the Dublin Natural History Society.[177] Kinahan sent some of the specimens to Britain to verify their identity, apparently half-suspecting that the animals might be greater horseshoe bats *Rhinolophus ferrum-equinum*. Some idea of the uncertainty at the time as to which bats occurred where in the British Isles, of the precise features to be used in their identification, and even whether certain species of bat actually existed, is given by the statement in his paper that "The pigmy bat has also occurred near Dublin, but this is generally now looked upon as the young of the pipistrelle".

Kinahan's enthusiasm was such that he travelled to Co. Clare in August of the same year to visit Ballyallia Cave himself, but found no bats. The journey from Dublin was not such a great undertaking as may at first be supposed, for the railway from Ennis to Ennis Junction, which was only a few minutes walk from Athenry, on the main Dublin-Galway line, had opened in June of that year. In March 1861 he returned to the county to scour the caves of the Burren district for bats.[178] Six out of the eight caves he investigated contained together a total of 35 bats, all of them lesser horseshoes, and, on checking Ballyallia Cave, he discovered a further 19. Of the 52 that he was able to sex, all but four were males. He concluded that the sexes occupy separate hibernacula.

I visited Ballyallia Cave in May 1969, but the only bat there was the decaying corpse of a lesser horseshoe on the floor. In January 1973 I found five hanging from the roof.[102] The above facts suggest that this cave may have been used continuously by lesser horseshoes for well over a hundred years, although probably only as a hibernaculum. The entrance was subsequently blocked by the local landowner in the interests of the safety of his livestock.[196] Happily, in a joint operation between Dúchas and the owner, it was reopened in March 2000 and the entrance is to be grilled. It is hoped that the bats will return.

In 1881 further specimens were obtained at Coole Park, Co. Galway, close to the border of Co. Clare.[217]

Lesser horseshoes were reported at Muckross, near Killarney, Co. Kerry, by a Mr Ray Hardy in July 1885.[176]

> Both males and females were common....flying about the old Abbey by hundreds....[In] the hay-loft....at Mr Herbert's stables....they hung from the beams above in great numbers. With my butterfly-net I could have taken ten at one stroke....

No further counties were reported to be occupied until 1929, when a J.E. Flynn found 36 hibernating in the cellar of an old castle at Glengarriff, Co. Cork.[134] In June 1932, he was passing an ancient beech some 100 m from the castle and

> ....on noticing the hollow centre I pushed a stick up it, and out flew a bat, followed at intervals by more, in twos and threes, till I had counted up to 53. I caught several which flopped on the ground....and found them to be Lesser Horse-shoes....The hole in this beech-tree extended upwards for four or five feet [1.5 m], and could not have been more than 18 or 20 inches [50 cm] in circumference at its widest part.

Lesser horseshoes roosting in a tree must now be extremely rare and I cannot find any other reference to it anywhere in the literature. The rarity of tree roosts in summer is odd. We know that summer roosts today are in buildings, but where did the bats go in prehistory when there were none? Temperature in Irish caves is generally low, presumably rendering them unsuitable as nursery roosts, unless perhaps if large numbers of bats clustered together. The only suitable alternative that I can think of is hollow trees.

The species was next recorded in Co. Mayo, when Major Robert Ruttledge, a centenarian as I write and probably the most distinguished Irish ornithologist of the twentieth century, took a total of 13 over the summers of 1942-43 at various places in the vicinity of his home at Cloonee.[288,289]

The species was not found in Co. Limerick until November 1984, when Mr Roger Forster, then gamekeeper on the Adare estate, discovered 18 in the cellars of the manor house there.[138]

Other published records prior to the National Bat Survey are negligible. The distribution revealed by the Survey is shown in Fig. 17. Since then the species has been recorded at various other sites - although many records have not been published - but the essential range has remained unchanged: in Ireland it is limited to the western halves of the chain of six maritime counties extending from Mayo in the north to Cork in the south.

All the other Irish bats, with the possible exception of Nathusius' pipistrelle, which is probably a recent arrival, are apparently widespread. Why, then, is the lesser horseshoe confined to the south-west? As bats are highly mobile, other parts of the country would surely be colonised if suitable.

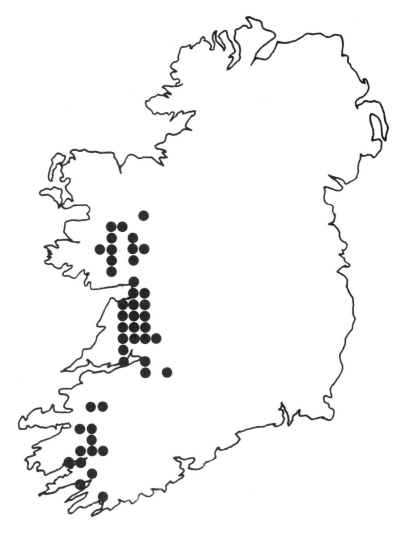

Fig. 17 The known distribution of the lesser horseshoe bat in 1988 by 10 km squares of the Irish National Grid.

In continental Europe the lesser horseshoe is a southern species, recorded south of a line running from about the north of France, through the middle of Germany and close to the southern border of

Poland (Fig. 18). In Britain it is now confined to Wales and the west of England, with a few older records from further east. So in both Britain and Ireland it occurs only in the south-west and is living further north than anywhere on the continent, in Ireland at the extreme northern limit of its world range. So one might argue that, because the climate is colder in higher latitudes, the geographical range in general, and that in Ireland in particular, is directly limited by a cool climate, perhaps because cold weather prolongs or otherwise interferes with hibernation in some way. Unfortunately, this will not wash, as winters in central and eastern Europe are much more severe than in the British Isles. Moreover, winter temperatures where the bat occurs in Ireland are similar to those in at least as large an area where it does not. Any argument for distribution in Ireland on the basis of summer temperatures, which might influence insect numbers, or other individual climatic factors is just as difficult to sustain.

It has been suggested that the availability of caves for hibernation is the limiting factor. But there are caves elsewhere in Ireland, even close to but not within the range of the lesser horseshoe: in east Cork/west Waterford, and particularly in Sligo/Leitrim/Fermanagh. For the present the factors which determine the distribution in Ireland must remain a mystery, but they are probably complex.

The National Bat Survey located 157 roosts of the lesser horseshoe, several in excess of 100 bats and two of more than 200.[261] With the restricted geographical range of the species, it is clear from this that, in those districts where it occurs, it is probably a very common bat. There have been two more recent investigations of smaller parts of its range.

In 1995 Ms Conjella McGuire, of Dúchas, undertook a major survey of the bats, particularly of the lesser horseshoe, in an area 225 km^2 in the northern part of Co. Clare including the Burren, where regional drainage is mainly underground and there are many cave systems.[208] She recorded 990 lesser horseshoes in 24 summer roosts (in attics or under the roofs of derelict buildings) and 165 individuals in 15 hibernacula. She also found a further 1,503 bats of other species during the summer in 52 roosts, of which 1,256 were pipistrelles. Lesser horseshoes, because they suspend themselves openly, are more likely to be seen than vesper bats. Nevertheless, roosts of the latter may be betrayed in other ways, for instance by droppings, and McGuire is an experienced bat worker. Even though she made a special effort to find lesser horseshoes, her data strongly suggest that in the area the species may be only less common than the pipistrelle.

Fig. 18 The geographical range of the lesser horseshoe bat in Europe.

The McGuire study concentrated on summer. In February 1994 Kate McAney organised a survey of potential subterranean hibernacula in Co. Clare, in the Cong-Clonbur district in Co. Mayo and at a few other likely places in Co. Galway.[197] Although this ran for only one week, it was carefully planned and had adequate personnel. Information on possible hibernacula was culled from the literature and also sought from members of the public via newspapers and the radio. All potential sites were then mapped and landowners asked for permission to visit them before the survey started. The work was undertaken by a group of 14 (operating as three teams) comprising herself, nine members of Bat Groups in England and Wales - all of whom experienced in speleology and identifying bats - a further British caver, two members of the Dublin Bat Group and a naturalist from Galway. In addition, various others, mainly from Dúchas, joined the hunt on particular days.

Altogether the group examined 95 sites comprising: 52 caves, 15 souterrains (man-made subterranean structures of the Iron Age), nine fissures (snake holes, sink holes and simple passages in limestone cliffs), nine tunnels, eight buildings and two mines. There were bats in at least some of all of these different types of potential site. Overall, 52% were occupied with a total of 562 bats, 97% of which were lesser horseshoes. A further 20% of sites contained bat droppings, which were probably mainly from horseshoes. The other bats found included long-eared, Natterer's, Daubenton's and a lone whiskered. There were no pipistrelles at all, doubtless because of their liking for confined spaces, and no Leisler's. The study not only underlined the abundance of lesser horseshoes in these districts but, as some of the hibernacula could be entered with comparative ease, the vulnerability of the individuals hibernating in them.

During the survey some days were cold; on others it was relatively warm; but even on the coldest days, with air temperatures at 2-3°C and those at night presumably lower, it was never less than 5°C in the hibernacula, and many were warmer. Interestingly, during the warmest days it was actually cooler in the hibernacula than in the open air, demonstrating that they were effectively insulated from temperature variation without. Regardless of the lengths of caves, the bats were all within 20 m of the entrances, most likely because it is no warmer further in. On the other hand the distance from bat to cave floor ranged from a few centimetres to 5 m. There was some evidence that lesser horseshoes move between hibernacula. Thus, in one cave visited on Monday, there were 50 individuals, whereas on the Saturday following there were only 35.

The survey work undertaken by the Dúchas and McAney leaves little doubt of the abundance of the lesser horseshoe in the districts in which it occurs in Ireland, especially in Cos Clare and Kerry.[151] O'Sullivan estimates the total population in Ireland to be 12,000,[261] although any such figure is bound to be very approximate. Whatever the number, the

Irish population is certainly faring better than anywhere else in Europe. Under the European Habitats Directive, a roost of lesser horseshoes containing more than 100 bats in summer or 50 in winter is considered to be of international importance. In Ireland there are now known to be at least 18 of the former and 12 of the latter.[52]

Everywhere else in Europe the lesser horseshoe is in retreat and, except in Britain, it has disappeared from the northern part of its geographical range. Reductions of up to 90% have been reported in populations throughout the continent since 1950, although there is no clear single cause for the decline. Habitat or land use change may have reduced available food, and disturbance in hibernacula may be a significant factor.[338] In 1993 Ireland signed the European Bats Agreement which aims to promote protection for bats by guarding key sites and habitats.

Under the auspices of the Dúchas and The Vincent Wildlife Trust, many underground hibernacula have now been fitted with steel grills, which keep intruders out but allow free passage to the bats. McAney, in continuing conservation work for the Trust, has identified the main problem in summer as colonies being forced to leave their traditional nursery roosts in large, often disused, slated buildings, for smaller ones roofed with corrugated iron, where they are more prone to disturbance. There are two main reasons for this: deterioration, so that traditional roosts are no longer suitable as nurseries, and renovation (particularly in recent years with the economic boom) or even the raising of buildings to the ground, both of which effectively remove the roofs from over the bats' heads. The Trust has been involved in the conservation of lesser horseshoes since 1991, initially in repair work to secure, maintain and improve roosts. More recently it has bought buildings. One, which had never been fit for the bats, was modified in early spring to furnish an alternative home for a colony which in summer had formerly been using rather tatty quarters nearby. By mid-summer half of the bats had moved into their new home. The Trust has also purchased four sites as reserves for lesser horseshoes and, at the time of writing, is negotiating to create three more. As a result of this conservation work, it is reckoned that over a thousand individual lesser horseshoe bats have benefited.[203]

In 1998 the National Heritage Council acquired and appropriately refurbished a derelict farm outbuilding at Ruan, near Dromore Nature Reserve in Co. Clare, which had been a major nursery roost. It now houses what is probably the largest nursery colony of lesser horseshoes in the world. In 1998 numbers there exceeded 400. This is a most judicious and farsighted development. For far too long in Ireland, the idea of "heritage" has been narrowly defined by many to be confined to things cultural and man made.

Destruction of roosts is not the only way in which lesser horseshoes may meet an untimely end at the hands of man. I visited the upper story of an outbuilding near Killarney in April 1974 and found a mummified horseshoe bat on the window ledge. On enquiring, I learned that the window had been open in the previous year and that a colony of bats had set up home there. Subsequently it was closed, imprisoning the animals.[259] Vesper bats usually go in and out of their roosts by small holes, and the closure of a window must rarely affect them. Horseshoe bats living in the upper parts of a building in reasonable repair are likely to pass in and out through an open window and are doomed if it is shut when they are within.

Cats may prey on lesser horseshoes in Ireland, especially because the latter roost openly. As they may emerge through windows, a cat can lie in wait for them on the sill. At one nursery roost McAney saw cats thus ensconced several times, trying to catch the bats as they passed.[200] In Britain one cat ate 22 bats, including greater and lesser horseshoes and long-eareds within a few hours. They were not even of any ultimate benefit to the brute, which regurgitated the lot.[337] Irish barn owls certainly catch lesser horseshoes occasionally. Of the remains of 3,900 items of avian and mammalian prey taken from pellets collected from a barn owl roost at Killarney, 27 were bats (0.7%), two of them lesser horseshoes.[326]

Since bats are pretty strictly nocturnal, anyone wishing to study them must venture out at night, and may often have to stay out until dawn. To be a bat worker, therefore, demands both dedication and stamina. Time crawls if nothing much is happening; it may rain, and it is nearly always cold in the small hours; one grows tired and is bitten by insects. But, apart from gathering information, there are partial compensations - the magical quality of twilight before dawn on a clear summer morning when the mist lies low on the grass, and a burst of euphoria when the dawn finally arrives. (Anyone who watches bats overnight is struck by dawn twilight. The well-worn saying that the darkest hour is before the dawn is nonsense.) I hasten to add that personally I have not stayed up overnight in this way very often. In contrast, when Kate McAney and Caroline Shiel were pursuing their postgraduate research, they spent around half of all nights in some summers in this way. Furthermore, when they arose on the following afternoon, the first task was to transcribe all of the data from tape-recordings made during the night, sometimes against a stop watch, for example if a precise time of emergence on the previous evening was required for each bat.

McAney concentrated her postgraduate research in the summers of 1983-85 around two nursery colonies of lesser horseshoes in Co. Clare, both in the roofs of north-west facing, two-story farm buildings, which were largely undisturbed. At these she made many accurate counts of the individuals leaving in the evening and returning in the morning, timing each emergence or entry to the nearest second. She sat up through the night, watching with the bat detector switched on to see what happened. She also went into the roosts from time to time during the day to observe the animals briefly or to check for the presence of young - this required care to avoid disturbance - and sometimes, over the period from births until the young flew, she also entered the roost after the adults had emerged so that she could see the young more clearly.

At Ballynacally, which lies beside the estuary of the River Fergus, the bats were in the roof spaces of two rooms in derelict staff quarters annexed to a farmhouse. From here they gained access to the lower floors via trap-doors and emerged through open sashes and missing window panes into a farmyard. The roost at Newhall was in an old outbuilding at Newhall House, outside Ennis, above the ceiling boards of an upper room, which were incomplete and extended only along part of its length. The bats could move freely into an adjoining room and from thence outside, via a staircase and unglazed window on the lower floor, giving onto a walled meadow. The window was densely shaded by two sycamores growing on either side.

Although most of her work was at these two sites, she also visited the roost at Ruan (the one since purchased by the National Heritage Council) mainly because it was easy to enter during the day without apparently upsetting the bats. (With hindsight, and many additional years of experience, she now believes that she may have disturbed them slightly, just the same.) Unfortunately, because they emerged via both a hole in a gable and from windows all around the ground floor, it was impossible for a single observer to see all of the beasts as they came out. Because they clustered tightly together, counting them in the building was also difficult, except early in the season when there were few of them.

The bats took up residence in the roosts in April, the colonies enlarging throughout May and June, and births were in the last week of June or in early July. As insects are most abundant in July and August, foraging is easiest then for the mothers, which have to produce milk over some 5 weeks for their rapidly-growing offspring. Numbers emerging in the evening generally peaked about 3 weeks after the latest births, at about the time when all of the young would have first flown. The maxima she recorded were 49 at Ballynacally, 95 at Newhall and about 100 at Ruan. Thereafter the colonies declined and the roosts were finally deserted by late September or early October.

Towards the end of the season, and at the beginning in April, numbers tended to fluctuate markedly over periods of a few days. For example in 1985 on 9 April there were no bats at Newhall; on 16 April there were four; 7 days later there were 20; and after a further 8 days only two. The rise corresponded to a period of warmer weather, the subsequent fall to a cool period. There was a similar fluctuation at Ruan at the same time. No observations had been made at Ballynacally. In fact, both early and late in the season there was an obvious general correlation between numbers in the roosts and daily air temperature. Presumably some of the colony were returning to their hibernacula when it was cold and coming back into the nursery roost when it was warm. Paddy O'Sullivan had ringed lesser horseshoes in a hibernaculum at Edenvale Cave and some of these were seen during the summer at the Newhall roost. In this case at least, the hibernaculum and summer roost were less than 1 km apart, so it would have been little effort for the animals to flit back and forward between winter and summer quarters depending on the prevailing weather.[200] Work elsewhere in Europe indicates that hibernacula and summer quarters are usually within 5 km.[151]

A bat "ring" is illustrated in Fig. 19, the only type permitted in the British Isles, which is available in several sizes. It is horseshoe-shaped, has no sharp edges and is made of a hard but lightweight magnesium-aluminium alloy, which cannot be roughened by biting. It is marked with a serial number and the words "LOND ZOO". The ring is attached loosely to the bat's forearm. Properly applied it should cause no damage, although not all species of bat can be ringed safely. The procedure requires training and a license. Applying a ring to the delicate wing of a bat is an entirely different matter from ringing the horny leg of a bird. Unless there is a specific reason for doing this - in the case of the Edenvale Cave, to determine where the bats spent the summer - or it is part of a well planned programme of research, I believe that it is best avoided.

The lesser horseshoes in the Ruan, Ballynacally and Newhall roosts generally clustered close together beneath the roof, presumably to prevent heat loss. However, when McAney visited the roost at Ruan on 12 July and the Newhall roost on 12 and 14 July 1983, the animals were hanging widely separated from one another. The summer had been a good one and 12-14 July were the hottest days of the year, with maxima at Shannon Airport (Co. Clare) of 29.5-30.6°C. Beneath the slates in the roosts it must have been considerably hotter, so the bats doubtless found themselves in the unusual situation of trying to keep cool.

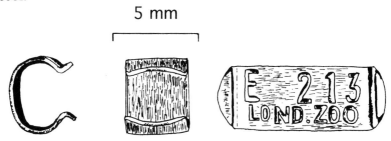

Fig. 19 A bat ring. In the illustration on the right the ring has been flattened to show the lettering and identification number.

It is late May or early June in the farmyard at Ballynacally.[199] We have arrived early, at least 45 minutes before sunset, and made a few notes on the weather, including whether it is overcast or clear. We have set the bat detector on the ground a few metres from the roost and settled ourselves on chairs near it. At sunset we read light intensity from the sky with an exposure meter and then we wait. The insects are beginning to bite already. The sun sets and soon there is a slight but perceptible decrease in the light. Birds gradually fall silent until all that is heard is the occasional querulous caw from the nearby rookery. In the dim interior of the lower story of the old annex building, we can see bats flying round and round, and, for the first time, the bat detector relays the characteristic long, warbling pulses of the lesser horseshoe. After a time - and this was very variable, from almost immediately until almost half an hour after circling started - a bat flies out of one of the windows and, in a couple of seconds, flies back in again. We called this an out/in, using the term strictly to mean such an excursion of less than or equal to 5 seconds. Before dark it is rare for a bat to stay out for *more* than 5 seconds if it is going to fly in again; on average this happens less than once each evening. Such an event we call an "early return". Shortly thereafter - the timing is variable but it never happens before sunset - the bats begin to leave in earnest, although, for a short time, the out/ins continue. There is no obvious pattern to the emergences, which seem to be more or less at random. The bats never come out in parties.

At Newhall events are similar.[199] The bats circle for some time by the window in the lower story before emerging and we can hear the long pulses on the bat detector. But there is a striking difference. There are fewer out/ins (on average 8.5 per night as compared with 11.3 at Ballynacally) and many more early returns - an average of 10.5 a night.

What does this peculiar behaviour signify and why the differences in the numbers of out/ins at the two roosts? Why were early returns common at Newhall but rare at Ballynacally? As explained in the last chapter, a bat flying in daylight is at a greatly increased hazard from being killed by birds. However, the earlier it comes out, the more flying insects there will be. So there must be a trade off between risk and advantage and there is probably a critical light intensity at which a bat will leave to forage, although this may differ both between species and between individuals of the same species. In the early evening at Ballynacally and Newhall the bats flew down to the lower stories and then circled near the windows. This was to allow them to assess the light. When a bat considered that it might be dark enough to leave it darted out to see whether this was indeed so. A moment was often long enough to appraise the situation and, if it was still too bright, the animal flew back inside: hence an out/in. At Ballynacally an out/in was nearly always enough for an accurate evaluation and so early returns were rare. However, at Newhall the entrance to the roost was shaded by the two sycamores. In the early evening a bat emerging there would have had to fly rather further before ascertaining that, clear of the shade, it was still too bright to set off. So early returns were commonplace there.

The time at which the first individual flew out and away was very variable, from a few minutes to over half an hour after sunset. When the last bat left it was late twilight. The total time taken for the whole colony to emerge was anything from about 10-40 minutes. Although showers had little affect on this, heavy rain slowed it to anything up to nearly an hour.

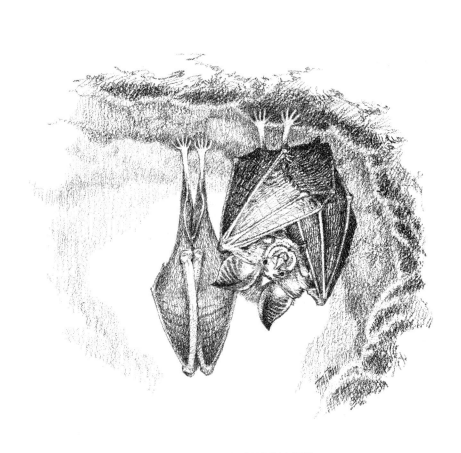

LESSER HORSESHOE BATS

McAney calculated the average time of emergence, for all the individuals in the colony, in seconds after sunset for each evening's observation. (For this purpose, out/ins were ignored.) This is a useful statistic indicative of the emergence for the colony as a whole with respect to sunset. It too was very variable but was strongly correlated with light. On bright evenings the bats came out late; on dark, overcast evenings, they emerged early. Once more, light intensity is the determining factor. Superimposed on the variation from night to night was a longer-term, seasonal pattern. The bats tended to come out sooner after sunset at the beginning and end of the season. The longest delay was around the summer solstice. Yet again, light intensity is responsible. As everyone knows, it gets dark sooner after sunset in winter than in summer, and the longest delay in nightfall after sunset is during the long days of summer, around the solstice, when twilight lingers on and on.

Comparison between all of the data from each roost showed that the overall average time of departure was about 4 minutes earlier at Newhall than Ballynacally. Naturally, the explanation for this might be expected to lie in some difference in the light at the two sites. Ballynacally and Newhall are so close together that sunset at both, calculated from an almanac for latitude and longitude, is virtually instantaneous. However, on the spot this is not so. The sun sets in the west. To the west of Ballynacally the land is flat for 2 km, whereas the ground immediately to the west of Newhall rises gently to some 100 m. So in practice sunset is slightly earlier there: it therefore grows darker earlier and the bats come out sooner.

After the last bat had departed there was sometimes, although not always, a quiet period of up to an hour when no bats returned to the roost. Even then lesser horseshoes were heard on the bat detector feeding in the vicinity. Thereafter bats entered and re-emerged throughout most of the night, and fed intermittently nearby, although there were often breaks in activity of up to about half an hour up to June.

Sometimes during the night at Ballynacally bats were found hanging under the staircase leading from the ground to the first floor of the roost building. Beneath, there were numerous insect wings and legs, indicating that this was a temporary night roost where the bats landed briefly to enjoy a newly caught prey. The leavings included wings of the small tortoiseshell butterfly *Anglais urticae,* which must have been taken while it was at rest, as butterflies are strictly diurnal.

In the late 1960s and early 1970s, Mr Neil Stronach often caught lesser horseshoes which had come into his home during the night at Milford House, near Kilmaine, Co. Mayo.[259] At one stage they were regularly entering through a window to roost for a short time above an airing cupboard, an ideal place to keep warm between flights. I saw them there myself once, apparently unconcerned by the people below.

Despite the comings and goings during the night at Newhall and Ballynacally, and although some bats came home to stay early, the majority of each colony were still abroad well into dawn twilight. There was always sufficient light to record the times at which these individuals eventually went to roost and from these data an average time of return before sunrise was calculated for each roost on each morning, the return period as a whole lasting anything from about 10-50 minutes. On only one morning did any individuals actually enter a roost after sunrise. The earliest arrivals flew repeatedly about outside emitting long echolocation pulses. Latecomers flew directly in, apparently usually producing only a single pulse.

Just as average time of emergence after sunset tended to be greatest around the summer solstice, the average time of return before sunrise was also greatest, and for essentially the same reason. Dawn twilight is longest at this time of year; it is lighter earlier, and the bats do not like being out when it is too bright. It is yet again a matter of light intensity.

All of this means that, when the nights are shortest, the bats come out later and go to roost earlier, thus further curtailing the limited time available for foraging. This must be inconvenient to suckling mothers, whose energy demands are greatest of all, and who may be nursing at any time from towards the end of June to early August. Fortunately there is a partial compensation, for insects are most abundant in high summer. If births were later, say toward the end of July, insects would still be plentiful during lactation and the longer nights would provide more time in which to catch them. However, this would be offset by the young having less time to perfect their hunting, and both young and mothers having a shorter period over which to build up fat reserves, before hibernation.

Besides her nursery roosts, McAney located four roosts of lesser horseshoes, in cottages and in a barn, which were occupied intermittently by a few individuals - usually only one. These bats were probably males although, because her policy was never to handle any lesser horseshoe lest she upset it, this was unconfirmed.

During the summer of 1984, from May to September, McAney tried to investigate the activity and habitat preferences of lesser horseshoes in the field on 15 nights within a radius of 1 km of Ballynacally and 14 nights within the same radius of Newhall. She first decided on the types of habitat that were typical of the terrain and then selected a few sites representative of each to investigate, to a total of twelve sites in each area. The habitats *inter alia* included trees (woodland strip and windbreaks), roadside hedgerows, farmyard, pasture and tree-fringed water.

To determine possible changes in activity through the hours of darkness, she divided each night into three equal segments: respectively post-dusk, mid-night and pre-dawn. Of course as the length of the night varied through the summer, so did lengths of the segments.

She monitored each site by walking up and down it for a 15 minute period, called a "run", with her bat detector tuned to 113 kHz and her tape-recorder switched on. She assumed (as I, her supervisor, did too) that only lesser horseshoes would then be picked up.

Unfortunately, there were no valid results from all of this work. Although her basic technique was sound - it has been used subsequently for Leisler's bat (Chapter 10) - the assumption that only horseshoes could be heard with the detector tuned to 113 kHz was faulty. Subsequent research has revealed that several species of bat in Ireland, although producing most of their sound at decidedly lower frequencies, also often emit a small proportion of it as high as 113 kHz. Moreover, because of the highly directional nature of the pulses produced by the nose-leaf, unless a lesser horseshoe is flying more or less towards the detector, it is unlikely to be picked up. Besides, because the sounds are high pitched, they do not carry far.

In the summer of 1998, Miss Rebeccah Cogan, then an undergraduate at NUI Galway, followed the lesser horseshoes from the big roost at Ruan as they departed in the evening to hunt.[52] As the bats left, she pursued them on foot with some help from a bat detector as far as she could (which was not far) and marked the spot where she was left behind on a large-scale map. The following evening she stationed herself at this point and repeated the process, thus mapping their route further, and on yet other evenings if necessary. In this way she determined all of the routes from the roost for a few hundred metres in each direction. These all lay along linear landscape features. Hedgerows were the most popular, but also tree lines and ivy-covered stone walls. Where the linear features were linked by scrub, the bats crossed this as well. The animals usually flew alongside the stone walls and, if crossing a road or track, this tended to be at at a point where the canopies of trees on either side touched. They never flew directly over open fields.

This suggests that lesser horseshoes dislike crossing open spaces. Work outside Ireland has revealed that many other bat species also use linear landscape features, or "flylines", when commuting to foraging grounds. Flylines not only provide insects for snacks *en route* but also cover from predators. It would be much easier for a bird to home in on a bat flying above a field in late evening or around dawn than one travelling along a hedge, alongside a wall or beside trees.

Cogan also discovered that the bats often perched for a time in a sycamore standing close to the building at Ruan and that other trees along the flylines were also used for such temporary halts. It will be clear from this, and from what has been said above regarding temporary night roosts, that, when lesser horseshoes are away from the main roost, they are by no means continuously on the wing. It has also been shown elsewhere that they may "perch feed": hang up somewhere and then make brief sallies to catch passing insects. So, not only are hedges important to the bats as flylines, but large hedgerow trees are also of some significance, and should be conserved. Provision for hedgerow trees in the future, when the present ones die or are felled, should be considered. If hedges are trimmed low and level between existing trees, as those along roadsides now are all too often, eventually there will be no hedgerow trees at all.

Studies in Britain show that not only do lesser horseshoes catch insects in flight, either by working a beat or by perch feeding; they also glean prey from leaves and even from the ground, so they are versatile feeders.[151] McAney analysed monthly samples of droppings extending from June to October from a total of eight roosts (total droppings analysed 630).[201] Prey included small night-flying Diptera, with midges predominating (13%) but rather more larger night-flying Diptera (24%), which included crane flies and window midges (Anisopodidae). The bats caught few diurnal Diptera. Moths and butterflies (19%), probably predominantly the former, were also important. These are usually impossible to differentiate in bat faeces, but the antennae of butterflies, which have a knob at the end, were recovered sometimes, and butterfly wings were found amongst the droppings at roosts, where they had been discarded by the bats. Butterflies are almost certainly indicative of gleaning. The bats regularly ate caddis flies (18%) and lacewings, also known as alder flies (Neuroptera 13%), which prey on other insects, particularly greenfly. So bats do not take exclusively insects that we might regard as injurious. The diet also included much smaller quantities of spiders, beetles (Coleoptera), wasp-waisted insects and bugs. The bats consumed a lot more smaller insects in June than in July and August. This is probably another instance of optimal foraging, discussed in regard to Daubenton's bat in the previous chapter.

Chapter 8. Leisler's Bat

Twinkle, twinkle little bat!
How I wonder what you're at!
Up above the world you fly,
Like a tea-tray in the sky.
Alice in Wonderland. Lewis Carroll.

Leisler's is the sportiest of the Irish bats and has long, narrow wings adapted to fast flight but with reduced manoeuvrability. Narrow wings usually mean high wing loading: a greater weight per unit area of wing, and in Leisler's wing loading is over two and a half times greater than that of any other Irish bat.[245] High wing loading means that stalling speed - the velocity above which the animal must fly if it is to remain airborne - is also relatively high. All of this applies equally to the specification of aircraft. The wings of a jet fighter are narrower and of lesser area than those of a piston aircraft of comparable size, which not only flies slower but can remain aloft at a much lower speed. The same is true of birds: compare the long narrow wings of swift-flying falcons with the broad ones of the buzzards, whose flight is leisurely to a degree.

Leisler's bat is also by far the biggest Irish bat. So the observer, with a little experience, may identify it in the air on a combination of its size and strong, steady direct flight, often in a straight line. This, with occasional dives, is typical of the beast as it commutes from roost to feeding place, at heights of up to 100 m,[308] well above the tops of the tallest trees. Not for this bat the judicious passage via flylines alongside hedges and walls. Commuting flight is thus quite different from the more fluttering progression of our other bats, especially the pipistrelle. Because Leisler's emerges earlier in the evening than all the other Irish bats, sometimes before sunset, the process of identification is made even easier. The echolocation sounds are also lower pitched than those of all our other species. The pulses are mostly frequency modulated from about 50-25 kHz, with most of the sound energy at the lower end, and emerge as a distinctive "chip-chop" through the bat detector. In certain situations the pulses may also be of almost constant frequency at around 25 kHz. They are particularly loud and Shiel could pick them up on a detector as far away as 100 m. While it might be imagined that a big, fast-flying bat, with relatively poor manoeuvrability and coarse echolocation would be unable to catch small insects - remember that low frequency means longer wavelength and less accurate perception - this, as we shall see, is not the case.

In the hand the most diagnostic feature is the tragus which, unlike that of any other Irish bat, is broader than long (Fig. 15). The base is narrow, the upper part wide and the overall profile is that of a mushroom. The fur is dark brown, but with the tips of the hairs characteristically much lighter than their bases. The hair extends thickly onto the wing and along the forearm, accounting for the former common name of "hairy-armed bat".

Whereas Leisler's bat is widespread throughout Europe, it is scarce and records are sporadic. Only in Ireland is it abundant. If there is a bat that is typically Irish, then it is Leisler's. Moffat thought it the third commonest species, after pipistrelle and long-eared,[231] and in the National Bat Survey it was shown "to be numerous and widely distributed" with several colonies of over a hundred individuals. The Northern Ireland Bat Group recorded it as the second commonest species, with one colony in Co. Fermanagh of 300 individuals counted in late July.[3] The biggest recorded Irish colony - also a world record - was in a nursery roost in a gate lodge at Castletownshend, Co. Cork, in 1974-75. As varying estimates of maximum numbers in this colony have been published elsewhere, it is worth setting down the facts, which were kindly furnished by Paddy O'Sullivan, who had by far the most experience with it. He did a count inside the roost in the mid 1970s in summer and found 600-620 bats. Allowing for males, which were almost certainly somewhere else in the district, this points to a local population of over a thousand head. Notwithstanding, the record number of Leisler's bats *in a single roost* must stand at something over 600, rather less than some published figures. In subsequent years the colony broke up, with various groups settling in other houses in the village. By 1978-79 the original building housed only about 300 individuals and by 1999 it was down to 135-150. From his observations on Leisler's bats over many years, O'Sullivan has concluded that once colonies reach a critical size, usually between 400-500, they will split.

In continental Europe, Leisler's is regarded as a woodland bat. During the summer it roosts in hollow trees there, although sometimes also in buildings and in bat boxes. Hibernacula on the continent are in hollow trees and in cavities in buildings.[297] It is indeed paradoxical that what is in essence a bat of woodland should be commonest in Ireland, where deciduous woodland covers less than one half of one percent of the landscape and all woodland cover is only around 6%. As mentioned in Chapter 6, commercial conifer plantations here are usually of limited use to bats, although there are exceptions, as described below.

One explanation of the abundance of Leisler's in Ireland may be the absence of a closely related species, the noctule *Nyctalus noctula*, which overlaps the geographical range of Leisler's everywhere else in Europe but is missing from Ireland.[338] The noctule is the bigger, weighing in at 15-49 g compared to Leisler's 14-20 g. It too is a fast flyer with narrow wings and uses relatively low frequency echolocation. Although both species are regarded as typical of woodland, they do not forage amongst the foliage like the long-eared. Instead both hunt above the treetops, as well as over meadows and water.[297] They apparently take similar prey but, because the noctule is bigger, it can tackle larger insects. Perhaps Leisler's is at a disadvantage where there are noctules. If so this might explain its success in Ireland. This notion of the absence of competition with the noctule as an explanation for the prevalence of Leisler's here goes back to Moffat,[228] who remarked that it

> is certainly far better known in this country than it is in England, where it may be a good deal 'kept under' in point of numbers, by the yet larger and more powerful Noctule.

Unfortunately the theory has never been proven and would require a study of both species living in the same district, which would obviously have to be *outside* Ireland.

In Ireland Leisler's bat roosts both in hollow trees and roof spaces in summer, apparently mainly in the latter, although a colony in a house is more likely to be noticed than one in a tree

The first record of Leisler's in Ireland was in 1858 - a specimen from Belvoir (pronounced "Beaver") Park, Co. Down, then some kilometres from Belfast. In the same year a man knocked down a second one with a fishing rod in Blackstaff Lane, Belfast. He gave this to a Mr Darragh who kept it as a pet, but eventually it found its way - presumably as a corpse - to Kinahan, who exhibited it at a meeting of the Belfast Natural History and Philosophical Society in April 1860.[21,22]

The next find, this time of the animals in quantity, was by R.M. Barrington (1849-1915), a naturalist of wide interests, who trained as a barrister but became a land valuer and whose private means allowed him to indulge his interest in Ireland's flora and fauna at will. On 28 June 1868.[22]

> ....when wandering through the lofty beech grove belonging to the Duke of Manchester at Tandragee [Co. Armagh]....my attention was attracted by a chirruping, clicking sound which apparently proceeded from a hole about twelve feet [3.7 m] up the trunk of a large tree. After some trouble I managed to get up to the hole, and as I did so the noise greatly increased. Within I saw a moving brown mass, and I thought I could distinguish bats. Cautiously putting my hand in among the chirrups and clicks, I made a grasp. They were bats....for I pulled out eight or nine, but, as they struggled so violently, all escaped but three or four. The bats in the hole were so much alarmed that they commenced flying out of the hole into the sunshine, and continued doing so for several minutes, tumbling and scrambling over one another in a ludicrous manner, so eager were they to reach the entrance. Probably eighty or a hundred bats thus flew out; some appeared to have young ones adhering to their teats, but I will not speak positively on this point....many bats still remained far up in the hollow trunk....It is not a little remarkable that the twelve specimens [obtained]....were females.

This was evidently a nursery colony. Two individuals were presented to the British Museum and later caused confusion when they were there initially misidentified as juvenile noctules.[1]

In February 1874 Barrington wrote to his brother-in-law, who lived near Tandragee, to see if further specimens might be procured, but the latter was unable to find any. However, on 18 May he located a roost in another of the old beeches containing "considerable numbers" of bats, which emerged in panic as they had for Barrington; 4 days later he tried again but the tree was, not surprisingly, deserted. After a further search he heard "squeaking....high up in another" and a colony was located "perhaps thirty-five feet [10.7 m] from the ground" in a hollow branch. He procured a ladder and climbed up.

> The cavity had two orifices, one in the trunk, the other in the bough itself. A bag was nailed over the lower opening, and the

bats were poked out from above. In this way sixteen were
taken.

It is quite possible that the same group of bats figured in all three
incidents.

Leisler's bat was found so often during the next 20 or so years that
in 1897 Jameson was able to list 10 counties where it was know to
occur, including Louth, where he shot specimens himself at
Killencoole and Branganstown.[167] By 1938 the list of counties had
expanded to 17.[231] But records from only a further three had been added
before the National Bat Survey commenced in 1985.

Alcock was the first to study the behaviour of Leisler's bat,
particularly in flight.

> At first it commonly flies at a considerable height, in open
> country taking long sweeps and wide zigzags, often being
> seen but once in an evening. Near woods and in favourable
> localities it will often remain for some time near one spot,
> flying at an altitude of 30-40 feet [9-12 m], with a faster and
> less irregular flight than the Pipistrelle....Later on it flies
> nearer the ground, very commonly shrieking loudly, and I
> have observed two bats at this time chasing one another
> exactly as two butterflies do, both flying very fast and
> screaming.

Evidently suitable concentrations of insects were found both high and
low. The "one spot" where the bat lingered may be interpreted as a
particularly satisfactory feeding place, where there were plenty of
suitable insects.

Alcock tentatively concluded that Leisler's bat flies for about an
hour after sunset and then retires to its roost for the rest of the night.[1]
At first Moffat agreed with him and particularly remarked that
individuals returning from such excursions shot by Alcock were as
"hard as a cricket-ball - crammed....to utmost capacity with insect prey".
It was probably one of them whose stomach contents was analysed by
G.H. Carpenter (1865-1939),[1] then working in the National Museum in
Dublin, but later appointed Professor of Zoology at the Royal College
of Science for Ireland. In one stomach he identified moth scales, caddis
flies and Diptera, including a midge and a yellow dung-fly, findings by
no means incompatible with a major investigation of the diet in Ireland

almost a century later.[302]

Over the summer of 1900, Moffat made a fuller study of when Leisler's bat flew.[218,228] On 22 July at Fassaroe, Co. Wicklow, presumably whilst visiting Barrington who resided there, he saw several Leisler's bats on the wing between 2.56 and 3.36 am, clear evidence that they might also fly in the small hours of the morning. By the way, in 1900 Greenwich Mean Time prevailed throughout the summer and sunrise would then have been around 4.15 am.

In August he set about watching Leisler's bats near his home at Ballyhyland, Co. Wexford, taking his station beside a roost in a hollow ash tree in a pasture field. On 10 August, when sunrise would have been at about 4.50 am, he saw and heard the bats at 3.35 am in the open "hawking and sporting above the level of the tree tops" and at 4.02 one darted into a hole in the ash "about seven feet from the ground". During the next 10 minutes a further three entered the tree by a hole further up. On the next evening five left shortly after sunset, and on the evening following all five issued forth, again around sunset. He noted that they skimmed close to the grass as they came out, but this was doubtless because it was essential for them to lose altitude on emergence in order to pick up speed; remember that the stalling velocity is high.

It being too dark to see when they went out for their morning flight, at midnight on 12 August he fixed a net over the hole and watched in the early hours of the following morning, "some brilliant Perseid meteors relieving the monotony of the vigil", and "at 3.15 am heard a bat gently flop into it". Repeated observations overnight convinced him that the species had both an evening and a morning flight of somewhat over an hour each and that the animals came home in the morning around half an hour before sunrise. Between sunset and sunrise, flying insects are most numerous at dusk and in the early hours of darkness, but there is also a minor peak towards dawn, which Moffat's bats might have been exploiting. These observations stood until the 1990s when they were greatly augmented by Shiel's research, also in Co. Wexford.[303,308]

In the 1940s Richard Faris (1911-84), a solicitor and - then something of a rarity - a naturalist in Co. Cavan, watched Leisler's bats near Kilmore on several occasions.[123]

> ....at least one lived in a gap in the masonry of the house and we often saw it setting out on its evening flight, direct and purposeful for the first 100 yards and then beating to and fro (while deciding on the best haunt?) and once more the direct

flight out of sight. This little hesitation in an otherwise direct flight is, I think, quite characteristic of the species....

The "hesitation" is most probably explained by a minor concentration of insects, at which the animal lingered for a bite, before continuing on its way to a larger meal elsewhere.

Dr Caroline Shiel began her work on Leisler's bat whilst a schoolteacher, during the summer holidays of 1991-92, when she was able to acquaint herself with the animal at leisure, to locate roosts, to time the bats as they emerged in the evening at some of them, and to collect droppings for analysis. In April 1993, therefore, she resigned her post and, well-primed and with some lines of enquiry well under way, began full-time research. She decided to concentrate her study, which continued until 1995, on a nursery roost in the attic of a two-storey cottage, with a tiny orchard at the back, on the outskirts of Baldwinstown, a hamlet a few kilometres west of Bridgetown, Co. Wexford. The bats emerged through two holes between wall and facia board on either side of a west-facing gable fronting a third-class road. The maximum number of bats she counted out was 207 in July 1991. One of the holes was elongate and wider than the other, so that several bats could pass through it simultaneously. Often a number lodged in it for a while, even before the first bat had flown out. A trapdoor led from the upper storey into the attic, so she could look at the animals inside as well. Unfortunately entering the attic tended to alarm them, especially the adults when there were young present. So from early in her work she only visited the roost immediately after the colony had departed in the evening, both to see when the young were born and, at the beginning of each month in 1991-93, to remove and replace polythene sheeting on the floor on which to accumulate droppings for analysis.

During the summer of 1993, she recorded the times at which the bats emerged on many evenings and watched throughout the night outside the roost 13 times. She also investigated activity and habitat preferences in the field with a bat detector. All of this work involved much the same procedures as those followed by McAney in the previous chapter. During this work Shiel discovered a second Leisler's roost in the district, at Rathangan, another hamlet, only 2.5 km distant from Baldwinstown. She had noticed that some of the bats passing overhead in the early evening were coming from the wrong direction for Baldwinstown. So she traced them further and further back on successive nights - rather as Rebeccah Cogan did with her lesser horseshoes, but in reverse - and eventually discovered the Rathangan roost. This was in a sealed chimney of a two-storey farmhouse, so there was no possibility of an observer entering it, but it was still possible to count and time the bats as they came out in the evening, through a small hole between wall and facia on a west-facing gable. Also, some idea of when there were young could be had from the distinctive calls which they make when their mothers are away. She included Rathangan in her study from then on, but did rather less work there than at Baldwinstown. The maximum number of bats she counted out was 175 in July 1995.

During the summers of 1994-95 she continued her observations at both roosts on some evenings, and, aided by her brother Robert (who is now a vet) during his summer vacations from school and university, she also radio-tracked bats from the roosts. A radio-tagged bat had to be pursued immediately it departed, so it was impractical on such an evening also to record the time at which each individual in the colony left. However, it *was* possible to note the time at which the first bat emerged. Her data from evenings when she was able to time all the animals out showed that there was a very close correlation between the average time of emergence for the whole colony - the statistic used by McAney - and the time when the first bat left. So she could use the latter as an index of the overall time for the whole colony, which meant that she had a reliable figure, representing time of emergence, for many more evenings than those in which she timed out every individual.

In 1993 and 1994 the bats first took up residence in the roosts towards the end of April, but in 1995 some had settled in at the beginning of the month. This was probably because the first fortnight in April 1995 was exceptionally warm. Furthermore, as many readers will recall, the summer of 1995 was also a scorcher, in fact the warmest in Wexford since 1959. In all three summers numbers of adult bats in the roosts increased, with some fluctuation from night to night, to June and then stabilised. The first births were judged to occur a few days before 13 June in 1993 and 1994, but a week earlier in 1995, because of the fine summer.[304,305] Readers will recall from Chapter 6 that for bats a warm summer means more insects, less daily torpor and earlier births.

Over the period that the young were born, a few adults remained in the roost during the night. These may have been giving birth. Thereafter the babies were left on their own at night, except when their mothers returned to nurse them. The newly-born bats were naked and pink with their eyes closed. At about a week old they darkened as a greyish coat of fur began to develop. The first few to arrive hung separately and were mostly torpid, whereas when all had been born they huddled tightly together in groups, presumably to keep warm, and were evidently active. This is important as torpidity would slow growth.[305]

The young first flew at about 4 weeks, in July, and all had taken to the air by about a week later; the number of bats emerging in the evening then peaked. From counts of bats when numbers stabilised in June and also when all of the young were flying, it was possible to calculate the approximate fraction of the adults in the colony which gave birth; this varied from year to year from about one third to two thirds. Of course a sizable proportion of the adult females in the roost would have been born in the previous year and presumably would not yet have mated, although this has yet to be verified for Leisler's bat.

That mother bats continue to suckle their offspring after the latter first take to the wing is important. For young Leisler's bats evidently need time to perfect their flying technique. At Baldwinstown a telephone wire lay across the flight path of emerging bats, but it was rare for an adult to collide with it. The juveniles often did. Whereas the adults took off directly from the exit holes, the juveniles sometimes crawled about a metre down the wall where they paused for up to 10 minutes, echolocating constantly, before apparently summoning up enough courage to launch into the air. For about two weeks after the young first flew, they foraged in the vicinity of the roost: at Baldwinstown over the orchard behind the house and the road in front, at Rathangan above the farmyard and a lane at the front of the house. At both roosts they sometimes landed on the gable wall and hung there for 10-15 minutes before taking off again. Maybe they were having a rest. When returning to roost they often settled on the wall a metre or so below the exit hole before crawling up and in. Returning adults alighted on the edge of the hole itself.

When the last of the young flew it is likely that the mothers of the first born had finished nursing. At any rate, numbers in the roost thereafter declined sharply as the adults gradually left for good, and after them the young, until by mid September in 1993-94, and the latter part of August in 1995, no bats whatsoever remained.[305]

Although squeaking was clearly audible from both roosts during the day, it grew louder and more continuous for about half an hour before the first bat flew. As adult Leisler's bats sally forth from the roosts of an evening they present some striking contrasts with lesser horseshoes. A lesser horseshoe flies out more or less on the level; the Leisler's bats dropped by 2-3 m, presumably to pick up airspeed and prevent a stall, before gaining height again and flying off. Because of their poor manoeuvrability and high stalling speed, anything like an out/in to test the light would have been out of the question. Indeed during emergence counts on a total of 130 evenings, on only five evenings did a bat come out, circle and then re-enter. All were at Baldwinstown and in three the animal in question was the first to come out. That this was to check light intensity seems unlikely. The explanation is probably that on each occasion the individual was among a group gathered in the large exit hole and was jostled out prematurely.

A further difference from lesser horseshoes is that Leisler's bats leave in groups: several individuals flying out in quick succession, or at Baldwinstown even simultaneously through the larger exit hole. Then a break; then another party. Outbursts of this kind may have some survival value. A predator, for instance a bird loitering with intent outside, might be confused by so many bats flashing past together and lack the presence of mind to select an individual victim quickly enough.

Outbursts are especially beneficial in this regard as Leisler's bat comes out early. At Baldwinstown the first few animals often, and over half of the colony exceptionally, had left before sunset. The earliest that any bat departed was over 23 minutes before sunset. However, the average time of emergence at Rathangan was 9 minutes later than at Baldwinstown and only occasionally did bats leave ahead of sunset, although the record was 16 minutes before. When Shiel and her brother watched both roosts on the same evening, Baldwinstown was nearly always the earlier colony to emerge. The reason for this may be the large exit hole there. The bats lodging in it would have been better able to test the light and, as there were a number of them, the chances of one setting off and initiating general departure, which followed more or less directly after the first bat had left, would have been enhanced.[304]

Leisler's bats are in fact the earliest abroad of all the Irish bats and this is almost certainly because they are also the fastest flyers. They thus have a better chance of outstripping hostile birds and can afford to set out early to take advantage of the greater numbers of insects flying before nightfall. Risk, however, is probably not entirely eliminated. Pat Smiddy tells me that one evening near Youghal, Co. Cork, in good light, he saw three Leisler's bats and about a hundred swallows *Hirundo rustica*, which were going to roost. The swallows occasionally chased after the bats, as they would after a bird of prey. There are also descriptions in the literature of the closely related noctule, which also emerges early and is an even faster operator, being attacked on one occasion by a sparrowhawk *Accipiter nisus* and on another by a herring gull *Larus argentatus.*[304]

Even during the hours of darkness, it appears that Leisler's is at less risk of being killed by owls than any of the other Irish bats are. All our species of bat (if we count the three pipistrelles as one) have been recovered from the pellets of Irish owls, except for the whiskered, presumably because it is so scarce, and Leisler's which, although common, may usually be too quick for an owl.

While showers had no affect on emergence, during heavy downpours the bats tended not to come out at all, or to wait for breaks in the rain, or sometimes they returned to the roost within minutes of departure. Rain probably interferes with echolocation and wetted fur would increase wing loading and heat loss.

The residence of both colonies over the summer is best considered in four periods: *prebirth*, before any young were born; *lactation*, when the mothers are suckling their offspring; *postlactation*, the short period between the end of lactation and when all of the adults had departed; and *juveniles only*, when these were the only bats left in the roosts. As far as was known, all of the adults in the colonies were female; the juveniles were, of course, of both sexes.

The time at which both colonies departed with respect to sunset varied from night to night, but at Baldwinstown was generally 15-20 minutes earlier during lactation, in June and July. This is almost certainly because producing milk for the young, which have to grow almost to full size in a matter of weeks, requires the consumption of a lot more food, and the increased risk in coming out earlier is offset by a better-stocked larder of flying insects. Such behaviour is quite the reverse of that of lesser horseshoes; they came out latest when the nights were shortest. However, Leisler's bats are faster fliers and are therefore less vulnerable. Moreover, lesser horseshoes can glean insects and so do not entirely rely on those on the wing as Leisler's has to. Surprisingly, the bats at Rathangan did not come out any earlier over lactation. This is difficult to explain but may have had something to do with the large exit at Baldwinstown allowing the colony there to assess light more precisely.

Once the emergence data were adjusted to take account of the earlier departure during lactation, it was evident that, even though the bats came out sooner than lesser horseshoes, light intensity affected their emergence in just the same way: they left earlier on dark, overcast evenings than on bright clear ones. Astonishingly this correlation was strongest during lactation at Baldwinstown, when the bats came out earlier anyway! But the risk of predation would be greatest then, so accurate assessment of light is most critical. All of this emphasises the knife edge on which the actions of the bats turn and how the hazard from predators is nicely balanced against the advantage of more flying insects, especially during lactation when energy demands are greatest.[304]

Observations overnight at Baldwinstown during prebirth revealed that bats returned to the roost throughout the night, but many within 2-3 hours after emergence. So most had only one flight. A few took a second flight in the latter part of the night. During lactation the animals came and went throughout the hours of darkness: they had both to forage hard and to return periodically to suckle young. In postlaction there was a return to the prebirth pattern as suckling had ceased. Note the similarity of the results for postlactation to the two flights Moffat observed in August 1900.[218] His few animals may have been adults, either females, which may have left the nursery roost, or males. Once the adults had deserted Baldwinstown, most juveniles tended to stay out all night, perhaps because, being less experienced in hunting, they had to spend longer at it to get enough to eat. As will be seen below, radio-tracking the bats usually confirmed and greatly expanded these conclusions.[304]

Often when a bat came back during the night it did not land at the hole immediately and crawl in. Instead it flew up almost to the entrance and then swung down and away as far as 6 m, usually repeating the swinging motion several times. Up to 27 such swings were counted during such a performance and up to five bats were seen at it together. As the bat approached the hole the echolocation pulses became progressively shorter and more rapid, as of course they do when a bat approaches anything.[305]

Another peculiar ritual on return was "swarming", which was commonest with the last group of bats to enter the roost before dawn. From 5-60 flew in circles, perhaps 10 m or more in diameter, at high speed, sometimes chasing each other, from above the roof of the cottage out over the road in front. This was accompanied by a characteristic call on the bat detector - a series of greatly shortened pulses of equal length, quite different from those made by the bats when foraging.[305] Swarming is impressive, for me one of the greatest spectacles in all of wild Ireland, ranking alongside a pond of breeding natterjack toads (Chapter 4). Presumably swinging and swarming, which have also been noted at another roost of Leisler's bats in Ireland,[195] are of some social significance, but this has yet to be worked out. Maybe the bats are advertising their ownership of the roost.

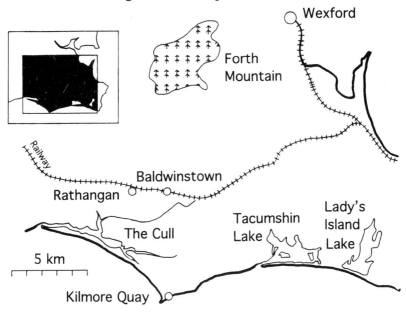

Fig. 20 A map of the area in Co. Wexford over which Leisler's bats were radio-tracked. Top left: the south-east corner of Co. Wexford to show the location of this area.

Fig. 20 is a map of the district around Baldwinstown and Rathangan. This is almost all flat and below 30 m above sea level. The land south-west of the roosts has mostly been reclaimed from the sea, and there are many partially tidal drainage canals there, mostly 10-15 m wide. These discharge into the Cull, an enclosed, mainly tidal estuary with wetland and saltmarsh. The canals are in turn fed by channels, not more than 2 m wide, which, for the sake of clarity, will be referred to as drains below. There are two slightly brackish lagoons to the south-east, with reed swamp and marsh: Tacumshin Lake and Lady's Island Lake. The south coast is bordered by sandy beach throughout its length and there are dunes from the western end of the Cull east to the village of Kilmore Quay (population 424), by far the largest settlement in the area. Almost all of the land is agricultural, with pasture predominating, and with no woodland, but to the north Forth Mountain rises to 237 m above sea level. Most of the mountain, and the land immediately around it, is planted with commercial conifers.

As already mentioned, over the summer of 1993 Shiel investigated habitat preference and activity in the field with a bat detector using the same approach as McAney.[303] Shiel choose 18 different sites representative of the habitats available, each of which she monitored for echolocation pulses of Leisler's with a bat detector for a 15 minutes "run" on each of 18 nights. Like McAney, she divided each night into three equal segments: post-dusk, mid-night and pre-dawn. Because Leisler's bats are fast flyers, she chose her sites up to 5 km from Baldwinstown. In fact, her radio-tracking in later years showed that they went even further.

No Leisler's bats were heard in 42% of the runs. There was little clear cut difference between habitats, and results from two or more sites of the same type often differed considerably. Perhaps, by accident, she had chosen some sites that happened to be atypical of the available habitats as far as the bats were concerned. The most consistent sustained foraging was at street lights, which, of course, attract insects. Nevertheless, the individual site the bats used most intensively was on pasture heavily grazed by farm animals, where there was abundant dung and doubtless numerous insects that had bred in it.

Average minutes with echolocation pulses for the three segments of the night were: post-dusk 1.33, mid-night 0.41 and pre-dawn 0.60. This matches the activity of the adult females observed overnight at Baldwinstown roost over the summer, apart from lactation, and quite possibly that for adult males, which might be expected to operate in the same way as females not suckling young. Thus the greatest activity is in the early part of the night, when all of the bats are out, followed by a reduction as the animals returned to the roost, and a minor rise towards dawn corresponding to a lesser peak in flying insects and some second flights. In fact the average figure for pre-dawn conceals a great deal of variation. On some nights Shiel picked up no echolocation pulses at all from Leisler's bat in pre-dawn. On others there was more activity than in post-dusk. However, unlike post-dusk, the extent of activity in pre-dawn was strongly correlated with air temperature. There was also a weaker correlation with air temperature during mid-night. A higher air temperature means more insects on the wing. These results may therefore be interpreted as follows. In the evening the bats will not have fed since before the previous dawn. In addition this is usually the part of the night when flying insects are most plentiful. Therefore all the bats come out to forage in post-dusk irrespective of temperature. During mid-night flying insects are scarcest but, depending in part on how warm it is, some bats may find it profitable to continue hunting and possibly others, which have returned to the roost, may come out again. In pre-dawn there is a minor peak in insects on the wing but this is also correlated with air temperature. The higher the temperature, the more bats will come out again. Of course, during lactation, females with young are active throughout the night, but this represents a relatively small part of the summer and is masked in the overall picture. The only clear trend in activity from month to month was a fall off in September.

To see whether there were any Leisler's bats flying in winter, for a total of 18 mild evenings from September to April in 1993-94 and 1994-95, Shiel monitored a few of the sites where she had recorded most activity in summer and others favoured by radio-tagged bats in April-May 1994. To maximise the chances of hearing something, she did this only from sunset until not later than 3 hours thereafter. She detected Leisler's bats up to the end of October, but never picked them up between November and March, although she heard pipistrelles both in December and March.

If the Leisler's bats were indeed still in the district during the winter, it is surprising that she failed to hear any whatsoever. Combined maximum counts from the two roosts in 1993 and 1994 were respectively 230 and 267 animals. A similar number of males might be expected to be around somewhere, bringing the overall total in both years in the area to over 400. The explanation, which would also account for the fall in activity in September, could either be that hibernation begins early, is prolonged and that the animals hardly ever forage in winter, or that most Leisler's bats leave the area and that maybe the move begins in early autumn.

The south-east tip of Wexford has exceptionally mild winters, with average daily temperatures in January of 5.5-6.0°C. Furthermore, the mean annual hours of bright sunshine, at 1600-1700, is higher than anywhere else in Ireland. There are no caves, so presumably any hibernating bats are in crevices in trees or in rocks. In south-east Wexford it may be that these are too warm for Leisler's bats to hibernate satisfactorily. So perhaps the bats may move elsewhere. Unfortunately in Ireland we know practically nothing about hibernation for Leisler's bat. In the literature there are several records of active individuals in October. Moffat noted a specimen captured in a bedroom at Cromlyn, Co. Westmeath, on 16 November 1894.[216] In 1975 Douglas Deane mentioned a colony hibernating in the roof space of the Charity School at Downpatrick, Co. Down, "a long time ago".[72] Unfortunately, as he did not give a date - not even a month - and the animals were evidently roosting openly, this must remain dubious. A few, roosting singly in winter, have also been recorded recently in Northern Ireland, but I have no other information about them.[285] There is only one record between December to March for a hibernating Leisler's for which I have details, for January 1996. Then Ger O'Donnell, of Connemara National Park, discovered one when the entrance kiosk there, which had been built of stone, was being demolished. He found several pipistrelles hibernating in crevices between the stones and a Leisler's bat on one of the roof beams. As the kiosk had been partly razed when he inspected it, the bat may have been hibernating in a crevice, was disturbed and had moved to the roof.

Shiel analysed the diet of Leisler's bats by collecting monthly samples of droppings from Baldwinstown over the summers of 1991-93 and also material from various months in the early 1990s from four other roosts, located in Cos Waterford, Cork, Leitrim and Donegal (total pellets analysed 1,195).[302]

Analysis of bat faeces for those recognisable parts of prey that have passed through the gut undamaged, is an invaluable technique. It yields systematic information on the diet of bats which is impossible to obtain otherwise, save by killing the animals and examining their stomach contents. It also gives the bat worker something to do during the winter, when bats are hibernating.

Bat droppings are often to be found below the entrances to roosts; unfortunately rain usually washes them away. The best place to collect them is on the floor beneath the bats in the roost and this is most easily done on heavy-guage, sheet polythene. To minimise disturbance, this should be laid early in the night after all of the inmates have emerged to forage. Monthly samples, which will provide an indication of any seasonal variation in the diet, can be gathered by replacing the polythene at the end of every month. The droppings should be carefully removed from the polythene, air-dried and then stored in screw-top coffee jars with a pinch of insecticide. The latter kills the eggs of those insects which feed on bat faeces. Material stored in this way will keep almost indefinitely. Failure to add insecticide may mean that when the bat worker comes to examine the material, it will have been chewed to a

LEISLER'S BAT

fine powder and crawling with maggot-like insect larvae.

There is nearly always an excess of material in each collection. So we usually take a sample of 40 pellets selected at random for analysis. The droppings are moistened overnight between two layers of water-soaked cotton wool. A single dropping is then placed in a small glass dish, a few drops of glycerine added, and gently teased apart with fine dissecting needles under a low-power microscope (magnification x20-x40). All bits which look distinctive, which are often jointed or divided into segments, are mounted on a microscope slide in glycerinated gelatine, which is a firm jelly prepared from gelatine and glycerine, with a little phenol as a preservative. This is quick to use, once heated to liquify it, and provides a permanent preparation if the slide is kept in a cool place.

Identification of the fragments demands patience and experience. A reference collection of potential insect prey is almost essential, as are illustrations in books used in the identification of whole insects. We have published a profusely illustrated manual, *Identification of Arthropod Fragments in Bat Droppings,* which greatly simplifies the process.[307] Even so, the beginner still needs time and effort to become competent. Sometimes a slide will have to be examined again and again before it is possible to put a name to everything on it.

A very small selection of the sorts of identifiable items that occur in the faeces is shown in Fig. 21. The antennae of male midges, which are densely clad with hair, are one of the simplest things to recognise. If a fragment is yellow, it may be from a yellow dung-fly. Among other bits of this insect which turn up are parts of the reproductive organs of the female, including the *spermathecae,* which store sperm from the male after copulation. These look rather like hand grenades. Shiel often encountered them before she was able to put a name to them. But when she discovered one attached to another part of the genitalia, the mystery was solved. Another instance of this sort was the "claw" found on the jaws (*mandibles*) of some scarab beetles. This she saw several times broken off, on its own, before coming across a jaw with the claw attached. Thereafter it was clear what it belonged to. The wings of moths and butterflies are covered with overlapping scales, which come off if rubbed with the fingers. Under the microscope these are objects of beauty and are readily recognisable. When a bat eats a moth, some of the scales persist in the gut long after the rest of the remains have been voided and are thus found in the faeces for some time thereafter. So, if the importance of moths in the diet is not to be overestimated, it is best

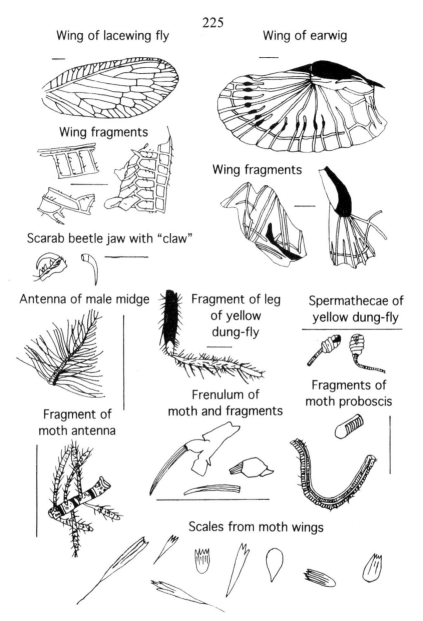

Wing of lacewing fly

Wing of earwig

Wing fragments

Wing fragments

Scarab beetle jaw with "claw"

Antenna of male midge

Fragment of leg of yellow dung-fly

Spermathecae of yellow dung-fly

Fragments of moth proboscis

Fragment of moth antenna

Frenulum of moth and fragments

Scales from moth wings

Fig. 21 Some identifiable fragments of prey found in bat droppings. The line scale given with each drawing (except the moth scales which are minute) is 1 mm.

not to consider a dropping as containing moth if the evidence is scales alone. Other diagnostic bits of these insects should be present, such as fragments of the *proboscis*, which the moth uses to suck nectar from flowers, of the *frenulum,* which couples the front and hind wings together, or segments of the antennae.

Shiel found the diet of Leisler's bats at the various roosts to be much the same.[302] The single most important item was the yellow dung-fly, overall forming 22% of the diet. It may be recalled from Chapter 6 that this insect is essentially diurnal, but that a minority will remain on cow pats for an hour to an hour and a half after sunset, before flying off to settle on vegetation for the night.[149] Shiel came across them whilst radio-tracking and they even flew into her car during the first hours of darkness. Leisler's bats, being built for speed with restricted manoeuvrability, cannot have gleaned these flies. Instead they must have caught them when on the wing. As the bats emerge around or before sunset and are often predominantly active in the early hours of the night, these insects must be particularly vulnerable to them.

Yellow dung-flies are, of course, associated with pasture, and so are dung beetles, which belong to the Scarabaeoidea, or scarab beetles, the latter contributing 13% of the diet overall. As the amounts of yellow dung-fly and scarab were correlated in the various collections of droppings - where one was higher or lower, so was the other - one could deduce that the bats were generally catching both at the same sort of place and that this was almost certainly over pasture, where dung from cattle, sheep and horses provides food for both kinds of insects. At least some of a number of the other types of insects in the droppings may also have been caught over pasture, for instance lesser dung-flies (Sphaeroceridae) at 7% overall, but they also occur on various other habitats and it is therefore impossible to say what proportion of them was taken over pasture and how much elsewhere. However, it might not be far wrong to speculate that in excess of half of the food of Leisler's bat in Ireland is made up of pastoral insects.

In view of the above, one must consider the widespread use of the drug Ivermectin as a wormer for cattle. Studies in Britain show that in cow pats it may debilitate or kill yellow dung-fly larvae. So dosing cattle with it may have implications for the conservation of Leisler's bat.

It is no surprise that night-flying Diptera, both large (22%) and small (9%), were also an important component of the diet of a bat that specialises in catching flying insects. Two thirds of the small Diptera were midges, although the amount was rather variable from roost to roost, actually reaching 17% of the diet in the summer of 1993 at Baldwinstown. Midges and other small flies dart about, and tiny active insects seem improbable prey for Leisler's bat with its low frequency, coarse echolocation and limited manoeuvrability. As it scarcely seems practical for a Leisler's bat to echolocate midges individually, it probably echolocates swarms instead, and midges and the other small Diptera swarm. Not only is there a mass of bodies in a swarm, which as a whole are presumably easier to spot with coarse echolocation than an individual midge, but, with so many midges together, at any instant some at least should be flying so that each presents the whole of the front or back of its body, plus its outspread wings, to the bat, thus maximising the target area for echolocation. Such an individual would clearly furnish a much greater surface to interrupt the sound pulses than, say, an isolated midge flying away from the bat.

Caddis flies (11% overall) were also an important component of the prey, and indicate feeding near water, as may midges, although these may swarm in other places.

Over the summers of 1994-95, Shiel and her brother radio-tracked 35 bats from the Baldwinstown and Rathangan roosts on a total of 122 overnights and 6 half nights.[308] It is a common practice in radio-tracking to give each animal a name in order to distinguish it more easily in the researcher's mind. This work was no exception: Amanda and Belinda were the first bats tagged and the other names followed in alphabetical order, after Z beginning at A again.

So as not to harass the bats, Shiel and her brother never radio-tagged more than one in an evening. They caught the animal in a hand-net as it emerged and sexed and weighed it. She then placed it on her brother's knee, with a piece of cloth over its head to quieten it, and trimmed the hair between its shoulder blades with scissors. The radio tag was attached with "Skinbond", a surgical glue. The animal was kept in a cloth bag until the glue had dried and then released.

The radio transmitters were basically of two types, weighing 0.7 g and 0.5g, each consisting of a button of epoxy resin containing the electrics and a wire projecting behind as an aerial. Fig. 22 shows the transmitter attached to a bat. She activated and tested each transmitter during the afternoon preceding tagging. It is an accepted rule, commonly called "the five percent rule", that transmitter plus adhesive should not exceed more than 5% of the weight of the bat. Shiel also avoided tagging heavily pregnant bats, which already had enough excess baggage to carry. Because of the 5% rule, and the weights of the transmitters, Leisler's bat was the only bat species in Ireland that could be radio-tracked at this time, although even smaller transmitters have since been developed.

Fig. 22 A Leisler's bat with radio-tag attached.

The theoretical maximum battery life of the tags was about 3 weeks, but 10 days was the longest period over which she was able to track a bat, because the transmitter, if still functioning, had come off by then. If this happened in the roost at Baldwinstown, it was usually possible to find it. If it was still transmitting it could be located on the floor in the evening after the colony had emerged, using the receiver, and sometimes even fitted on another bat on the following evening. If it had ceased to transmit, it could still be recovered at the end of the summer, after all the bats had left, when the season's accumulated droppings were removed and searched. As each transmitter cost over £90 and the manufacturers would replace the battery for a nominal sum, they were worth trying to retrieve.

Two radio receivers were available, each with a directional aerial held in the hand. The direction in which the bat had gone could thus be found by turning the aerial to where the signal was strongest. The signal was a series of bleeps: bleeps use less power than a continuous tone and therefore prolong the life of the battery. Most of the tracking was by car and a whip aerial was mounted on the roof which received the signal from any direction. But this was used only when the signal was lost and it was necessary to drive around until it was picked up again. The terrain in south-east Wexford was ideal for radio-tracking bats because there are no hills to interrupt the transmission. Because Leisler's bat often commutes at high altitudes, certainly when flying out in the evening, this further enhanced the distance over which the signal carried: 1.2-1.5 km. Generally this was maximised when the directional aerial was raised above the surrounding terrain. Fig. 20 shows that the Waterford-Rosslare railway roughly bisects the region. Roads cross the line by humpbacked bridges, the parapets of which provided ideal places for an observer to perch with the aerial in a hand extended above the head. Fortunately there were such bridges beside both Baldwinstown and Rathangan roosts. So the direction in which the bat made off could usually be discerned right away.

By heading in roughly the same direction by car and checking the direction of the signal every few minutes, it was usually possible to keep in contact with the animal as it travelled to its feeding site. The process was facilitated by the bat apparently usually commuting in straight lines. The chief problem was to keep up with it. It was often possible to get within 20-30 m of the beast as it foraged, or even closer if it passed overhead when commuting. Occasionally it could even be seen.

How fast does Leisler's bat fly? Speed, calculated from feeding site back to the roost over distances greater than 5 km, sometimes exceeded 30 km/hour. Unfortunately such estimates of commuting speed are bound to be on the low side, as the bat always arrived back before the observers, who could only take their own time of arrival as the nearest estimate of the bat's. Besides, the bat might not always have travelled in a straight line; and it could also have dawdled briefly on the way, perhaps to snack on a few insects. Fortunately Shiel and her brother were twice able to track an adult female along a straight road, with the gain control on the radio receiver, which increases or decreases its sensitivity, set to minimum. This meant that the range of the receiver was minimal and that the bat was flying directly overhead. The speeds then were 48 and 56 km/hour (30 and 35 mph). This is well above the economic cruising speed for Leisler's bat calculated from its body weight and the size and shape of the wings using aerodynamic equations. But Leisler's bat commutes in the open, often well before nightfall, so speed is essential to avoid predators. Anyway, commuting fast in a direct line probably uses less energy than travelling along hedges to a distant foraging site, a much longer journey, even if it were undertaken at a more economic speed. Shiel and her brother also timed a juvenile male in the same way at 40 km/hour (25 mph).

Besides the nursery roosts at Baldwinstown and Rathangan, some of the radio-tagged bats used other, temporary roosts, in buildings and in hollow trees; she located 12 of these. Some were occupied only during part of the night, whereas at others the bats also stopped over during the following day, for up to three consecutive days. No animals ever day-roosted away from the nursery during lactation, when doubtless the mothers would have been unable to leave their youngsters unfed for even a day. Three bats tagged at Baldwinstown also day-roosted at Rathangan, and then returned home, which suggests that the two colonies were not mutually territorial. Why an adult should use a temporary day roost is difficult to explain. Of course it would save time commuting to a good feeding site distant from home, but some of the temporary day roosts were within a few kilometres of the nurseries which, as we shall see, is a negligible journey for a Leisler's bat.

On 24 July 1994, during postlactation, Kelly, an adult female, stopped at a temporary roost in a building from about 1-4 am, during which Shiel picked up the mating songs of male Leisler's bats in the vicinity on the bat detector. From late July to September in both 1994 and 1995 she also heard the songs elsewhere in the district, usually around mature trees, especially beeches. Evidently in Ireland Leisler's bat may mate, or at least indulge in courtship, onwards from late July, shortly after the young are weaned.

In order to interpret the remainder of the results from this study, the reader must have some idea of relative abundances of night-flying insects in each of the summer months. Unfortunately there is no such information for Ireland, but there are ample data from agricultural land in England, which it seems reasonable should reflect the situation here. In April insects are scarce; numbers increase in May and then rise sharply in June and again to a maximum in July, with usually a slight fall-off in August. The net effect of this is that during the prebirth period insects are generally scarcer than later on.[308]

When the number of flights which a bat took in a night is mentioned below, readers should bear in mind that the interval between each may have been spent either at the nursery roost or, much less often, at a temporary roost. Also, when a maximum distance from the roost is mentioned, this means the maximum distance measured in a straight line.

The activity of the bats varied in a similar way in the summers of both 1994 and 1995 and fitted in well with what Shiel had found out in 1993 during all-night watches at Baldwinstown and from her "runs" with the bat detector.

During prebirth (April to early June) even though insects are generally scarcer, energy demands on adult females are much lower than during lactation. So they spent on average under 2.5 hours flying each night. Often there was only the one flight, with a second in the latter part of the night if it was warm. This verifies the importance of increased temperature - and more flying insects - later in the night in justifying a second flight. Three flights were recorded on 3 nights, but on part of all three there was sustained rain, which probably forced the animals to enter a roost for shelter. Despite the limited flying time, the bats generally flew further away to forage during prebirth than at any other time: sometimes south-east as far as Tacumshin and Lady's Island Lakes or north right up to Forth Mountain (Fig. 20). The maximum distance that any bat commuted - 13.4 km - was recorded during prebirth and the average per night was around 7 km. Insects are not only relatively scarce but feeding patches - the places where they are numerous enough to make sustained foraging worthwhile - must also be relatively few, and therefore widely spread, so the bats often had to fly further to get there. When feeding at Tacumshin Lake and Forth Mountain they were most likely exploiting the clouds of midges which swarmed there in the early part of the summer. Any radio-tagged individual at Forth Mountain was not alone: it was obvious from the bat detector that several other Leisler's bats were hunting with it, right up to the summit. So prey must have been plentiful. It is amazing that such a small mammal as a Leisler's bat, with a brain size less than that of a mouse, clearly knows the best places to forage, even when these are relatively small areas many kilometres from home. Presumably it is a question of experience and memory and bats, of course, may live for several years or more. One bat might discover a feeding patch by following another but, judging from what Shiel saw in the evening at her nursery roosts, they did not usually commute in parties, even though, as already described, they emerged in small groups. It was quite obvious that the bats were deliberately selecting specific sites at this time and not trying here and there for the best place.

During lactation (mostly June - only females producing milk were radio-tagged) there was a marked change in behaviour. Even though the hours of darkness were minimal, the animals spent more time flying each night than in prebirth, on average some 3.5 hours. The young grow to near adult size in a matter of weeks and the mothers need to work hard to produce enough milk for them.

During lactation in 1995, Shiel tagged three bats on successive nights and, while she tracked one of these in the field with one receiver, her brother stayed beside the roost each night for a week with the other checking the three out and in through the night. The time all three spent flying per night increased through lactation to almost 6 hours a night. This is only to be expected: as the young grew bigger they needed more milk. Mothers took more flights each night during lactation than in prebirth: an average of three, although up to five were recorded in both years. This was because they had to return regularly to the nursery to suckle their babies. Shiel's results indeed point to one long round of foraging and nursing young throughout the night just before the young are weaned, even though flying insects are more plentiful than during prebirth. However, more insects mean more feeding patches, so at least the bats did not have to go so far to feed. The maximum distance that the animals flew from the roost was on average only a little over 2 km - under a third of that during prebirth. The greatest recorded distance was 7.2 km, much the same as the *average* during prebirth.

With postlactation (1994 July and early August, 1995 July) the maximum distance flown from the roost rose slightly, perhaps because the mothers no longer had to return and suckle young and could choose the best feeding patches without regard to their proximity to the roost.

Because of the five percent rule, Shiel tagged juveniles (both females and males) only when they were almost full grown, after the adult females had deserted the Baldwinstown and Rathangan roosts (1994 August, 1995 mid July to August). Whereas the number of flights by juveniles were much the same as those taken by the adults during prebirth - one or sometimes two a night, the time which the youngsters spent flying increased as they grew older, commonly to over 6 hours a night, with a maximum of over 8.5 hours. Such results dovetail those from the overnight watches at Baldwinstown in 1993. Such prolonged flying times might be at least partly explained by the juveniles' need to lay down fat for hibernation and because their hunting technique is probably less efficient than that of their elders. The maximum distance they flew from the roost also increased as they grew older and they eventually hunted at places not used by the adults since prebirth: at Tacumshin Lake and Forth Mountain. Furthermore, the cumulative distance flown in the course of the night from feeding place to feeding place, gradually rose until it was greater than that recorded for almost any adult.

As flying insects should have been nearly as plentiful for the juveniles as for the mothers during lactation, why did the juveniles fly so far from the roost when there were presumably good feeding patches closer at hand? The answer is probably that they were exploring the surrounding countryside, familiarising themselves with the home range to be used in future years. Juveniles also used temporary day roosts more often than the adults had, possibly as a preliminary step towards deserting the nursery roost. Some juveniles also made temporary long distance migrations before finally forsaking their nursery. Thus Frank, a male juvenile tagged at Baldwinstown in August, returned in the morning and spent the following day in the roost, but was lost 14 km north on the next night. However, he returned two nights later, moved to a temporary roost on the following night less than 2 km away, where he spent the next day, before coming home again, where he roosted for the next 4 days, when the transmitter failed. When they desert the nursery roost, some juveniles at least apparently quit the district entirely. Dennis, another male tagged at Baldwinstown, was lost 29 km to the north.

At the end of August 1995, Shiel tried to find out where at least some of the juveniles went after they had departed for good. At the time she reckoned that up to four of these could have tags which were still functioning: Dennis, George, Harry and Ingrid, all from Baldwinstown. She therefore hired a light aircraft from Waterford Airport, fitted a directional aerial to the wing strut on each side and searched the surrounding countryside for radio signals. With the plane flying at an altitude of 457 m (1,500 feet), the effective range of reception was enhanced to an astonishing 8-10 km. Once she had the approximate position of a bat, she then tried to pinpoint it on the ground.

George she never found. Harry she located 4 km south-east of Baldwinstown, but the signal was too weak to allow him to be pinpointed. Ingrid, and Dennis's tag, which had come off, she found in the attic of Monksgrange House, a mansion 35 km north of Baldwinstown, at an altitude of 180 m on the lower slopes of the Blackstairs Mountains. Ingrid was tagged 3 weeks after Dennis had been lost, so it is astonishing that both bats ended up at a place so far from their nursery. A few days later, on 23 August, Ingrid moved into a hollow beech stump, shared by other Leisler's bats, in the grounds of the mansion. Thereafter Shiel checked occasionally in the evening to see whether Leisler's bats continued to emerge from it, which they did on 26 August, 3 and 17 September and 5 October. No bats emerged on 10 November, nor on 5 January nor 4 April 1996. This evidence, slim as it is, hints that at least some of the Leisler's bats living at Baldwinstown and Rathangan may move inland for the winter, perhaps, as already discussed, because south-east Wexford is too warm for satisfactory hibernation. Notwithstanding, this is far from being proven. Sleeman reported a Leisler's bat caught in a hand-net early in the night on 4 October 1986 on the Great Saltee Island, which lies 5 km south of the coast of Wexford. He thought that the animal may have been migrating, although he did not suggest to where.[310] In continental Europe there are several records of long distance migrations by Leisler's bat, including one from Russia to Turkey of 1,245 km.[151] There is obviously a lot yet to be learned about what Irish Leisler's bats do in wintertime.

The percentages of the total recorded feeding visits to various habitats in 1994 and 1995 are shown in Table 18. Evidently the bats used feeding patches on a wide range of terrain, although pasture, pasture with a drain (the bats foraging mainly around the latter) and canal accounted for over 75% of feeding visits in both years. The intensive use of pasture readily accounts for the preponderance of pastoral insects in the diet, as does feeding over canals and drains for the significant proportion of insects associated with water, like caddis flies and midges. Nevertheless, these findings do not mean that pasture is a preferred habitat, for it accounted for almost all of the area around Baldwinstown and Rathangan. Canal represented only a tiny part of the total area, but around one fifth to a third of the feeding visits, so canal is preferred to pasture. The district where Shiel radio-tracked is rural and the total area illuminated by outside lighting minute, so the figures in the table for lights are relatively enormous. Lights are probably the most favoured habitat. The bats often took advantage of strongly illuminated places, such as the floodlighting at the Hotel Saltees at Kilmore Quay. They even hunted around the lights on the pier there. Studies in Sweden, where there are many more species of bat than in Ireland, indicate that bats which hunt most at streetlights are those adapted to fast flight and foraging in the open, a description which in Ireland applies to Leisler's bat more than to any other.[294]

Readers will recollect that in 1995 both early April and the summer as a whole were exceptionally warm. The effects show up in Table 18.

Several habitats which were apparently ignored in 1994 were exploited in 1995, probably because it was only with higher temperatures that there were sufficient insects flying above them to make them worth foraging over. Note that the bats also used pasture more often in 1995. During prebirth in 1994 the bats were never located feeding over pasture (i.e. pasture without a drain) at all. In contrast, during prebirth in 1995 pasture accounted for an amazing 36% of feeding visits. The warm period in early April had doubtless awakened the overwintering stages of the pastoral insects prematurely and advanced their breeding season. So there were enhanced populations of flying adults over pasture much earlier in the summer.

Table 18 - Percentages of feeding visits to various habitats by Leisler's bats in Co. Wexford in 1994 and 1995 as revealed by radio-tracking.

Habitat	Percentage of feeding visits in	
	1994	1995
Lights	8	6
Pasture	31	44
Pasture with drain	24	13
Canal	31	19
Lake (Lady's Island or Tacumshin)	4	<1
Stream	2	2
Estuary (The Cull)	<1	5
Beach	-	<1
Dunes	-	5
Marshy coastal pasture	-	2
Forest (Forth Mountain)	-	4
Total recorded visits	226	321

Among other pointers to the conservation of Leisler's bat derived from Shiel's work is the likely importance of hollow trees. It is evident that these are often used as temporary roosts and they might be important for hibernation. Mature or dead deciduous trees should therefore be managed sympathetically. They may be of particular importance in Ireland where there are no woodpeckers to open up cavities in younger trees, and coniferous woodland is usually felled before natural cavities develop.

Chapter 9. Hunt the Muskrat

Let anyone who meets a muskrat on foot on dry land watch out, or he may be bitten by an animal that has the equipment to bite with....Gentle old ladies, children, horses, cattle, even rolling tumbleweeds, may be attacked by desperate muskrats acting as if they felt cornered....Whatever else may be said of them, normal muskrats are not sissies, and, when scared in addition, they can be savage.
Muskrats and Marsh Management. Paul L. Errington.

This history of the muskrat *Ondatra zibethica* in Ireland is based on official records in the Department of Agriculture, on newspapers, on conversations taperecorded in 1980 with four of the men who took part in its extermination - Messrs Paul Burke, Joe Hogan, George Maher and Michael Whelan - and on other sources detailed below.

The muskrat is a rodent native to North America and lives in streams, lakes, ponds, swamps and marshes. It has a compact sturdy body clad in rich brown fur, with the belly hair silvery. The ears and eyes are small, the latter noticeably beady. The animal weighs 800-1,500 g and measures 25-35 cm from nose tip to the root of the tail, which is 20-28 cm long, naked, scaly, and markedly flattened from side to side. The hind feet are partially webbed.

The muskrat is primarily a vegetarian, preferring the leaves and stems of aquatic plants, but will sometimes take animal food if available, such as fish carcasses, dead or moribund birds and freshwater clams, especially if succulent plant material is in short supply.

To a muskrat home is either a burrow dug in a river bank, which may be extensive and has its entrances below water, or a house or lodge: a mound of aquatic vegetation, built in shallow water, some 1-1.5 m in diameter and projecting up to a metre or so above the surface. House building reaches a peak in late summer, the lodge providing both shelter and victuals for the winter, during which the rats burrow into it and gradually eat out the centre until only a shell is left. Eating one's house is a bizarre occupation, even for a rodent, and brings to mind Hansel and Gretel nibbling at the witch's gingerbread cottage.

Muskrats are pugnacious and fight among themselves. They are uneasy at any distance from water, where, as the quotation at the beginning of this chapter makes abundantly clear, they may attack humans unprovoked; and the two pairs of incisor teeth, the hallmark of any rodent, are formidable weapons. Except during droughts, a muskrat at any distance from water is usually low on the social scale and has not managed to stake out a territory for itself. However, if it wanders far enough, finds sufficient to eat, and is not eaten itself, it may come across suitable habitat uncolonised by others of its kind and be able to settle down.

In North America, mink are omnipresent in the life of the muskrat, for both dwell in the same habitat, and whereas a mink is usually prudent enough to refrain from tackling an adult in good health with its own territory, it can pick off young ones, or individuals that are sick, injured, or in poor condition (as often happens during droughts), or those which have no territory and are obliged to spend more time on dry land than they would wish. Mink predation on muskrats presents a dilemma for those who hunt mammals commercially for their pelts, for both species are important fur bearers. The fur of the muskrat is known commercially as musquash - the Red Indian name for the animal - or, when sheared and dyed, as "Hudson seal", for it is then similar in appearance to sealskin.

Muskrats were imported into several places in Great Britain for fur farming from 1927 until 1932. Escapees soon established populations in the wild so that, by 1932, there were feral colonies in 14 counties. Further imports were then halted by the Destructive Imported Animals Act,[151] for these rodents were indeed destructive. Their compulsive burrowing undermined banks and dams and caused riverside trees to fall. Vegetation cut adrift choked watercourses and diverted water onto surrounding land, effectively turning it into marsh. The Act was followed by a campaign of extermination, which proved laborious as the relatively mild climate in Britain and abundant food allowed an extended breeding season, from February to November, facilitating a potential of six or seven litters of eight young to each female. However, the animals were eventually eradicated by 1937, but only after some 4,500 had been killed.

And so the scene is set for the saga of the muskrat in Ireland.[105]

Details of the first arrival of muskrats are often cited incorrectly. The facts were kindly provided to me Mr R.J. Minnit, whose father Mr C.F. Minnit imported the beasts. In January 1929 the Minnits settled in the family home, Annaghbeg House, the local "big house" at Dromineer, Co. Tipperary, about 2 km from where the Nenagh River drains into Lough Derg. There they farmed the land attached to the property. (All of the localities mentioned in this chapter are shown on the map in Fig. 23.) Mr Minnit's wife wanted some occupation besides housework that would provide additional income, and thought that breeding chinchilla rabbits for the fur trade might fill the bill. However, as muskrat farming was then becoming popular in England, the couple eventually decided that this might be a more remunerative proposition. They made enquiries about how they should proceed, both in England and with the Department of Agriculture of the then Irish Free State. The Department expressed interest, gave advice on the wild plants that might be used as fodder, and wished the project every success. In view of ensuing events, this correspondence must subsequently either have been overlooked or lost.

In due course - still in 1929 - the nucleus of the breeding stock, almost certainly a male and two females, arrived at Annaghbeg in their travelling cages. The latter were unsatisfactory as permanent quarters, being too cramped and insufficiently reinforced to resist the chiselling incisors of their occupants. So the Minnits, father and son, set to work to build permanent, commodious, heavy-duty accommodation. Unfortunately, before this was finished it was one morning found that the rodents had gnawed through the travelling cages and escaped. The Minnits were concerned, not only at the loss of their valuable livestock, but for the welfare of the animals, which they supposed would soon be polished off by predators. However, after a few day's fruitless search, they had to give their charges up for lost.

In the summer of the same year, Professor Donald Griffin, whom readers may recall from Chapter 6 first demonstrated echolocation in bats, was a boy of 13 and was visiting Dublin with his father, an enthusiastic amateur student of family genealogy, who had come to Ireland to follow up his interest. In order to avoid boring his son, who even then was a budding mammalogist, Griffin's father encouraged him not only to visit the Natural History Museum but to seek out the mammal man on its staff. At that time the Museum employed only three biologists, one of whom, Eugene "Bugs" O'Mahony (1899-1951), dealt with mammals, birds and much else beside. Evidently O'Mahony was impressed by Griffin and was keen to encourage his interest. An exchange of study skins and a lengthy correspondence followed. In 1978 Professor Griffin kindly provided me with photocopies of such of O'Mahony's letters that he still had.

24 February 1930

....I am pleased to get the 'photos as I've been doing some work on the muskrat lately - the Forestry Dept. asked for information re the food as there is someone wanting to introduce them into this country for rat farm purposes & the Forestry people wanted to know if the rats would damage their plantations. I have heard nothing more about the idea since and it's a month ago since I sent in my report.

7 April 1930

....I told the Dept. what the muskrat fed on and I warned them as to the danger of introducing foreign animals into this, or any other country, particularly a rodent....the only enemy I can think of is the fox (*Vulpes vulpes* L.) and possibly the stoat (*Mustela hibernica* T & B.H) and I don't place too much faith on the amount of checking they may do.

So I've not heard anything more & it is possible that the project to import the rat has been squashed. I'm not trying to be funny, and I hope it was because the list of "introduced" pests is long enough in most places & we have, fortunately, got few....

Unaware that it was already too late, the civil service were clearly pondering the advisability of allowing muskrats into the country. Files from the Department of Agriculture confirm that from at least December 1931 there was a dawning awareness of the inherent dangers, and with it a hardening of attitude. Consideration was given as to how imports might be prevented and the solution adopted was to ban the landing of muskrats, along with certain other small mammals, under an order for prevention of foot and mouth disease. Delay was occasioned because it was uncertain whether muskrats could actually transmit it and the order was not made until 31 January 1933, the *Foot and Mouth Disease (Importation of Rodents and Insectivora) Order 1933.*

In the meantime two more muskrats had arrived and were being kept at a "silver fox farm" at Bahana, near Enniskerry, Co. Wicklow. Dr Patrick O'Connor, Keeper at the Natural History Museum in Dublin, and therefore O'Mahony's boss, drew attention to their presence in Ireland in a letter to the *Irish Times* of 27 October 1932 and urged legislation for their destruction. As O'Connor was a botanist, this may have been at O'Mahony's instigation. An internal letter in the Department of Agriculture files of 14 November discusses O'Connor's letter and remarks that the owner of the fur farm had imported muskrats and other small mammals; a note of the same date indicates that a further import, of beavers *Castor canadensis*, was projected.

The two muskrats at Bahana, being already in the country, would have been unaffected by any ban on imports and they caused the Department much anxiety. One novel solution mooted was the retrospective imposition of a prohibitive excise duty on the species, legislation enabling this being already on the statute books. Legally minded readers will be aware that such a move would now be impossible, as the 1937 constitution quite rightly prohibits retrospective acts.

That the official view was widely appreciated is clear from an item in the *Irish Independent* of 8 December 1932.

> All hopes of developing a musk rat industry in An Saorstat seem doomed as legislation is being prepared to restrict the entry of musk rats.....A. Rohu of Messrs Rohu and Son, manufacturing furriers, skin dressers and dyers, 2 Castle Market, Dublin....said that the man who introduced musk rats into this country would earn himself an unenviable reputation, and might run the risk of being lynched when the destructive habits of those animals became known....

Eventually a *Destructive Imported Animals Bill* was drafted, apparently in February 1933, no doubt inspired by a similar bill then undergoing consideration by the government in Northern Ireland and which was passed on 28 March, the *Destructive Imported Animals Act (Northern Ireland) 1933,* which dealt primarily with muskrats.

Events overtook the Free State's nascent bill in dramatic fashion with a report, in both the *Irish Independent* and the Nenagh *Guardian* of 15 April, of three muskrats shot in Co. Tipperary. These were killed by Mr Joe Hogan at Dromineer, at the mouth of the Nenagh River, and were undoubtedly the descendants of Mr Minnit's beasts, which had not perished at all, or not until they had reproduced themselves. When Mr Hogan first saw them in the water he took them for young otters. Once they were shot, however, their exotic character became apparent. Afterwards he caught a further three young ones in a net and kept them in a barrel, feeding them mainly on cabbage. These proved a local spectacle and attracted many visitors.

A wire from the Department of Agriculture was immediately

dispatched to the Garda at Nenagh to obtain one of the rodents in order to verify the species, and by 19 April the identity of a specimen sent to the Natural History Museum had been confirmed by O'Mahony. On 24 April the Ministry for Education, which was in charge of museums at the time, agreed to send O'Mahony to investigate; 2 days later he was in action in the district and had caught a muskrat at Killadangan, which was subsequently stuffed and is still on display in the Museum. Dr O'Connor joined him shortly afterwards and the upshot was a report to the Department of Education on 1 May, the most important passages of which read

> I am satisfied that, as indicated by the abundance and nature of burrows, well-used tracks, foot-marks, dung and "lodges" or winter houses, the Musk Rat is well established and thriving and rapidly multiplying at the mouth of the Nenagh River, on both its banks for a distance of two and a half miles from the mouth and in the neighbouring tributaries, drains and streams to a width of about three quarters of a mile [1.2 km]....some idea [of the numbers] may be got from an observation, which was made in one of the most densely populated regions, of fifteen colonies in a stretch of about one hundred yards [91 m] along one side of an open drain....The lakes of Poulawee, Clareen, Claree and Nagelane [see Fig. 23] and the associated streams and adjoining marshes were examined without any conclusive evidence of the presence of the creatures being found...

It is perhaps curious that the rats could have been around for so long and have built up to such numbers without anyone having noticed them, but this was also the case in Britain due to their elusive and nocturnal habits. In any case interest in the natural world among the general public in Ireland, and indeed in Britain, was much less then than it is today. There may, in fact, have been an intimation of the pest before the first animals were shot: Mr George Maher told me that some time before this the ground had collapsed under a man ploughing at Annaghbeg and that there was some discussion locally as to what sort of creature could have constructed the tunnels so revealed. There is what appears to be a brief retrospective reference to this in the Nenagh *Guardian* of 29 April 1933.

Evidently O'Mahony had no real idea of muskrat numbers and, in

MUSKRAT

view of subsequent events, he overestimated them. In a letter to Griffin dated 2 February 1934 he wrote:

> In spite of my warning, the Forestry people allowed the muskrat into Ireland....In April 1933 I was sent down, officially, to see if the beast was doing much damage and to find out its distribution. A large area is now inhabited by Ondatra and there must be at least 100,000 of them in the river basin - a conservative estimate of the numbers when I was there was 20,000.

Local animosity to the aliens was kindled almost immediately and muskrat bashing soon assumed the proportions of a regular pastime, as revealed by the following excerpts from the *Midland Counties Advertiser* of May 1933, which suggest that the rodents were also common around Nenagh.

> Mr Daniel Quigley....Nenagh, cycling through the town holding a musk rat which he had killed on the Nenagh River at Brook Watson [close to the confluence of the Nenagh and Ollatrim Rivers]....Encouraged by Mr Quigley's kill, a number of men with guns, dogs and sticks, travelled along the banks of the Nenagh River to the Shannon's edge, and killed five more musk rats....On Friday evening a musk rat, measuring 24 inches [61 cm], was killed on the edge of the Nenagh River by Mr Robert Morgan....The rat was first tackled by Morgan's dog, but eventually it was killed with a stick. It is five miles [8 km] along the River Bank to Dromineer....only 50 yards [46 m] from the town to where the latest rat was killed...The hunt for musk rats continues in Nenagh and is becoming a popular sport. The number killed so far is ten.

But antipathy in Nenagh was more than matched in the Department of Agriculture, where there was concern, not only about the undermining of river banks, but more especially over the possibility of the rats reaching the new hydroelectric installation at Ardnacrusha on the River Shannon, below Killaloe at the southern end of Lough Derg. This was by far the largest work of civil engineering in the infant state and had only been completed in 1929; so such alarm was understandable. Legislation was drafted by the end of May and the *Musk Rats Act 1933* was shepherded through the Dail by the Minister of Agriculture, Dr James Ryan, receiving royal assent on 24 July. This specified *inter alia* that separate licenses were required to import a muskrat, to keep a muskrat and to purchase muskrat pelts. This was important, and farsighted. Wildlife biologists today usually agree that, to exterminate or cull a pest species efficiently, it is essential that bodies should be of no value to the general public. Thus no one has a vested interest in protecting the pest. Under the Act, even when a license was granted for the keeping of muskrats, this could only be for scientific research. Finally, the Act decreed that "The Minister may take such steps as he considers necessary for the destruction of any musk rats at large". This he proceeded to do.

Actual extermination did not begin until September. Mr Michael Whelan believed that he started work as a trapper on the last Monday of the month, and therefore on 25 September 1933. This seems highly probable as the first reports of rats having been dispatched appear in the Nenagh *Guardian* of 30 September, and the *Midland Counties Advertiser* of 5 October mentions trapping as having started "a week ago". Considering the time required to draw up and enact the necessary legislation, and for the initial organisation of a large-scale trapping programme, the interval between the discovery of the infestation and the commencement of destruction was commendably short. In any event this unavoidable delay may have been partially offset by the fact that 1933 was the driest year recorded for the standard meteorological period 1931-1960 at the nearby station at Cloughjordan, droughts having particularly detrimental effects on muskrat populations.

Mr Thomas Garvey, an inspector in the Department of Agriculture, had been sent to Shropshire during the summer to study control measures there. He was in overall charge of the campaign, which had its headquarters in the old military barracks at Nenagh, where traps, protective clothing, spades and other implements were stored.

With Garvey there were two experienced trappers from Britain to instruct the others, one named Vallings (who may have been a Canadian) who returned to England in November, and Joe Wade, who was foreman. Local men were otherwise employed: there were five by 14 November and this was eventually augmented to ten, including: George Burgess, Paul Burke, Martin Cleary, William Fahy, Joe Hogan (who shot the first muskrats), Hugh McGrath, George Maher, George Waterson and Michael Whelan. At the beginning a leaflet with a full description of the animal and its habits was issued to one and all. The men were also supplied with forms on which to report their activities and on the numbers of rats caught. The trappers often operated more or less in pairs, each pair working around 100 traps. For transport there was a van and two or three boats, one being a motorboat, but bicycles were also regularly put to use.

Generally speaking the morning was always spent in checking or laying traps. The afternoon was sometimes also given over to inspecting terrain outside the known range of the animals for any signs of fresh infestation. Mr Hogan recalled travelling with Joe Wade as far as Clara, Co. Offaly, 66 km to the north-east to do this.

The traps were of the gin or steel type. Mr Hogan generously presented me with a rusted but still-serviceable example: an Oneida Jump No. 1, of United States origin and the standard type used for muskrats in North America. The trigger plate is a disc 5.6 cm in diameter and the jaws 9.3 cm in diameter. The latter are toothless, their purpose being to hold the victim with minimal wounding. If the trap was positioned properly, the rat then generally fell into the water in its struggles to escape and the weight of the trap was enough to keep it below the surface so that it drowned. If the trap was not placed so as to facilitate drowning thus, the animal was able to gnaw through its leg and escape. A few larger gin traps were also available, their advantage presumably being in giving a more substantial hold, either further up the leg or on the body itself, and greater weight in keeping the victim below water. But these traps were more difficult to carry in quantity and less popular among the Irish trappers. There is also a brief mention of cage traps in the government files.

A great part of Garvey's account of the extermination campaign[143] deals with the general habits of the musk rat and is intriguingly reticent on details in Ireland, perhaps because it was published only a year after the rats had been exterminated, and civil servants are discreet about divulging information of recent events that may prove controversial. He specifically mentions only two "sets" for laying traps. The "ditch trap", used in ditches and drains, consisted "of a platform attached to which are four cylindrical leads" which "guide the muskrat to the platform on which the traps are placed". The "floating island" was a plank or log moored to the shore bearing traps hidden with vegetation. Anyone with experience of trapping mammals will appreciate the time and effort required to assemble these, let alone transport the platforms and planks by boat or across fields. They seem unnecessarily laborious in an intensive campaign to exterminate a rodent. The trappers I spoke to used them only occasionally. As they also considered the muskrats easy to trap, at least in the early stages of the campaign, such elaborate measures would usually not have been needed.

The procedures the trappers described as most common included setting traps on the shelving banks of streams and drains, where the rodents would emerge to sit and feed on the vegetation that they had cut in the water. Sometimes excavation by a trapper, for instance in producing a secluded hollow in a bank, could improve a trap site. However, it was always essential to stake the chain attached to the trap so that the animal would fall into the water in its initial struggles to escape, and so drown. Burrow entrances (which were, of course, submerged) were also favourite sets. Traps so placed may also have had the potential advantage of an absence of any human scent on them. A further excellent site was around the bases of the muskrats' houses. Particularly mentioned by trappers in the construction of houses were "flaggers" - yellow flags or irises - and Mr Maher recalled the leaves of a rhubarb-like aquatic plant cut into lengths of about 6 inches (15 cm), almost as if measured. Another trapping stratagem was to narrow a stream at one point by packing bushes or sods at each side and to place a trap in the narrow channel thus formed, or close beside it. In selecting a site for a trap it was important to bear in mind that availability of aquatic vegetation, the rats' staple food, nearby was essential. A "good dirty drain" was especially recommended.

Besides muskrats, the traps caught brown rats and various waterfowl: rails and ducks being particularly mentioned. One trapper approaching a trap by boat to retrieve what he thought was a muskrat, discovered at the last moment that it was an otter. It sprang at the boat which, in the confusion, was almost capsized. Not the least of the hazards of muskrat trapping was the business end of the muskrat, and at least two severe bites were recorded, one through a wader.

The rats were caught in large numbers to begin with. The Nenagh *Guardian* (presumably receiving its information direct from headquarters) mentions cumulative totals of 59 on 30 September and 100 and 117 respectively on 7 and 14 October. Garvey gives an average of 30 for each of the first 9 weeks: about 270 by the end of November. The remaining figures, again cumulative totals, are from government files and are for pelts: "130 or so" on 20 October; on 18 November it was predicted that 250 would be available on the week following; and there were 300 on 10 January 1934. It is quite possible that not all bodies were in sufficiently good condition to make skinning worthwhile, and that the above are therefore minima for muskrats killed. Certainly, Garvey's account mentions that sometimes the bodies of trapped muskrats had been almost entirely eaten by rats.

The bodies were skinned at headquarters, the pelts stretched on boards to dry and then stored. They were eventually sold off to furriers for final dressing, the best at an average price of 4s 6d (22.5p) each, to the Hudson's Bay Co. and George Rice Ltd. in England, and to Peamount Industries Ltd. at Newcastle, Co. Dublin. Mr Burke mentioned dead and putrefying muskrats being found which had not been trapped and were useless for skinning. These may have been killed during fights amongst the animals themselves. The old military barracks, besides acting as headquarters, served a local social function. One correspondent, who was 15 at the time, remembers attending boy scout meetings there and being fascinated by the muskrat skins.

The exceptionally dry weather of the summer of 1933 continued into the winter and trapping was apparently unhindered by flooding. Capture rates decreased but Garvey was cautious, rather than optimistic, in interpreting this. He considered that the muskrats had gone to ground and judiciously submitted substantial estimates of expenditure for 1934-35 in February. However, after undertaking "an extensive survey of the River Shannon from Limerick City to Banagher, five miles on both sides" over 4 weeks in March and early April, he concluded that numbers were indeed well down. By 17 April only one or two captures were being made each week. The last muskrat was trapped in May 1934, bringing the total officially killed in the campaign to 487.

The number of local trappers was down to seven on 5 May, to three by October 1934 and, from April 1935 until at least a year later, Joe Wade operated on his own. The authorities were taking no chances especially as, even after the demise of the last rat, there were literally hundreds of reports of individuals sighted "from reliable sources". The trappers told me that Joe Wade had been guaranteed employment for life: a job trapping rabbits in state forests awaiting him when it was quite certain that the muskrats had been extirpated. This was a shrewd move on the part of the Department of Agriculture, for no one is enthusiastic about working himself out of a job.

Joe Wade's official "diaries" while working on his own have mostly been preserved and it is significant that on a number of occasions he came across what he considered evidence of feeding by muskrats only to conclude, after trapping, that he had been mistaken. Wade was the most experienced of the trappers and thus references to signs of muskrats elsewhere in government records should be treated with caution unless the animals were caught in the localities mentioned. The authorities had to take all suspicion seriously. In this context a memorandum from Garvey of November 1933 serves as an example, for there is a reference to "definite signs of infestation from Killaloe to Portumna". But, as will be seen below, all the muskrats were killed in roughly 120 km^2 including the Nenagh River, the contiguous part of Lough Derg and a single locality in Co. Clare.

While close on 50 years had elapsed between the campaign and my discussing it with the trappers, there was remarkably close accord among them on the area that had been colonised by the muskrats, shown in Fig. 23. Most of the Nenagh River as far upstream as the town and especially the surrounding drains held muskrats. Drains, peaty soil and aquatic vegetation were emphasised as particularly conducive to their presence. The rats were not obtained much further upstream than the town on the Nenagh River itself but on its tributary, the Ollatrim River, there were colonies at Lisbunney and Islandbawn. Further up, both watercourses were unsuitable, being rocky and lacking water weeds.

The muskrats had largely occupied Dromineer Bay, notably around the mouth of the river, including Goose Island (the tiny island in the Bay in Fig. 23), where lodges were particularly remarked. The animals continued abundant along the shore of Lough Derg, north to Luska Pier where, again, there were lodges. Muskrats were also taken in the inlet north of Cameron "Island" - in reality a peninsula - and a single capture on the stream flowing into it marked the most northerly point of the range. The rodent in question was secured near the small lough at the head of the stream. Mr Whelan particularly remembered this, as the rat escaped once, leaving a forepaw in the trap, before being caught again by a hind paw and drowned. Muskrats were in residence over much of the lower reaches of the Ballycolliton River west of the road bridge, including Lough Nagelene and the Black and Annagh Loughs. Poulawee Lough was heavily infested and Mr Whelan considered that the greatest densities, other than those around Dromineer, might have been here. It will be recalled that in April 1933 O'Mahony and O'Connor found no traces in Poulawee, Nagelene or in the surrounding district. Claree and Clareen Loughs, on the other hand, were still clear. An individual was trapped at Springmount, on the Ballycolliton River, marking another extreme of the range. To the south the rodents were found at an inlet of Lough Derg at Hazel Point and there were a few captures at Youghal

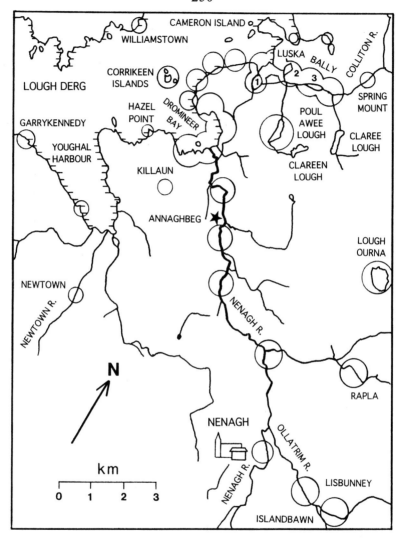

Fig. 23 Distribution of the muskrat in Ireland. A star marks the point of introduction. 1 = Black Lough, 2 = Annagh Lough, 3 = Lough Nagelane. Localities specifically mentioned by the trappers as being infested are approximately indicated by circles. Although few circles are shown on the Nenagh River, trappers considered that generally most of it below Nenagh, and the surrounding drains, was occupied.

Harbour and Garrykennedy. One was secured at Newtown on the Newtown River, and "three or four" in a "bog hole" at Killaun. Besides these isolated pockets, there were two others: at Rapla and Lough Ourna. The latter is interesting because there are apparently no watercourses which would have assisted migration from the main centres of infestation.

The only islands in Lough Derg yielding muskrats outside Dromineer Bay were the Corrikeen Islands. It was possibly from them that a rat reached the western shore of Lough Derg at Williamstown Harbour, in Co. Clare. There was only one muskrat trapped on the western side of the lough, probably in late March or early April 1934, and therefore towards the end of the campaign. This event was not only recollected by the trappers but is mentioned in government files. The animal was an adult female and a nest of seven dead young was discovered shortly afterwards. This serves as an excellent example of the potential for a single pregnant female rodent, introduced into new territory, to found a whole population. Rodents, of course, pay no attention to the laws of incest.

Chapter 10. The Rabbit

Burrowing is by the oldest right of prescription the rabbit's own profession, and the young ones from trying it in sport soon learn to earth the ground in good earnest.
Ordnance Survey Memoirs of Ireland. Anon.

Rabbits and hares along with the picas (which are confined to Asia and north America) all belong to the mammalian order Lagomorpha. Lagomorphs are fairly small, the largest species not exceeding 7 kg in weight. They are all similar in appearance, and although picas have fore and hind legs of equal length and relatively short ears, they sufficiently resemble the others as readily to betray their kinship. Lagomorphs are not rodents and are but distantly related to them. True, both groups of mammals have a pair of continuously growing incisor teeth at the front of both the lower and upper jaw. Nevertheless, lagomorphs have, in addition, a second pair of incisors hidden behind the first, and there are a number of obvious differences in the bones of the skull. Many rodents feed mainly on seed; no lagomorphs do. Lagomorphs are essentially consumers of green plant material - leaves and shoots - sometimes supplemented with bark and succulent roots, such as carrots. In terms of niche or profession, lagomorphs are a sort of miniature version of the larger herbivorous mammals, such as sheep, deer, antelopes and horses. Indeed, in the course of evolution it may have been competition with the bigger herbivores that not only kept lagomorphs small, but constrained their diversity and restricted the number of species, for in all there are only 69 extant.

Everyone knows the main differences between rabbits and hares: the former are smaller, slower and hop rather than leap. Although it is also often assumed that rabbits burrow and hares do not, this is not entirely true. Those species of rabbit which are good runners also spend their lives above ground and even our own rabbit *Oryctolagus cuniculus* - a notorious burrower - may sometimes do the same, especially where there is sufficient cover and the soil is difficult to mine, or where the animals are colonising fresh ground.

Originally the rabbit *Oryctolagus cuniculus* was confined to southern France and the Iberian Peninsula. Through human agency, wild populations now live over most of Europe and in many other parts of the world. Imported and released for the utility of its meat and skin, the rabbit has almost invariably proved an unmitigated pest wherever it has been naturalised. Strictly speaking, all the populations outside the original geographical range are probably not wild at all but feral - domesticated stock gone wild. The animals in the Iberian Peninsula today, which must represent the original strain, differ from those elsewhere in being somewhat smaller with proportionately longer ears. They are also supposed to have somewhat paler coats. Rabbits elsewhere are in all probability descendents of animals originally kept in captivity by the Romans, the differences having been brought about by selective breeding. It is possible that this period of domestication also imparted a fillip to the constitution of the beast, which has enhanced its hardiness and manifest subsequent adaptability.

Rabbits are probably the most familiar of all Ireland's wild mammals. Despite this, there has been little enough research done on them in the field here, although domesticated strains are commonplace laboratory animals. As our native mammals go, probably only the hedgehog and some of our bat species have had less scientific study here. The rabbit in Ireland is also unusual in being the only species of wild mammal to have been the basis of a significant industry, for in time past it has provided both meat and pelts in incalculable abundance. Of course, other species of mammal in Ireland have been killed for their skins, but in comparatively small numbers. The skins, suitably dyed and sheared, have often furnished cheap simulated versions of more expensive furs under the names of coney seal, beaverette, chinchillette etc. Probably at least as important in the way of commerce has been the use of fur itself, shaved from the skin, which has been used to produce felt for the manufacture of hats.

Because most people are so well acquainted with the wild rabbit, the variations in the colour of its coat are often remarked. Silver-grey, fawn and black varieties are not unusual in Ireland. The longer hairs on the back bear bands of sandy brown and dark grey along their length and variations in the relative proportions of these are largely responsible for the apparent differences in coat colour. On the mainland black rabbits are probably at a disadvantage because they stand out against vegetation, making them more liable to predation. On several Irish offshore islands where there are no ground predators, black rabbits are commoner. On the western end of Dunvillaun More, Co. Mayo, for instance, they make up 20% of the population and, against black banks of peat there, may even be at least partially concealed from predation by great black-backed gulls *Larus marinus*.[41]

Although rabbits prefer dry, sandy soil, they may be found in a wide range of habitat in Ireland but are usually absent from bog, probably because wet ground does not lend itself to the creation of snug burrows and suitable forage is scarce. (Their occurrence near peat banks mentioned above is curious.) Like cats, rabbits are reluctant to get themselves wet, although they have been known to swim on occasion.[92]

The mapping of a burrow system requires labour, although rather less than that for a sett (Chapter 3), because the tunnels extend over a smaller area, usually do not go as deep and are often in sandy soil. Only one has been fully excavated in Ireland, at Cappagh in Co. Waterford, by Major Gerald Barrett-Hamilton (1871-1914), probably the most distinguished Irish mammalogist of his generation.[21] He selected the site for investigation because the soil was stony and the tunnels therefore not expected to run to any depth. In the event none was more than 71 cm beneath the surface. In any case the work was also undertaken in 1910 when gentleman naturalists could rely on cheap labour if required. The plan is shown in Fig. 24. The depression in the centre had the appearance of having originated from subsidence caused by tunnels too close to the surface, so the galleries leading out of it may have been confluent at some time previously. The sharp turn in the passage towards the top left was probably because the animals had encountered hard ground there and were obliged to change direction. The two loop galleries towards the bottom of the diagram would both have allowed a rabbit pursued by a stoat to enter its hole and rapidly re-emerge and thus avoid being cornered. As there were only three chambers in the whole system, it seems likely that at least some of the animals used passages as resting places.

There were two short, blindly-ending tunnels, one the best part of a metre from the rest of the system. A doe sometimes digs such a burrow, known as a stop or scab, for her young and makes a nest of grass and moss at the bottom, lined with fur plucked from her own body. After her litter is born she visits the stop to suckle them, usually at night, and seals the entrance with earth when leaving. Of course litters may also be born in a burrow system. Sometimes a doe chooses more unusual sites and at least one litter has been found above ground in Ireland, in a depression in vegetation resembling the form or resting place of a hare.[21] I was once called to a builder's yard in Galway to identify two mysterious creatures in a hollow in a bag of cement. These turned out to be young rabbits, practically hairless, with their eyes

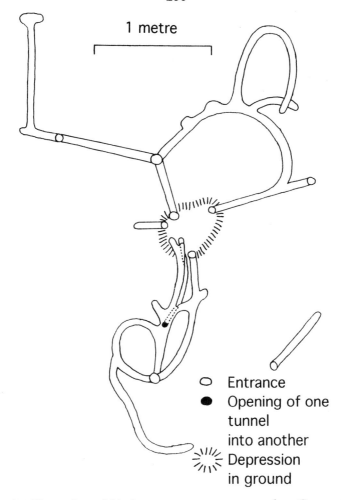

1 metre

○ Entrance
● Opening of one
tunnel
into another
Depression
in ground

Fig. 24 Plan of a rabbit burrow system excavated at Cappagh, Co. Waterford.

closed and coated in powdered cement. Their mother had made a bizarre choice for her stop. The only published information for Ireland on reproduction and development of the young indicates that they go abroad when about 3 weeks old.[21] However, even though we have no figures for Ireland, it is obvious that rabbits are prolific, although no more so than many rodents. In Britain, where the breeding season is mainly from January to August, does produce litters of three to seven at a minimum interval of 30 days.[151]

Unfortunately for rabbits, they do not always enjoy uninterrupted rights of occupation of their subterranean dwellings. Both mink and rats may move in unhindered, and also foxes, although the latter often have to enlarge the accommodation. All make uncomfortable, not to say lethal, companions. Rabbits living on the coast may also find themselves ejected by seabirds which nest underground. On Skellig Michael, Co. Kerry, for instance, puffins *Fratercula arctica* commandeer their burrows.[190] Manx shearwaters *Puffinus puffinus* on Lighthouse Island, off Co. Down, also evict the rightful owners for the same purpose.[133] As both birds have formidable beaks, doubtless the bunnies do not argue. Short-eared owls *Asio otus,* which occasionally occur as migrants in Ireland, usually nest on the ground and often settle on it. In 1835, during the first Ordnance Survey of the Parish of Magilligan, Co. 'Derry, it was noticed that these birds were taking shelter in rabbit burrows when disturbed. Sad to say, several were shot and one was trapped whilst emerging.[276]

In Ireland the diet of the rabbit has been determined in detail in Connemara in Co. Galway. Most of the district is bog and therefore unsuitable for rabbits, but here and there on the lower slopes of the mountains where the underlying rock is marble, the soil is richer and consequently there is often a colony. Judging by the distribution of the droppings at such sites, which at about 10 mm in diameter are readily distinguished from those of hares at nearer 15 mm, the rabbits confine their activities to the few hectares around the burrows. Studies in Britain have shown that, as in Connemara, most rabbits do not forage far from their holes.[151] Miss Siobhán Duffy investigated the diet of three such colonies in Connemara at NUI Galway by analysing droppings collected each month over a full year.[87]

The outermost layer of cells on the leaves and stems of plants forms a sort of skin, the *epidermis.* This has a waxy coating and, probably because of this, the epidermis resists digestion in the guts of herbivorous mammals. Naturally it is torn into tiny strips by their teeth, but pieces large enough to allow the cellular structure to be made out under the microscope survive. Fortunately the structure and

Fig. 25 (Opposite) The cellular structure of the epidermis of some plants. U indicates upper surface of leaf, L the lower surface. Not to scale.

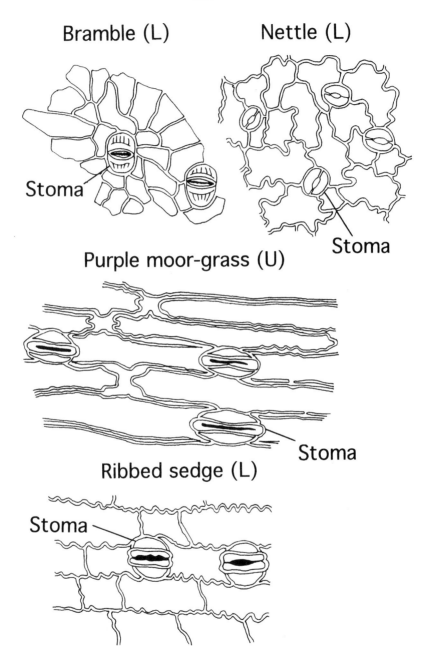

Bramble (L)

Nettle (L)

Stoma

Stoma

Purple moor-grass (U)

Stoma

Ribbed sedge (L)

Stoma

arrangement of the cells in it are usually characteristic for each species of plant, and so such fragments can be used to identify the plants eaten. It also contains holes, most commonly on the lower surface of the leaf, called *stomata* (singular *stoma*), through which the plant breathes. Each stoma is opened and closed by a pair of cells enclosing it known as *guard cells*. The form which these take often proves a useful additional aid in identification. Some examples of plant epidermis are illustrated in Fig. 25. In order to determine the food of a particular herbivore, one requires not only its faeces, but also epidermis for comparison from all the plants growing in the area in which it lives. So a full reference collection of plants must be assembled and pieces of epidermis painstakingly peeled off with a sharp blade, mounted on microscope slides and then drawn or photographed. This, and familiarisation of oneself with the multifarious shapes of the cells, often accounts for half of the labour involved in such a study.

A rabbit dropping is a compact mass of epidermal fragments and other debris glued together by mucous, but these can be separated by overnight immersion in a very dilute solution of sodium hydroxide (caustic soda, a strong alkali - drain cleaner is the cheapest source). A sample of pellets is treated in this way, the alkali afterwards neutralised with dilute acid, the material filtered to remove the tiniest particles (which are unidentifiable and get in the way later) and bleached (with bleach) and mounted on glass slides. Several areas of the slides are then viewed under the microscope and scored for presence or absence of epidermis from the various plants. This technique gives a very good indication of the relative proportions of the various plants that have been eaten.

The results from the three warrens were fairly similar and are summarised in Table 19. Overall, grasses were the most important food, which is not at all surprising, for in the many studies of rabbit grub outside Ireland grasses have usually been found to predominate. When one considers the rabbit as a pest, damage to crops is perhaps the first thing that springs to mind, but by grazing grass, the rabbit is at least of equal economic consequence in depriving farm animals of their food. Although less plentiful than some other grasses around the warrens, the rabbits ate fescues more often in Connemara than other species, probably because these are more nutritious. On the other hand matt grass, which was abundant at all three sites, was consumed in smaller amounts than the other common grasses, doubtless because it is less attractive to herbivores in general. The rabbits ate more purple moor-grass in summer than in winter, for this species is most palatable when young. Cattle and sheep will eat the fresh leaves, but ignore it otherwise.

Table 19 - Summary of the main plants eaten by rabbits at three warrens in Connemara. Figures are percentages. + = present.

Food item	Warren		
	Mweelin	Cur	Illion
Ferns	+	+	+
Mosses	2	2	2
Rushes	5	5	3
Sedges	7	3	6
Grasses	68	73	75
Other but mainly unidentified monocotyledon	5	6	6
Dicotyledon	13	11	7

Grasses, sedges and rushes all have long, narrow spear-like leaves and belong to one of the two major divisions of flowering plants, the *monocotyledons*. Besides the fragments of monocotyledonous plants which Duffy identified, there were others which she could not because they had been damaged or were too small. Although these are entered against unidentified monocotyledon in Table 19, they were almost certainly mainly from grass. So grass probably provided more than three-quarters of the food and monocotyledons as a whole accounted for around 85-90%.

The rabbits ate a little moss - rather more in winter than summer, because it continues to grow overwinter and so provides live tissue which is then otherwise scarce.

The remainder of the food consisted almost entirely of *dicotyledons,* the other major division of the flowering plants, which have broader leaves. One would, however, be mistaken in inferring that the relative unimportance of dicotyledons meant that the bunnies concentrated on only a few kinds. Duffy identified over 50 species in the droppings. Almost all were eaten in tiny amounts and the animals were probably very dainty in this respect, most likely selecting only the tenderest leaves and shoots. Although shrubs and trees, such as heather, bramble, hazel, willow and oak evidently took the bunnies' fancy occasionally, most of the dicotyledons eaten were herbs, and these they consumed more often in summer, when fresh greens would have been readily available, than in winter, when most of them would have withered.

Dr Alan Wolfe, of University College Dublin, analysed rabbit droppings from North Bull Island, Co. Dublin, and, like Duffy, he obtained samples from each month over a full year. The terrain on the island comprises saltmarsh and a little aldermarsh, dune grassland and a golf-links and so provides much better rabbit habitat than Connemara. Nevertheless, Wolfe's overall findings were remarkably similar. The overall diet comprised 1% mosses, 2% rushes and sedges, 85% grasses, and 12% dicotyledons. The bunnies were also taking full advantage of the superior grazing on the island as over half of the diet consisted of fescues.[385]

Not only do rabbits usually concentrate on the most nutritious forage in the form of grasses, and thus compete with the farmer's stock. They also graze the grass much closer to the ground than farm animals usually do, and so inhibit its regeneration. In turn this promotes the growth of thorny and poisonous species and generally plants which are distasteful to other herbivores - a further headache for the poor farmer. The area around a well stocked warren has the appearance of a snooker table, with the grass trimmed almost to the ground, but with many noxious plants impudently erect and unharmed.

It is sometimes argued that, since rabbits provide meat - and delicious lean meat at that - they should be considered a viable economic alternative to cattle and sheep. Whereas rabbit meat, in terms of protein content, is comparable to that of sheep, rather less of the body is edible: about 41% as compared to 50%.[348] A more significant factor is the relative amounts of forage consumed by the two species to produce the same weight of animal. A full-grown rabbit weighs about 1 kg, a sheep perhaps 35 kg. An assessment from Yorkshire from the the late eighteenth century - that ten rabbits will eat as much as one sheep - seems to be still fairly widely accepted today as a rule of thumb,[300] although the equivalent number of rabbits is sometimes said to be even smaller. So sheep are at least three times more efficient than rabbits in converting graze to the substance of their own bodies. Besides, rabbits are much more difficult to manage and, if a warren is close to a neighbour's land, its inhabitants will pilfer from it willy-nilly. Although rabbits were an important industry in Ireland until the advent of myxomatosis, it is only on marginal land, unfit for domestic stock, that they are a valid economic alternative to sheep and cattle. This generally means sand dunes or small islands, where it is too much trouble to transport farm stock back and forth. The subject of islands is considered further below. Prior to myxomatosis, many rabbits were shot, trapped and snared on agricultural land and sold, because the landowners were glad to be rid of them, but almost every one went to market owing a farmer money.

The reasons why sheep are so much more efficient than rabbits in converting forage to meat are simple. Green plant material is hard for mammals to digest and much of the process is brought about by bacteria in the gut. Rabbits *refect,* in other words eat their own faeces, so the food passes twice through the digestive tract for improved extraction of nutrients. Nevertheless, this is less efficient than the ruminant system of chewing the cud, in which the food passes back and forward from the stomach for repeated mastication. An even more important factor in the efficiency of sheep over rabbits is the differences in their sizes. Small size means relatively greater surface area and so a rabbit must burn a greater fraction of its food as fuel to maintain its body temperature than a sheep. The latter, being larger, can also carry a much longer coat, is thus better insulated than a rabbit and so loses less heat per unit area.

Only rarely can the rabbit's intensive grazing be regarded as beneficial, although it has been on the National Trust's Nature Reserve at Murlough, Co. Down,[371] an area of sand dune and heath rich in wild plants and associated animals. The reserve can boast of 22 species of butterfly alone. The Murlough dunes were used as a commercial warren up until the arrival of myxomatosis in the 1950s, such management probably dating back to the thirteenth century. When the Trust acquired the land in 1967, the rabbit population had recovered from the myxomatosis outbreak of the 'fifties, but not before there had been marked changes in the character of the reserve. Sea buckthorn, a spiny, impenetrable shrub, had established itself over large areas and there were more self-seeded trees. There was also a smaller variety of plants; remember that rabbits only nibble at most dicotyledons, but trim down the grass that would compete with them. Much of the heather, untrimmed by rabbits for years, had grown old and shrubby. Before myxomatosis, therefore, the rabbits had played a major part in maintaining the diversity of plant and consequently of animal life on the reserve. In 1976 a drought and several outbreaks of a severe strain of myxomatosis caused a major decline in rabbit numbers. By the early 1980s they were found only in one small part of the reserve and there was some concern over further changes in the habitat. More rabbits were a priority. In 1985 91 were brought in, released and cossetted during their first winter with a supply of root vegetables. In 1986 another 112 were put down and a further 89 in 1987-89. Resurveying in the early 1990s revealed that the animals were well established in several places and seemed to have at least halted the undesirable changes. Unfortunately myxomatosis yet again devastated the population and at present additional grazing is being provided on fenced areas by Exmoor ponies which do not, of course, catch myxomatosis!

Many readers will have had personal experience of rabbits raiding kitchen gardens and crops, but published details in Ireland are few. Up until the arrival of myxomatosis, rabbits were the most serious of all pests to forestry and their depredations on young trees meant that all new plantations had to be expensively fenced with fine mesh wire netting.[247] Nor are older trees immune. Besides attacks on seedlings and saplings, it is fairly widely known that the pests will also bark trunks when snow blankets the ground. Barrett-Hamilton noted that they may still do this in districts where winters are rarely severe and instanced such damage to ash, blackthorn, laurel and ivy in the mild climate of south-east Ireland.[21]

Surprisingly, the rabbit's menu is not exclusively vegetarian. Heaps of gnawed snail shells have been discovered several times at the mouths of burrows. One naturalist found so many at some entrances in the dunes at Portrush, Co. Antrim, as to almost block them. While such evidence for rabbits actually having eaten snails is circumstantial, another naturalist observed rabbits actually dining upon them in Co. Donegal.[21]

Observations of other kinds of behaviour by rabbits in Ireland are sparse but Moffat made two.[223] He remarked on the sentinel-like stamp of the hind foot, given as an alarm signal. This he interpreted not as a direction to cut and run, but as a warning to others to be on the *qui vive*. Moffat watched rabbits as well as bats at night and noticed that the slightest crackle of a dry twig or rustle of leaves was sufficient to make the first one to hear it sit up, listen and stamp. The stamp was followed, not by a general scamper, but by a profound silence in which the bunnies listened with all their ears. Moffat also considered that another danger signal, the white tail of the rabbit, bobbing up and down as its owner dashes off, was by no means a call to "follow me". Indeed following the first rabbit to do this would be to run toward the one nearest to danger. The bobbing tail is to be interpreted simply as "Run", and each individual bolts by the most direct route to its usual place of refuge.

There has been only one attempt to estimate rabbit densities in Ireland, on the Mullet Peninsula in Co. Mayo, apparently in August 1971, by a group from the student Zoological Society at the University of Reading, who spent several vacations researching there from 1969 to 1972.[156] Judging by the variety of work which they subsequently published, the society at the time must have been an unusually vigorous one. Field trips by undergraduate societies tend to run to holidays rather than serious research, and findings more often than not remain unpublished and are lost. The students investigated an area of 169 ha, consisting of a central patch of 55 ha of high dunes, with much exposed sand, but also long marram grass, surrounded by 114 ha of flattish grassland, well colonised by vegetation, but providing less cover. They selected 30 areas of 30x30 m and counted the numbers of burrows in each. They then worked out the percentage of burrows containing rabbits by watching 20 with binoculars for 2 hours each. In addition they also smoothly sanded the entrances to 60 and checked them for tracks after a day. Finally they counted the numbers of rabbits entering and leaving selected holes, sometimes stirring the bunnies up by showing themselves or firing a shotgun into the air. They thus had estimates of the numbers of burrows per unit area, the percentage of these occupied and the average number of rabbits using a burrow. They came up with an overall figure of 237 rabbits/ha - 383/ha in the dunes and 91/ha in the surrounding grassland. The higher numbers of both rabbits and burrows on the dunes was probably because of the greater availability of cover, but the bunnies also prefer to dig into a slope, so that the spoil does not accumulate so readily outside the hole. My only criticism of this study is that the students did not say how they dealt with the possibility of individual rabbits using more than one of the burrows where the counts were made. If this happened, the densities estimated would be too high and, from the comfort of my armchair, I am inclined to think that this might have been the case. The figures do, however, give some indication of the enormous population of rabbits possible in a small area, something that was almost certainly commonplace over much of Ireland before myxomatosis.

Rabbits have been introduced into many of the islands off the coast of Ireland, presumably because they provided a ready source of meat and required little tending. As they rarely swim, and then not far, they cannot escape. Poaching is also minimised. There are often no ground predators to speak of and, as the soil is often sandy, it is admirably suited for burrows. It is even possible that some island populations are healthier, as would be the case if the founders did not happen to carry some of the parasites to which every individual is at risk on the mainland. The list of islands colonised in Table 20[21,34,67,133,209,243,253,275] may not be comprehensive, but it is certainly impressive. Besides these, there are no fewer than 16 islands named Rabbit Island and nine Coney Island, either offshore or in inland lakes. These too presumably must have supported the beasts at one time or another.[6]

Because conditions are so favourable on islands, numbers often build up to plague proportions and there are many references to this in the literature. For instance this occurred in the early years of the twentieth century on Lambay Island, Co. Dublin, where there are records of the animals being caught for sale as far back as 1749. From September 1904 to December 1906, 24,000 were killed in efforts to control them.[17] As Lambay is almost 2.5 km^2, this represents almost 100/ha. On Copeland Island, off Co. Down, (1.2 km^2) rabbits eventually rendered the growing of root crops futile and during the winter of 1950-51 a trapper killed 11,000, or 92/ha.[74] When I visited the south coast of the island in August 2000, the bunnies were more numerous than I have seen them anywhere since myxomatosis, and the sward was grazed flat as a pancake.

On some small islands it is at least theoretically possible that numbers may grow so great, and fodder be so depleted, that the population dies out. During a seabird survey of the Blasket Islands, in Co. Kerry, in 1988, no rabbits were seen on Beginish Island. On a visit during the following year, Pat Smiddy found rabbit skeletons but no living rabbits, nor did he spot any during several subsequent trips to the island during the 1990s. So it seems that the animals became extinct there early in, or just before, 1988.

One of the most protracted campaigns against rabbits on an island, and certainly the most colourful, was on the Great Saltee, Co. Wexford, by the owner, the self-styled "Prince Michael" alias Michael Neale, the son of a Wexford farmer.[282] He also came into conflict with the County Council for failure to pay rates on his kingdom, although that is another story. In 1949, after his attempts to afforest part of the Island were frustrated, at least partly by the bunnies, he declared war on them. His

Table 20 - Offshore Irish islands on which rabbits have been recorded since
1900 listed by county. The animals may since have become extinct on some.

Co. Antrim
Rathlin
Co. Cork
Bere, Bird, Cape Clear, Dursey, Eyeries, Furze, Garinish
Co Donegal
Aranmore, Cruit, Doagh, Gola, Inishirrer, Inishleane, Inishtrahull, Owey, Rathlin
O'Beirne, Roaninish, Tory
Co. Down
Copeland, Lighthouse, Mew and, in Strangford Lough, Dunnyneill, North Rock,
Rainey, Round and South.
Co. Dublin
Lambay, North Bull
Co. Galway
Eddy, Finish, Hare, High, Inishmaan, Inisheer, Inishmore, Inishturk, Mweenish,
Omey, Tormore
Co. Kerry
Beginish, Great Blasket, Great Skelllig, Hogs Head, Inishnabro, Inishtearaght,
Inishtooskert, Inishvickillane, Puffin, Scariff, Valencia
Co. Limerick
Foynes, Greenish
Co. Mayo
Achill, Caher, Clare, Duvillaun More, Horse, Kid, Inishbiggle, Inishbofin, Iniskea
North
Co. Wexford
Great Saltee, Little Saltee, The Keeraghs (which individual islands unspecified)

---

first act was to release a couple of ferrets, but these died during their
first winter. He then airlifted in a dozen foxes but, as these had been
kept together for some time before being taken to the island, they were
all infected with distemper and died. One report of the affair gives
mange as the cause of death.[71] In any event, the Prince was by no means
beaten. He now procured 46 cats from an animal shelter and flew these
to the Great Saltee. This created a sensation when it became generally
known. The public was appalled. Letters of protest streamed in to the
Irish newspapers, on the one hand from those concerned with cruelty to
cats, and on the other from bird lovers, alarmed at the potential

slaughter of nesting seabirds by the pussies. Questions were asked in Dail Eireann. In due course a number of Gardai, a vet and a neutral observer were dispatched to determine how the cats were managing, but were unable to find any. In fact they were doing remarkably well, allegedly killing rats and puffins and begging from picnicking parties. However they died out within some 8 years, probably because few of the kittens survived the winter. The rabbits are still there.

Whereas rabbits living on the mainland are today always at risk from a fresh outbreak of myxomatosis, those on an island have the advantage of isolation. The disease is likely to die out on an island and the animals are then safe until it is re-introduced. This has led to problems in recent years. For instance the natives of Inishmore, one of the Aran Islands in Galway Bay, had rabbit troubles in the 1980s, when farmers complained that the pests were competing for grazing with their cattle. The grass runways of the airport there were in danger of being destroyed and the burrows had to be gassed with cyanide pellets, resulting in an estimated kill of some 5,000.[129,209]

Because of the advantages, some of the earliest warrens in Ireland may have been on islands. Although the Romans had domesticated the rabbit more than a thousand years before, it was not brought to Britain and Ireland until Norman times. For some centuries warrens in Britain were afforded a great deal of care. Sometimes earthworks were constructed for the inhabitants to tunnel in, and holes made with a special spade to encourage the animals to start burrowing themselves. Lacking such encouragement, they might hang about for months before getting down to serious excavation. The does were in some places furnished with hutches in which to rear their litters. The warren itself was usually enclosed by a ditch, bank, wall or fence. This forestalled visits by predators, four-footed and otherwise, and kept the citizens of the warren from escaping. Not only were the rabbits highly valued but, if allowed outside, they would have laid waste to neighbouring fields and woodland. The warren was also often actively managed to encourage suitable food plants.[300]

From the coming of the agricultural revolution in the eighteenth century, much of the land formerly considered useless for anything but rabbits was gradually improved and put to more profitable uses. However, small numbers of rabbits at least were tolerated on estates as an adornment to the landscape. As the practice of game preservation and stocking spread (Chapter 5), rabbits were encouraged to provide variety in shooting and they multiplied as keepers mercilessly persecuted their predators, which tend to prevent any increase in the rabbit population when numbers are low.[300] Incidentally, this is why, despite their proverbial rate of breeding, rabbit numbers are slow to recover after an outbreak of myxomatosis. Paradoxically, although foxes were preserved, this sometimes benefited rabbits, for fox coverts often made ideal sites for warrens. Although the rabbit had no legal protection outside the warren, landlords could forbid people living on their land from killing them. So many tenant farmers ended up not only paying one rent to the landlord, but a second in kind to the rabbits. The Ground Game Act in 1880 allowed tenant farmers to destroy them, irrespective of the wishes of the landlord. However, by this time the animals were well established in the wild in both Britain and Ireland.

Up to about 1700 the adult rabbit was known as a coney, the term "rabbit" being reserved for the young. Thereafter the names became interchangeable until eventually coney was regarded as archaic. In the older literature a warren is referred to as a coneygarth, conyger or conigree, all corruptions of the words "coney" and "earth". The Irish word for warren is "coinicér" (cunickere).

The history of warrens and their management in Ireland, although probably similar to that in Britain, is much less complete. However, there is little doubt that from Norman times they became widespread and the frequent mention of them in legal documents indicates the value attached to them.

There are many references to rabbits in Irish place names, which are easily recognised, the word *coinín* (a rabbit) appearing in terminations like *-coneen, -nagoneen* or *-nagoneeny,* as in Kylenagooneeny (the wood of the rabbits) in Co. Limerick or Carrickcooneen (the rock of the rabbits) in Co. Tipperary. It is sometimes corrupted as in Lisnagunnion (the fort of the rabbits) in Co. Monaghan.[173] There are also many townlands with names which suggest the presence of a warren, although in some cases the "Warren" may only have been the name of the local landowner. The ones which I have been able to trace, 52 townlands in all, are listed in Table 21.[6] In addition the "co" has apparently disappeared from some anglicised names, of which Nicker, in Co. Limerick, and Nickeres, in Co. Tipperary are examples.[173] There are still further localities, below the level of townland, with names indicating former warrens, such as Kinnegar, a district close to the shore in Holywood, Co. Down. (Kinnegar townland in Table 21 is in Co. Donegal.) Furthermore, on the south coast of Ireland several areas of sand dunes are known as "Burrow" or "The Burrow".

Table 21 - A list of townlands in Ireland with names which suggest the former presence of rabbit warrens.

Ballyconnogar Lower (1)	Cunnicar (1)	Warren and Bog (1)
Ballyconnigar Upper (1)	Cunnigar (2)	Warren or Drum (1)
Coneyburrow (6)	Conywarren (1)	Warren High (1)
Coneygar (1)	Doogan's Warren (1)	Warren Lower (2)
Coneykeare (1)	Greaghawarren (1)	Warren Middle (1)
Conicar (2)	Kinnegar (1)	Warrenpoint (1)
Conicker (2)	Old Warren (1)	Warrensfields (1)
Conigar (2)	Rabbitburrow (1)	Warrensgrove (1)
Coniker (1)	Rabbit Park (1)	Warrenstown (6)
Connigar (1)	Rabbitpark (1)	Warrentown (1)
Cunnaker (1)	Warren (4)	Warrentown Mountain North (1)
		Warrentown Mountain South (1)

The earliest reference to rabbits in Ireland is in a deed c1185, which reads[58]

John, son of the King of England....granted Alard son of William....near Waterford, the land of Carreckenard, the land of Carrec Eghon [etc.]....He also granted him....the hunting of stag, doe, pig, hare, wolf and rabbit.

The first reference to a warren may be in a in a charter of 1204 granting rights *in warrennis cunigariis* with lands to Hugh de Lacy in Connaught.[192] The next mention appears to be in a murage tax - one exacted to pay for the upkeep of city walls - of 1234 in Waterford, which includes imposts on the sale of the skins of rabbits, hares and foxes. This date approximately matches that of the earliest rabbit remains recovered in recent archaeological excavations in the city, which are from the thirteenth century.[204] There is a brief reference to foxes devouring an unusually large number of rabbits in a warren at Kilcosgrave, Co. Limerick, during the reign of Edward I (1272-1307).[369] In 1282 skins of 20 rabbits from Ballysax, Co. Kildare, were priced at 1s 4d (6.7p) and 100 "great coneys" from the same locality in 1287 at 13s 4d (66.7p),[21] surprisingly high prices in view of those in succeeding centuries given below. It is probable that rabbits in the late thirteenth century Ireland were still a novelty, and a dish only for the rich man's table. There are further references to warrens in the district around Dungarvan, Co. Waterford, in 1234 and 1298 and another to warrens at Kinsale, Co. Cork, in 1299.[370]

In 1324 the profits of hunting the "cunicularium" at Rosslare, Co. Wexford, a locality famous for its sand even today, formed a portion of the returns of the lands of Aymer de Valence.[21] In 1333 the rental of the Manor of Lisronagh, in the Clonmel district of Co. Tipperary, refers to a warren on the estate which is valued at 10 shillings (50p) and, confusingly, is expected to provide an annual crop of 24 rabbits at one penny each - not many as rabbits go.[174] Perhaps the 24 rabbits were intended to form part of the rental. In 1349 Agneta Cassel made a grant of land to her bother John Rathcoul with a warren at "Dondalch".[155] This was probably Dundalk, for a further document, of 1412, has survived relating to another grant of lands at Dundalk by what was perhaps a relative, Juliana Racoule, and this too specifically includes warrens. They also appear in a quit-claim of 1416 by Alice White of her rights in lands at Callstown and Glaspistol, Co. Meath.[155] Of the same date is a transfer of property by Alice Archer and her son Simon MacCarrowyll to Richard D. Heydane, Archbishop of Cashel, which contained warrens in the tenement of Rathsax.[174] Yet another reference in a legal document appears in a grant of 1461 by Richard Felan, chaplain, to Patrick and Nicholas de Launde of lands in Co. Laois including warrens "near the Monastery of Leys" and at "Dungarven".[124] The poem the *Libel of English Policie*, written about 1430 on the exports of Ireland, already mentioned in Chapter 5, includes "Felles of kydde and conies grete plenté".[21]

In the following century rabbit skins continued to be exported to Britain, rising in price from 2/6 (12.5p) per hundred at the beginning to 5s (25p) towards the end. In 1588 45,000 were sent to London alone.[193] The earliest estate map in Ireland, of Sir Walter Raleigh's lands at Mogeely, Co. Cork, in 1589 shows an area of 70 acres (28 ha) as "the warrene" and an adjoining 50 acres (20 ha) as "the warrene close".[42]

In his pamphlet *A briefe description of Ireland* (1598) Robert Payne includes as an addendum "The copie of an instruction for a Warrane".[271] Whether this was designed for use in Ireland, or, indeed, whether the author had any real experience of warrening, is not clear. He planned a combined rabbit-run and orchard all within 2 acres (1 ha) which sounds impractical, and his ideas on keeping the bunnies from barking the trees seem fanciful. The whole was to be surrounded by a ditch and a wall built from the earth dug from it. In the middle there was to be "a little lodge", containing eight hutches, with a buck rabbit chained in each and 64 does at liberty. The animals were to be pampered with

....the shortest and sweetest haye you can get; you may give the[m] grass, mallowes, coleworts [cabbages], sowthistles, the tops of carrots, scarates [?], and any yearbs or weeds....

The article ends with an estimate of the probable profits from the scheme.

In 1593 the Earl of Ormonde made a grant of part of his property in Cos Carlow and Laois to William Harpole for 21 years, a condition being that William was to supply 12 couples of rabbits every week for the provision of the Earl's houses.[155] Rent in rabbits for land containing warrens was not unusual. For instance in 1608 in a lease granted by Thomas Fitzgerald to Richard Eustace of land at Gorteenvacan, Co. Kildare, payment took the form of £2 and "twelve cupple of good fat connies", the money to be paid at the Feast of St Michael and at Easter in equal portions, and the rabbits to be delivered to Fitzgerald's House at the Feast of All Saints and at Christmas.[130] Professor Nicholas Canny also told me of a supply of fresh rabbits to Lismore Castle, Co. Waterford, in 1637 as part of a rental agreement with a warren keeper.

In about 1600 Fynes Moryson refers to "great plenty....of conies" in Ireland and around 1635 Sir William Brerton noticed, near Wexford, "abundance of rabbits, wherof here there are too many, so as they pester the ground".[21] Sir George Rawdon, writing to Viscount Conway in 1668 regarding the latter's estate in Co. Antrim on the eastern shores of Lough Neagh (on which Arthur Stringer was huntsman some years later), mentions that "We have two burrows made for rabbits by the lough side". It would therefore appear that workmen there had dug these initially to start the warren off. The trade in rabbit skins went on through the 1600s. Fortunately a summary of exports from Ireland in 1665 is given in the *State Papers*,[210] during which a total of 103,200 rabbit skins were shipped out.

Exports of skins in the eighteenth century were recorded not by number but by weight and sometimes fur alone appears in the records. From information on the weights of skins kindly provided me by a furrier, I was able to estimate the equivalent in rabbits and came up with an impressive 114 million during the period 1697 to 1819.[109]

RABBIT

The Royal Dublin Society's *Statistical Survey* of Co. Meath in 1802 illustrates one way in which the bunnies could be confined to the warren, the one in question extending along the sea shore from the River Boyne to the River Nanny, a distance of some 5 km.[349]

> The rabbits burrow in a heap of sand, blown off the sea-shore by the easterly winds, and feed on a salt-marsh parallel to it, being prevented from going on the uplands and corn grounds by broad drains, which are constantly full of water; they are taken by pass-nets, placed between them and the burrows, on their hasty return from feeding at night, being alarmed by the barking of dogs, kept for the purpose.
>
> They are all disposed of in Dublin market, the skin being generally more valuable than the flesh; and they are sold by the warreners, at one shilling and sixpence [7.5p] to two shillings [10p] the pair. I have been informed this warren is worth three hundred pounds per annum....and the ground, so employed, is not valued at one shilling per acre.

Arthur Young in his *Tour of Ireland* (1780) mentions being told of a warren at Horn Head in Co. Donegal "25 miles [40 km] long" but no details of this seem to have survived.[388] It is unmentioned in the Ordnance Survey Memoirs.

Probably the most productive area of warrens at the time, and certainly one of the largest, was in the Parish of Magilligan, Co. 'Derry,[184] an area of some 53 km², about half of it a great triangular promontory of dunes and flat sandy ground jutting into the mouth of Lough Foyle. The warrens here have been better documented than any others and must have provided the chief business of the district, at least from the late seventeenth century until well into the nineteenth. Even those farmers who were not full-time warreners often supplemented their income from rabbits because of the generally poor quality of the soil. Much of the land was fit for little else, a survey of 1768 describing some of the farms as "good for Nothing but rabbits and Swine".[184]

The *Downe Survey* of the 1650s mentions "Cunny-warrens and Sandy hills" in several of the townlands and a Hearth Money Roll of 1683 includes "Hugh O'Cahan, warrener". Raising rabbits for market rather than for the pot seems to have been established as a regular occupation by the 1680s. When all of the rented holdings were relet in

1713-14, the most expensive appear to have been those with the largest areas of warren. Rent was often specified in rabbit skins, or their equivalent in money. One of the largest individual warrens, in the townland of Scotchtown, prior to 1721 had a rent of 30 dozen annually. When the lease was renewed in 1732-39 it was raised to 40 dozen and by the late 1750s to a staggering 120 dozen, which seems to imply a growing appreciation of the profitability of the industry. However, a farmer's overheads in the enterprise were minimal, apart from the wages of cottiers, who did the work. In 1761 the Earl of Abercorn's agent

....wrote a gentleman at Magilligan who had a large warren to agree with a person to come and kill rabbits at Barrons-court but can get no-one from thence as they are all cottiers kept chiefly for that business, and have neither ferrets or nets of their own.

Leases on some farms stipulated the open season, so rabbits could only be caught from mid-November until mid-February, the precise dates varying from lease to lease. One of the latter required that the nets in which the animals were caught should be delivered up on 2 February on penalty of a fine of £5.[184] The open season almost certainly reflects the period over which the fur is at its best, as one would expect in winter.

Arthur Young (1780) commented on the value of the industry in the district as follows.[388]

At Magilligan is a rabbit warren, which yields on an average 3000 dozen per ann. last year 4000 and 5000 have been known. The bodies are sold at 2d the couple; but the skins are sent to Dublin at 5s. 7d. to 6s. [30p] a dozen, selling from 1500l. [£1,500] to 1800l. a year.

Dublin was the main commercial centre for rabbits at this time, both for exports and manufacture of hats. A bad season in 1780 saw the hatters demanding that the Irish Parliament impose a swingeing duty on the export of rabbit fur, but this was wisely resisted.[184]

The Royal Dublin Society's *Statistical Survey* of Co. 'Derry (1802) points out that the value of the land as warren was very variable, an acre in some places yielding rabbits worth £10 *per annum* and "in other places, 30 acres will not contain the value of £1".[295] The "swelling grounds" were said to be the most valuable "because rabbits will not burrow in the low flats". Whereas this is not strictly true, the animals do prefer to dig in a slope, as the students from Reading University found on the Mullet. The total land under warren was estimated as exceeding 1,500 acres (6 km²) although, as will be seen below, this was probably too low a figure. The annual production of rabbits was said to be around 2,000, but this probably refers to 2,000 dozen in view of Young's figures and those for later years. The prices fluctuated between 8s (40p) and 12s (60p) per dozen. Sales began in November with the beginning of the open season.

> The flesh of these rabbits is either consumed in the farmer's household, or, in still greater proportion, is carried through the neighbouring towns and country. The carrier hangs the rabbits over his horse's back; he sells them from 6d. to 10d. per couple, skinning or *casing* them at the same time. For the skins he is accountable to the proprietor of the warren.
> It is well known, that warren rabbits are greatly inferior to those of demesnes as to flesh, but they are greatly superior as to fur. Those at Magilligan, which are fed on moss and bent, have the longest fur.

An account of 1812 gives the annual production as 2,500 dozen, the skins selling for 13s (65p) per dozen and carcasses at 4d (1.7p) each, resulting in a total return of £2,100 (!) *per annum.*[184]

The Ordnance Survey Memoir for Magilligan, compiled in 1835, deals with warrening there at some length, the surveyors having obviously taken some trouble to examine what "constituted so large a proportion of [the local] productive economy".[5] It is therefore probable that the area stated to be under warren - 3,000 acres (12 km²) - is more reliable than that in the Statistical Survey, although it might have increased during the intervening years. Production was then about 3,600-3,700 dozen pelts annually. However, the upper figure is too low because it was the practice to reckon a deficient skin only as a half. Such faulty items were called *racks* or *libbocks* and sold off cheaply to pedlars. When these, along with rabbits stolen or otherwise disposed of than at a public sale, were taken into account, the surveyors reckoned that the actual annual production might have been as high as 50,000. Unfortunately the increase in the output of rabbits over the early part of the century was not matched by prices, these having gradually fallen to

3s 6d (17.5p) per dozen. Those carcasses which were not consumed locally, where they were bought at 2d a pair, were sold in towns in the county at a penny more the couple. The season for rabbits was still in "the 3 winter months" and many of the labouring masses must then have been thoroughly sick of the sight of them on the dinner table. It was not unusual for servants to stipulate that they should not be fed them exclusively. The following grace was commonly repeated to the surveyors by such persons.

> For rabbits hot, for rabbits cold,
> For rabbits young, for rabbits old,
> For rabbits tender, rabbits tough,
> We thank thee Lord, we've had enough.

The Memoir again mentions the great variation in the numbers of rabbits produced per unit area, just as the Statistical Survey did, but, with some perspicacity, observes that it is an error to encourage warrening on ground capable of being put to tillage, not only because crops are more remunerative, but also because of the erosion of sandy soil by the burrows.

The warrens were often started by digging the first holes with a rabbit spade, a labourer cutting a groove in the surface of the ground and then roofing it over with sods and earth. The bunnies were said to be attracted by the fresh diggings and, finding partial accommodation awaiting them, were disposed to complete the job.[5]

The skins were sold by public advertisement, such a one appearing in the *Belfast Newsletter* in January 1790, informing readers that a deposit of fifty guineas (£52.50) was, as usual, required immediately after the sale. Another in 1797 warned that only cash would be accepted as payment.[184] The Memoirs state that two men, known as *casters,* were sworn to act as independent arbiters between buyer and seller. The purchaser bargained on the price per dozen for the quantity he required and then paid in cash. Unsold skins were used to make glue.[5]

The Memoirs also refer to a smaller area of warren in the adjoining Parish of Aghanloo, also surveyed in 1835. Although no figures on annual production are given, numbers had been falling because too many rabbits were being killed; others were being eaten by rats in the traps.[5]

There are several references in the period 1747-66 to the rabbit warren at Murlough, Co. Down, in the papers relating to the Downshire

Estates.[277] It is clear that the skins of at least some of the rabbits were sold to hatters in Dublin. Nets, brass wire to make snares, and expenses for powder and shot are all mentioned, so the animals were taken by a variety of means. It is nowhere stated unequivocally how many rabbits were dispatched in any year. However, a letter of 4 January 1755 mentions "60 dozen" (720) "already caught", which probably refers to the preceding December. An account dating from the following December details a total of only 3,614 killed between 10 January and 1 April. The "flesh" from these sold for £15 6s 6½d. Expenses, including wages for the warrener and his three partners, powder and shot and coarse linen to make bags in which to bring the skins to Dublin, totalled £15 1s 5d, so that the net profit was a mere 5s 1½d (less than 26p). However, the skins were presumably sold separately and, even if this had been 6s per dozen (the price given above by Arthur Young in 1780), should have grossed £90. But either prices were higher than this or the estate warreners were working at less than full capacity, because several letters over the period offer rentals for the warren ranging from £90 to £140 *per annum.*

The subsequent history of rabbits and the rabbit industry in Ireland is sparsely documented. Estate game books sometimes list numbers shot or trapped, including what may have been a world record by Sir Victor Brooke and four companions on his property at Colebrooke, Co. Fermanagh, who slaughtered a total of 4,426 rabbits in 4 days.[74] The export of skins continued to be important and also, surprisingly, of rabbit meat. In the 1870s, for instance, most of the rabbits sold in meat-markets in Manchester came from Ireland.[300] From the perspective of today, it is hard to appreciate just how plentiful the animals were in the nineteenth century and in the first half of the twentieth. All of this was to change with the arrival of myxomatosis and with it the collapse of the trade.

Myxomatosis was first recognised by scientists when it killed off laboratory rabbits in South America. It was subsequently proved to occur in the native rabbit of Brazil *Sylvilagus braziliensis,* in which it is a mild disease,[348] resistance presumably having been built up over a long period during which the rabbit and virus eventually came to co-exist. The subsequent appearance of the disease in populations of wild European rabbits around the world was often brought about unofficially and has a murky history.

Although most people in Ireland bewailed the "terrible cruelty" of introduction of the virus here, it proved a blessing to farmers in almost eradicating arguably their most serious outdoor mammalian pest. Moreover, whilst the swellings caused by the disease are unsightly, the amount of pain an infected animal suffers is unclear. It is certainly insufficient to prevent a rabbit from feeding and moving about.

Few official records of the arrival and spread of myxomatosis in Ireland have survived. I have therefore been forced to rely on newspapers and the files of the late Douglas Deane, who alone seems to have retained notes of the spread of the disease in Northern Ireland, some of which he published in a newspaper article.[64]

Myxomatosis first appeared in Britain in 1953 and in Ireland in the summer of 1954, the first outbreaks, in Carlow, Kildare and Wicklow, being reported in the *Irish Times* on 20 July. Concern was expressed in the newspaper on the future of the rabbit industry, reckoned to gross over a million pounds annually, with half of this coming from exports alone. On the following day, the Society for Prevention of Cruelty to Animals was said to be organising volunteers to put infected rabbits out of their misery. The Society urged farmers not to spread the disease but to "take proper steps" to control rabbits instead, something that was easier said than done. The next day the paper divulged that sales of rabbits in the Dublin markets, reckoned at as many as 1,000 a day, had ceased. A Dublin hospital with a standing order for two batches of 100 carcasses each week had cancelled it. The paper deplored the loss of jobs in the export trade, which, with packing of carcasses and preparation of skins, was labour intensive.

According to the *Irish Farmer's Journal* of 24 July, there had been a conspiracy to bring in the virus in the previous May, a meeting having been called in Dublin to discuss ways and means. A representative had then gone to England to assess the situation there, and to satisfy himself that the virus was both harmless to anything but rabbits, and effective enough to make it worth the trouble of introducing it to Ireland. The article then detailed its release at various centres round the country "by the end of May". The *Journal* naturally approved. The *Irish Times* of the same date encapsulated the polarisation in public opinion.

> The introduction of myxomatosis to Ireland has given the rabbit a new status. On the one hand the humane type of citizen is raising his voice against the cruelty of condemning the animals to a lingering death; on the other the farmers have welcomed with unalloyed joy this new weapon against one of their oldest enemies.

In between the two are the folk, mainly boys, who are building up a nice little trade in diseased rabbits.

It seems clear that diseased rabbits were indeed changing hands for cash, the most valued animals being those in the early stages, which, it was reasoned, would continue to disseminate the infection to their fellows for the longest time, before passing on themselves; corpses were considered to be useless. The vector of the disease in Ireland may be assumed to be the same as in Britain, namely the rabbit flea *Spilopsyllus cuniculi*. On wild rabbits several of these insects may often be seen attached to the exposed skin around the ears. When the rabbit dies, the fleas do not leave until it cools. So a moribund rabbit, or one which has just expired, might still have proved entirely satisfactory in starting up the disease in a fresh area.

If an item in the same newspaper 3 days later is accurate, some kind-hearted farmers had been killing myxomatous rabbits and were commended for same by Our Dumb Friends League. But myxomatosis marched on relentlessly. By 29 July there were reports of it in Wexford, Cork, Mayo, Westmeath and Clare, so, within days of its first being noticed, it was widespread in the Republic.

The first occurrence in Northern Ireland[64] was said to be on 1 August near Gortin, Co. Tyrone, although Deane did not receive this information until nearly 3 weeks later. The second report was from Ballinamallard, Co. Fermanagh, on 10 August. On 20 September the virus was at work at Castleward, Co. Down, and on 20 October at Bushmills, Co. Antrim, and Limavady, Co. 'Derry. By Christmas it was established in at least 14 different areas within the six counties. Deane was in no doubt that both the arrival of myxomatosis and the rapidity with which it spread were largely the result of human agency. According to the *Irish Times* of 23 July, the Ministry of Agriculture of Northern Ireland took no measures to prevent the smuggling of infected rabbits across the Border - any would have been futile anyway - and regarded the prospect of rabbits being "wiped out" as a "good thing", the damage they did more than balancing the benefit from skins and meat.

The pattern of appearances of myxomatosis, some localised and others more widespread, has continued in Ireland since, although these have been almost completely undocumented. However, rabbits now appear to be making something of a comeback. The history of the disease is probably similar to that in Britain. Most of the original strains of myxomatosis killed almost every rabbit that they infected, but this was of no advantage to the virus. Fewer hosts meant reduced chances of spread. Less virulent strains therefore tended to do better and the overall severity of the disease diminished, so that more rabbits survived in subsequent outbreaks. The incidence of Grade I of the virus, which kills 99% of rabbits, had decreased dramatically in Britain by the early 'sixties and it may now be extinct. In addition, as those rabbits which have any inherited immunity are more likely to survive and pass this on to their young, it was predicted that the rabbit population as a whole should tend to become more resistant. Such inherited resistance was first detected in Britain in the 1970s.[125]

In recent years there have been outbreaks in farmed and domestic rabbits in various parts of the world of a new sickness, rabbit haemorrhagic disease, which initially gave rise to speculation hailing it as the "new myxomatosis". This has proved unfounded. Examination of blood samples from wild rabbits collected in Britain and Ireland has revealed high levels of antibodies to this sickness, indicating that milder strains were already present in many populations of wild rabbits in the British Isles. Exposure to these confers significant immunity to the more severe variety.[352]

The horror with which some people view myxomatosis is tied up with the cuddly childhood image of the bunny rabbit: soft, reassuring, innocent, comforting and carefree. It is unlikely that such compassion would have been afforded if a less endearing species, such as the brown rat, had been the victim, but then rats are "not nice". Anyway the life of a rabbit is far from comfortable and serene. Even if Peter Rabbit had eluded Mr Macgregor, the odds were stacked against either he or any of his siblings surviving to have families themselves. It is inevitable that the great majority of the offspring of any species of animal so efficient at reproducing itself must perish young. Besides disease and starvation there also are a number of predators in Ireland which are delighted to assist bunnies in departing this life.

The partiality of mink, stoat and badger for rabbit has been dealt with in earlier chapters and the otter and probably the marten take them occasionally. Foxes eat a lot of rabbits. In the 1960s I examined the stomachs of 340 adult foxes and 163 cubs with food in them and 45% and 59% respectively contained remains of lagomorphs. Partially consumed cadavers around earths showed that foxes caught both rabbits and young hares.[111] As recounted in Chapter 12, rats are sometimes also serious predators. Domestic dogs and cats both kill rabbits, the latter sometimes becoming accomplished specialists. Such partiality is reflected in the fact that of the four cats from which I have obtained fleas, only one yielded cat fleas *Ctenocephalides felis;* the other three carried rabbit fleas, these having doubtless transferred to pussy from their previous hosts on the principle of "any port in a storm".

Several species of bird also prey on rabbits. The staple item in the diet of the buzzard *Buteo buteo* in Ireland is usually said to be rabbit, but actual published instances of the birds taking them are scarce. The only one that I can find refers to the stomach contents of a buzzard shot at Glenarm Castle, Co. Antrim, in 1873.[36] The buzzard was virtually exterminated in Ireland at the end of the nineteenth century, but a pair nested in Co. Antrim in 1933. Four pairs were breeding there in 1953 and in 1954 ten, after which numbers fell away, this being attributed to shortage of prey caused by myxomatosis.[291] In recent years there has been a remarkable recovery and this is probably at least partially explained by the increased availability of rabbits. Many birds, especially owls and hawks, regurgitate undigested parts of their food as pellets. Lagomorph remains have been found to be common in pellets collected from around nests of hen harriers *Circus cyaneus* in Ireland.[86,298] Further evidence of rabbits, young or old, as prey of birds, native or visitors to Ireland, either from pellets, direct observation of kills, or of their remains at nests, is available for short-eared owl,[4,94] golden eagle *Aquila chrysaetos,*[65,148] sea eagle *Haliaetus albicilla,*[350] peregrine falcon *Falco peregrinus,*[355] gyr falcon *Falco rusticolus,*[357] kestrel *Falco tinnunculus,*[114,350] merlin *Falco columbarius,*[49] rough-legged buzzard *Buteo lagopus,*[350] raven *Corvus corax*[28] and great black-backed gull.[38]

Besides predators, wild rabbits have also to contend with parasites. An investigation of those of 22 rabbits killed in Co. Cork[39] turned up most of the ones that one finds in wild rabbits in Europe:[151] on the outside the rabbit flea and louse *Haemodipsus ventricosus* and the pasture tick *Ixodes ricinus;* on the inside, in the gut, three species of roundworm and three of tapeworm. Because rabbits eat mainly grass and defaecate on the ground, the probability of them acquiring such worms is high. For two of the roundworm species *Passalurus ambiguus* and *Graphidium strigosum* it is a simple matter of the eggs from the parasite passing out, contaminating the grass, being accidentally consumed by another host and hatching in the intestine. The eggs of the third species of roundworm *Trichostrongylus retortaeformis* actually hatch on the ground to give larvae, which, with luck, are then ingested by a new host. The three tapeworm species,

Cittotaenia pectinata, Cittotaenia denticulata and *Cittotaenia ctenoides,* have more complex life cycles. Their eggs must be eaten by mites living in the pasture. Inside the mite the larva emerges from the egg and forms a cyst around itself. Nothing further happens until the infected mite is consumed by a rabbit along with its forage. The parasite then arouses in the gut, develops and goes about its business. *Cittotaenia ctenoides* is not large by tapeworm standards, but a big specimen, which might measure as much as 80 cm in length, is long enough for a rabbit.[335] Twelve liver flukes lurked in the liver of one of the rabbits. Although this parasite is usually associated with sheep and cattle, where it is of great economic importance in many of the wetter parts of Ireland, it can occur in other hosts. The life cycle of the liver fluke is complex. The eggs pass out in the faeces of the host and each gives rise to a larva which must find a particular species of snail, *Limnea truncatula.* The larva then enters the snail and multiplies in a series of complex stages (which many undergraduates in zoology classes have difficulty in remembering) until eventually other larvae emerge, encyst on vegetation and wait to be eaten by a cow, a sheep, or perhaps even a rabbit.

A study of 179 wild rabbits at Queen's University Belfast, produced a similar range of parasites. In addition the livers of the rabbits were examined for the microscopic single-celled parasite *Eimeria stiedai,* the causative agent of the disease coccidiosis, which may be fatal. This turned up in almost 10% of the animals.[18]

Chapter 11. Hares

Begorra! Mrs Cafferty, yer lepin' like a hare!
Phil the Fluters Ball. Percy French.

There are only two species of hare in Europe: the brown or European hare *Lepus europaeus* and the Arctic (sometimes called varying) hare *Lepus timidus,* the various races of which have other common English names. The race in Ireland is known as the Irish hare, that in Great Britain as the mountain hare (N.B.). The Arctic hare is smaller than the brown and has a more rounded outline and shorter ears. If the brown hare's ears are folded forward they extend to beyond the end of its nose. Those of the Arctic hare scarcely reach the nose tip when stretched toward it. Most books also state categorically that the Arctic hare's tail is entirely white, whereas the upper surface of the brown hare's is black. As the brown hare tucks its tail down when running, the white underside is then usually invisible. This feature ought therefore to provide a means of distinguishing the two species from the rear when one accidentally starts a hare. In Ireland, however, as we shall see, things are not that simple.

The geographical ranges of the two species in Europe are shown in Fig. 26. The brown hare is also found further south, into Turkey and Israel, and east into central Asia. The Arctic hare has an even wider distribution, living up to its name by occurring right around the northernmost parts of Europe, Asia and North America. One might infer that the brown hare is essentially a species of temperate climates and the Arctic hare of colder, more severe conditions. This is borne out by the fact that the Arctic replaces the brown on the Alps and on much of the high ground in Britain, and it is for this very reason that the race there is commonly called the mountain hare. In this context I particularly recall a naturalist colleague during a lecture showing a slide of a fence on a Scottish estate which, he remarked, served as the boundary between the two species, with brown hares occurring only below it and mountain hares only above.

The generally colder climates in which the Arctic hare dwells are reflected in its more rounded shape and shorter ears, both of which reduce its surface area, and readers will by now be well aware of the advantage of reduced surface area in conserving heat. A further adaptation of the Arctic hare to the cold is that, unlike the brown, it turns white in winter over most of its range. This, of course, greatly enhances concealment when there is snow. The transition from the darker summer coat to the white winter one, and *vice versa,* is brought about by the moulting of the one in early winter and of the the other in spring.

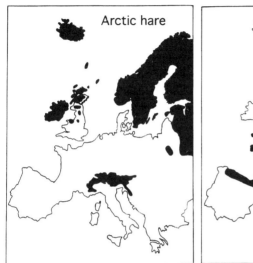

Fig. 26 The distributions of Arctic and brown hares in western Europe.

Whereas there is some slight overlap in the geographical ranges of the two species, there is only one large temperate region where the Arctic hare is to be found living from sea level to the tops of the highest mountains - Ireland. Here it is, of course, known as the Irish hare, a distinct race or subspecies, *Lepus timidus hibernicus.* It is unfortunate that in recent years some people - and not all of them in Britain - have fallen into the habit of referring to the beast as the Irish mountain hare, which is confusing, for it is by no means restricted to high ground. A further anomaly is the curious fact that the hare on the Irish predecimal threepenny piece, judging by the relative length of its ears, was not an Irish hare but a brown one!

The attention given to the Irish hare by zoologists and others from when its distinctness was formally recognised in 1833[350] until the present, mirrors the extent of published research on Irish mammals in general over this period. In Victorian and Edwardian times there was considerable interest in the animal, partly due to its being uniquely Irish and partly because it was a game animal and was coursed; and in those days field sportsmen were much more inclined to commit their observations on the natural history of their chosen quarry to print than today. There followed a long interval when almost nothing further was learnt, until interest was renewed in the latter part of the twentieth century. In the 1990s there were no fewer than three PhD theses on the animal: by Dr Alan Wolfe at University College Dublin, who did most of his work on North Bull Island in Dublin Bay; by Dr Rebecca Jeffrey, at Trinity College Dublin, who studied mainly a population on farmland on the Wexford Wildfowl Reserve on the Wexford Slobs (which is also Ireland's only statutory hare reserve), but did some additional work on a mixed farm in Co. Meath; and by Dr Karina Dingerkus, at Queen's University Belfast, who operated over most of Northern Ireland.

The brown hare is, of course, mainly brown, but so is the Irish hare. However, the fur of the latter is of a rich, russet hue, the difference being immediately obvious when skins of the two species are laid side by side. The long guard hairs, which give the Irish hare its colour, are black at the tips and when the latter are worn away the animal may appear almost as red as a fox.[21] The underside is lighter, the belly in particular being sometimes almost cream. The summer coat of the mountain hare, on the other hand, is of a much greyer hue, being dusky rather than tawny. When the animal is changing from summer to winter pelage, or *vice versa*, the intermingling of the white and grey gives rise to a tint referred to by some fanciful people as blue, this having given rise to the alternative name of blue hare.

Not all Irish hares are red-brown. Sooty individuals sometimes turn up, some described as "perfectly black" and others with the back black, fading to grey on the flanks. A well-marked buff variety was noted in Galway in the closing years of the nineteenth century and was shortly afterwards found to be fairly common along the east coast, from Malahide to Balbriggan.[21,378] Douglas Deane claimed to have seen such animals at Castleward, Co. Down, in the 1970s.[70]

Snow which lies for any significant part of the winter is rare in Ireland, except on the highest mountain tops. Therefore winter whitening, a boon to the Arctic hare over most of its range, would here seem a liability. In fact many Irish hares do not whiten in winter. Some certainly do, although this is nearly always only partial. Even on individuals otherwise snow-white, there are usually patches of brown on the back or on the top of the head, and the tips of the ears often remain dark coloured. In addition there is almost always a sprinkling of russet hairs elsewhere amidst the white. In a given area occupied by a number of Irish hares, it sometimes happens that some will turn almost completely white, others will be half-and-half and still others may only whiten a little or not at all.[21,231]

It must be emphasised that, as white hares tend to stand out against vegetation free of snow, they are more readily spotted than those which have not whitened. There is therefore a tendency to overestimate the proportion of the population which acquires even a patch or two of white. Indeed the net result of partial whitening would appear to be to render the animal more conspicuous whether there is lying snow or not. Still, it might be advantageous where light falls of snow are common and vegetation is only partially blanketed.

Most of our knowledge on winter whitening in the Irish hare stems from numerous casual observations in the nineteenth century. There has been no systematic study of the phenomenon and one is long overdue. Furthermore, it has been demonstrated that the mountain hare does not have two annual moults, but three: brown to brown (early June-mid September), brown to white (mid-October to January) and white to brown (mid-February to late May).[151] So it is also highly probable that there are also three in the Irish hare, which is so closely related, although this has apparently never been explored.

The onset of the seasonal whitening in the Irish hare is erratic, although it apparently rarely starts before December. An individual may whiten in patches until the process is complete, or the moult may be evenly spread, so that the animal appears to be undergoing a gradual bleaching. The spring moult is even more haphazard and may begin as early as the end of January or as late as early May, so its timing is not much different from that of the mountain hare.[21]

There is an appreciable body of evidence, again mainly in the form of casual observations, that whitening of the Irish hare is commoner in severe winters, when, of course, it should prove most advantageous.[21] There is even some indication that a spell of cold weather, or perhaps lying snow, may actually trigger the moult. Alexander More (1830-95), formerly Keeper at the Natural History Museum in Dublin, whilst being pushed about the streets of Dublin in a bath chair in his declining years, noted that, after only one week of snowy weather about the middle of January, the number of hares hung up in poulterers' shops showing white patches was conspicuously greater than at the beginning of the month.[217,222] There is also a modicum of evidence to suggest that hares living on mountains are likely to whiten more often and more completely than those on lowland.[21] Studies in Scotland lend considerable indirect support to the above tentative conclusions, because there mountain hares at higher altitudes whiten earlier and more completely than those living lower down, and both air-temperature and snow-lie affect the progress of the winter moult.[151]

One characteristic of the pelage of the Irish hare has been discovered only recently: that, unlike other Arctic hares, the Irish hare's tail may not always be entirely white. All of Wolfe's hares on North Bull Island had white tails while in their winter coats. Between May and October, however, all of the animals that he radio-tagged at some stage developed black fur to a greater or lesser extent on the upper surface of their tails. Later he spotted this on other Irish hares elsewhere, in Cos Dublin, Kildare, Meath, Wicklow and Westmeath. Since he operated in the eastern part of Ireland, it is entirely possible that such darkening of the tail is more widespread.[381] This, of course, means that the tail cannot alone be used to distinguish Irish and brown hares, which are also present in Ireland as a result of deliberate introductions. We shall return to this point at the end of the chapter. In the meantime we are concerned only with the Irish hare.

The heaviest adult Irish hare on record was a non-pregnant female of 4.3 kg which Wolfe caught at Mosney Holiday Camp, Co. Louth, in the winter of 1994-95. It had been alleged that hares had been ravaging flowers planted there, and so Wolfe and others, to prevent the beasts from being destroyed, netted some 22 and transferred them to North Bull Island. From his, and other published figures, it is probable that the range for adults is about 2.3-4.3 kg and the length of the head and body is 521-559 mm.[21,103]

Like most of our mammals, the Irish hare is primarily, though by no means exclusively, nocturnal and, unlike rabbits, essentially a surface dweller, relying on concealment in ground cover or on its speed to evade predators. Commonly when at rest it lies in a *form:* a hollow in ground vegetation made by the animal's body. It might be argued that many of the resting places are not true forms but simply convenient nooks among vegetation of even amongst stones or rocks.[222] Notwithstanding, it is well established that Irish hares sometimes go to ground if a convenient hole is available. Arthur Stringer regarded this as unexceptional. Barrett-Hamilton once witnessed a coursed hare save its life by vanishing down a rabbit burrow and on several occasions saw *leverets* (the young of hares) voluntarily doing the same thing or concealing themselves in the trunks of hollow trees.[21] A naturalist living near Donaghadee, Co. Down, watched what he believed to be the same hare habitually enter a rabbit hole and several times saw leverets use rabbit burrows in the same district.[268] Deane was told that a hare had probably given birth under a meteorological hut at the edge of Aldergrove Aerodrome (now Belfast International Airport) in 1948 and 1949. At any rate, leverets were often observed to emerge from beneath the hut, only to re-enter on the approach of a member of staff.[60] In 1896 a pseudonymous correspondent to the magazine *The Field* wrote

> On many Irish mountains the Hares take to natural fissures in the rocks, or to natural water-courses, called by the natives water-breaks, formed by the percolation of the water through the peaty formation overlying the rock or other hard subsoil, often to a depth of several feet. In many localities, as for instance the Bannermore chain in Donegal, where there is little covert, the Hares become nearly as subterranean in their habits as Rabbits. In these holes or crevices they seek safety from their enemies or shelter from the bad weather, coming to the entrances of their "burrows", if such they may be termed, to bask in the sun, their "seats", as they are termed, being clearly marked.[14]

Sometimes the hare may even go as far as a little excavation to furnish itself with a snug resting place. During the winter of 1911-12 Barrett-Hamilton found that several near his home in Co. Wexford had scooped out hollows for themselves in an arable field of clay soil, and crouched in these with their backs sheltered by some 15 cm (6 inches) of excavation. Similar scrapes were reported in the mud flats beside North Bull Island in January 1955.[234] Whether the animals ever go the whole hog and dig full-blown burrows for themselves is dubious, despite an unverified description of such activities on the Mullet Peninsula in Co. Mayo dating from 1840. As the burrowing of rabbits was destabilising the dunes there, the landowner allegedly determined to exterminate the rabbits and introduce hares

....which, he knew, or thought he knew, would not burrow; but here he was mistaken; for the animal soon found that it must leave the district, or change its habit; for if, on a winter night, it attempted to sit in its accustomed form, it would find itself buried, perhaps twenty feet [6 m], in the morning, under blowing sand....Accordingly, the hares have burrowed; they choose out a thin and high sandhill....Through this Puss perforates a horizontal hole....with a double opening; and seating herself at the mouth of the windward orifice she there awaits the storm; and as fast as her hill wastes away she draws back, ready at all times to make a start, in case the storm rise so as to carry off the hill altogether.[267]

Those readers who have started a hare from its form will have found that often, after a brief initial dash, the beast will take its time or, as Moffat put it

When disturbed - if there are no dogs among the party - the hare ambles off in an amusingly leisurely fashion, and with an appearance of the utmost unconcern - generally sitting down after the first few paces to look round at her disturbers, but soon renewing her retreat - which is not in a particularly straight course - without any appearance of alarm or hurry, such as a rabbit would present in the short rush to its burrow.[222]

The hare most likely feels secure in relying on its swiftness to escape almost anything, once a decent initial distance has been established between itself and potential danger. Although greyhounds can outpace a hare on unbroken ground, most dogs have no chance of catching one on rough terrain once it is ahead by that critical distance. Indeed a dog which bothers to continue the chase in such circumstances simply makes a fool of itself. I have witnessed this myself, the hare in question frequently stopping and sitting bolt upright to look back, almost as if to tease its pursuer. Although a hare in its form must be aware of an approaching man, it may often wait until he is almost on top of it before moving off, doubtless counting on its speed should a break become unavoidable. It is, however, also noteworthy that although a hare in the open will see a moving man at a few hundred metres, it may fail to notice a motionless observer, even if he is standing in the open, and may approach him quite closely.[21]

How fast a hare can run must vary considerably with local conditions. Barrett-Hamilton, both a mammalogist and a keen follower of coursing, estimated a speed of 48 km/hour (30 mph). Mr J.L. Desmond, current Secretary to the Irish Coursing Club is more conservative and estimates the top speed at approximately 32 km/hour (20 mph). Barrett-Hamilton also mentioned that a noted steeplechase rider many times "turned" a hare on the Curragh, Co. Kildare, after about a quarter of a mile's [400 m] gallop on a chaser. Unfortunately he failed to mention how far the hare was ahead when the race started, although it may be deduced that a hare is rather slower than a racehorse.[21]

Barrett-Hamilton also investigated the hare's leaping ability in the following ingenious manner. In December 1890, when there was a rare fall of snow at his home - Wexford being renowned for the mildness of its climate - a hare happened to cross his front lawn. With great presence of mind, he shouted and sent one of his dachshunds after it. The uproar from their united efforts sent the beast off at speed. He then examined its tracks and measured the lengths of successive leaps to the nearest inch (2.54 cm), obtaining the following figures, which I give in centimetres to 5 cm: 230, 115, 230, 115, 220, 105, 155, 110, 220, 120, 150, 305.[18,191] So the Irish hare is capable of at least a 3 m long jump.

Stringer was certain that, in the course of its rambles, a hare will avoid, as far as possible, unnecessarily wetting itself and, should this occur, will often lick itself dry.[341] So it is all the more surprising that the animal sometimes shows little reluctance to swim, which it does with its ears laid back as when running at speed. When frightened, hares have been seen both to cross the tidal estuary of the River Suir, Co. Wexford (nearly 100 m), and to the lake shore from the islands in Lough Inagh, Co. Galway (several hundred metres).[21] Thompson, when out one day with a friend collecting marine animals from the edge of Strangford Lough, started two hares which had been lying on seaweed-covered rocks far out from the tide line. He reckoned that one of them would have been cut off by the advancing tide within a few minutes, and would have remained so for 12 hours, unless it swam ashore. Another of his friends, whilst quietly fishing in a river, watched a hare, which, quite undisturbed, entered and swam across a pool although, by going only a short way lower down, it could have gone over bone dry. On the other hand, a gamekeeper told him of hares often crossing a millrace by running over a wooden pipe that spanned it, rather than by swimming, or jumping it, which they could easily have done.[350] The most remarkable reported instance of swimming was of a doe which, in order to attend to the needs of her three leverets, each night swam to them on an island in an artificial lake in a demesne on the outskirts of Dublin.[78]

The Irish hare has a prolonged breeding season. During the winter of 1972-73 various coursing clubs in Cos Galway, Clare and Limerick, kindly allowed me to take away for dissection a total of 51 hares killed at their meetings.[103] The bucks were fecund from December onwards and at least some does were either pregnant or in breeding condition by January. I was also able to estimate the litter size at two or three from four females. Barrett-Hamilton took it for granted that hares bred from at least March to June, but in the district around his home he also found either young, or suckling or pregnant females, in January (including "rabbit-sized" leverets towards the end of the month), February, August and October. He also concluded from his observations that there were normally either two or three in a litter.[21] Thompson considered that the usual number was three, although some gamekeepers informed him that there were rarely four or more.[350] Wolfe tells me that his dog once found a litter of four in short grass. Stringer reckoned that the animals bred from February until November. To summarise, therefore, it appears that young may be born as early as January until autumn, when births probably tail off in October, or possibly November. As gestation in the mountain hare is 41-44 days,[151] and presumably of comparable length in the Irish hare, and as leverets may be born in January, actual mating can occur as early as December. So breeding, either as mating or giving birth, may extend from December right through to at least the following October. The usual litter size is evidently two to four.

The young in a litter are not always found together, sometimes being dispersed by their mother to individual forms. Leverets above ground are vulnerable and leaving each in a separate form means that only one will be lost through any single act of predation.[21] There they remain for most of the time during the next few weeks, the doe visiting to suckle them. As Stringer remarked

> The hare doth not lie or sit with her young, but comes to them and gives them suck, and so covers them with moss or fogg [a type of grass], her own fur, old grass, or what happens to be nigh the place where she hath them....In the open, plain country and no covert, they have their young in old caves, under hollow stones, in the bottoms of old walls, or any such

like hole or vault that she can find in February, March, April or May. In the other months she commonly breedeth or kindleth in meadows or high rank corn, fern or lying rushes growing up against hedge-bottoms.

Leverets, unlike young bunnies, are precocious. Thompson noted that young extracted from a freshly-killed pregnant doe in late term had their eyes fully open and "within the first hour of their untimely birth" were able to run about.[350] Barrett-Hamilton observed a captive doe to be energetic in defending her litter with both teeth and claws and twice saw a doe in a field chasing away crows, as if she resented their proximity to her offspring.[21]

The madness of the March hare is proverbial and in the Irish hare such apparent insanity takes the form of boxing, kicking, bucking, dodging, leaping sideways, mutual sniffing nose to nose and rushing madly round in a circle. Such behaviour, which is an expression of aggression, occurs between males competing for females in heat, and between the sexes, when the female is almost ready to mate....but not quite. It is perhaps more marked in spring, but, being associated with mating, it is by no means restricted to that season.[21,222]

Wolfe recorded one type of mating behaviour which had previously been undescribed for the Irish hare, or indeed the Arctic hare in general. This was *flehmen.* In many mammals the sense of smell is not restricted to the nose. There are also blindly-ending tubes elsewhere in the head, called collectively *Jacobsen's organ*, lined with scent-sensitive cells, which usually open into the roof of the mouth. Jacobsen's organ serves to sample odours breathed through the mouth or brought in by the tongue. Very often these are smells arising from the scent glands or urine of other members of the species, especially those odours which reveal to males that a female is on heat. When Jacobsen's organ is thus employed, there is usually a peculiar curling of the lower lip, called *flehmen,* as the odour is breathed or brought in with the tongue. Wolfe saw such behaviour seven times in male hares on North Bull Island. In each case a male appeared to be following a scent trail and had his head near to the ground. Every so often he stopped, lifted his head, turned down his lower lip and licked his upper lip and muzzle. All of these bucks were obviously particularly intent on what they were doing and sometimes came close to observers in full view without apparently noticing them. Wolfe was able to track three of them, all of which eventually tried to copulate with does.[380] Since completing his PhD thesis, he has also observed similar behaviour in Co. Westmeath.

"Madness" is most spectacular amongst large groups of hares and the Irish hare has a justified reputation for gregariousness. Thompson repeatedly saw from one to 300 in some demesnes in the north of Ireland. On one property in Co. Down they were so numerous, and thus caused so much damage, that they had to be culled and "on several occasions" were "sent into Belfast by the cart-load".[350] In the 1960s there was an impressive herd at Belfast International Airport, which then amounted almost to a tourist attraction. When the Mammal Society held its annual conference in Belfast in 1967, I led an excursion there where members counted over a hundred.[10] Deane saw over 300 in 1969 and states that there had been hares living on the grass between the runways since 1918.[70] In 1976 he asserted that numbers had dwindled, suggesting that this was because of a rise in the number of jet aircraft, and the consequent increase in noise.[73] Be that as it may, hares have been just as numerous there in recent years. Professor Ian Montgomery told me that he counted 178 while his plane was taxying before take-off in the middle 1990s, and, in answer to my enquiries in 2000, I was told by fire staff that there were "hundreds" of hares there. When Belfast was served by the older airfield at Nutts Corner, herds of upwards of 200 hares lived there too, but these dispersed after it was closed down in the early 1960s, and the beasts were no longer protected by the rigid restrictions on trespass necessarily imposed at airports.[70]

In 1979 the Irish Biological Records Centre (Chapter 1) reported the Irish hares as "common and widespread, occurring both on lowlands and mountains".[243] There have been several systematic attempts to find out just how common in different habitats.

The earliest work was on the windswept blanket bog and low hillsides of north Mayo.[366,368] From 1967 into the early 1970s, Mr P.J. O'Hare, of the Agricultural Institute (then An Foras Talúntais), directed research on grouse around the Institute's Peatland Experimental Station at Glenamoy, both on native bog and on areas of it that had been transformed into agricultural grassland in the 1950s by drainage, application of lime and fertiliser, exclusion of sheep, surface seeding and the planting of shelter belts of shrubs and lodgepole pine. The total area of study extended over some 100 km².

The staff, with the occasional help of some Scottish zoologists, counted grouse, and incidentally other birds and mammals, on the bog by flushing them with trained pointer and setter dogs. The density of hares there was low, averaging about 1/km². The hares on the grassland were conspicuous and could be counted from cars on nearby roads. Here numbers increased dramatically after treatment to 125/km², before settling down to around 40-50/km². Since the hares were not flushed, even these figures may have underestimated the population density. On the other hand, it seems possible that individuals from the surrounding bog may have been visiting the treated areas to feed. In which case these figures might have overestimated actual numbers. It is clear, nevertheless, that the artificially-created grassland suited the animals very much better than blanket bog.

Using binoculars, Dr John Whelan, of University College Dublin, counted Irish hares on a mixed farm near Celbridge, Co. Kildare. This he did at first light 4 days a month over the whole of 1982. So as far as possible to nullify the effects of adverse weather, which might have restricted visibility and discouraged the hares from emerging from cover, he went out only on dry mornings with the wind no more than a gentle breeze. However, the weather still apparently influenced the results, numbers being particularly low in January-February during a heavy snow fall, and in October in an unseasonably cold spell when the hares were still in their summer coats. Then many of them had probably retired to shelter by dawn. The average density based on all counts was 4.0/km². However, this underestimates numbers because there was appreciable variation between counts in the same month, from which one might deduce that, during the lower counts at least, some individuals were in hiding. Anyway, simple observation must be expected to give a lower figure than flushing. If we consider the maximum counts each month as a minimum estimate of population, then on March, April, May, August and December the density was at least 6.6/km². The highest count was in December, corresponding to a density of 8.1/km². It seems reasonable to infer that for most of the year the actual density was 7/km² or more. The animals' only apparent habitat preference was for short grass where there were no farm animals. Whelan never saw hares grazing the cereals grown on the farm.[372]

Rebecca Jeffrey,[169] who counted hares ten times on the farm in Co. Meath over a full 12 months in 1992-93, recorded densities ranging from 4.7-17.6/km² and averaging 12.4/km², figures not too dissimilar from Whelan's. Nevertheless, her main study area was on farmland on the Wexford Slobs which was 80-90% carefully managed pasture, much of it grazed in summer by livestock, with the remainder mostly barley and fodder-beet. Like Whelan, she also did her counts at first light with binoculars in good weather. During the summer she operated on foot, spotting the animals from vantage points, but also walking through tall vegetation to raise any that had might be hidden there. From October to April she was obliged to view the hares only from a four-wheel drive vehicle, so as to minimise disturbance to the Greenland white-fronted geese *Anser albifrons,* several thousand of which winter on the Reserve each year. Altogether, between June and October 1994 and between May 1995 and April 1996, she checked numbers 31 times. Densities varied between 24.5-129.4/km^2 and averaged 50.5/km^2, populations comparable to those on the treated areas at Glenamoy. Like Whelan, she found that the beasts preferred fields with short grass that were free of livestock.

In the mid 1970s it was estimated that there were 50-70 hares (17/km^2) on North Bull Island[251] which, as mentioned in the last chapter, lies in Dublin Bay and is mainly saltmarsh, dune grassland and a golf-links. The island is much used for recreation by Dubliners. The population increased to some 150 (51/km^2) in the mid 1980s, but over the period that Wolfe counted the hares there, numbers fell from 50-55 (17-19/km^2) in December 1990, to 15-20 (5-7/km^2) in April 1994. Even though the hares were reproducing, Wolfe saw very few leverets and found no evidence of juvenile hares growing to maturity and thus replacing adults which died, let alone augmenting the population. He considered that human disturbance, which alone probably had a greater impact than predation and availability of food and cover, was the main problem, and especially uncontrolled dogs. Some Dubliners even brought greyhounds onto the island to course the hares. Such annoyances probably tended to force does to deposit leverets on the least disturbed habitat, the saltmarsh, where they would be likely to have died from drowning or chilling.[382]

The above results on population density, all that are from the Republic, indicate that Irish hares are scarce on blanket bog, commoner on mixed farmland and commonest on areas of the best managed pasture, especially where there are no farm animals for at least part of the time. The highest densities were at Glenamoy, on experimental grassland where there were no sheep, and on the Wexford Wildfowl Reserve, which is stock-free for part of the year and is otherwise managed for the benefit of wildfowl and, to a lesser extent, hares. Perhaps hares, being of a timorous disposition, are uneasy in the proximity of any large animal, including humans.

These apparent habitat preferences are by no means the whole story, as demonstrated by a hare survey by Karina Dingerkus over Northern Ireland in 1994-96. This work was necessarily restricted to winter and spring, when cover was minimal and hares easiest to see.[80] Almost the whole area of Northern Ireland has been surveyed for land use and local terrain classified into one of 23 different types. To find out which of these land classes were most favourable for hares, she choose 150 1 km squares, randomly spread over the whole of the six counties, but so that each land class was represented by 5-8 squares. She then found out which squares contained hares and which did not. Like a lot of ostensibly simple research, this involved an enormous amount of hard work; fortunately she had a little assistance from volunteers. In each square her approach was two-pronged. First, she ruled a 1 km long line through the square on a map and then walked along this in the field checking for sightings of the animals themselves or indisputable signs of them, such as droppings, tracks and forms, which she became expert at recognising. If the landowner was certain that he had seen hares recently in the square, she also accepted this as evidence of their presence. Anyone who has tried to walk in a straight line across farmland in Northern Ireland, most of which consists of numerous small fields surrounded by hedges which are often thick, will appreciate the effort involved in this. Her second approach was to examine each field in each 1 km square individually, to note the habitat in it and to check again for the presence of hares. As this was an even more intensive form of survey, it is hardly surprising that it tended to turn up more hares, or signs of them. She made no attempt to estimate hare densities, but she saw no more than seven hares in any square.

Of the 150 squares, she found hares in 97 (65%). The overall results on habitat preference are most easily appreciated if the 23 land class types are amalgamated into just three; these, together with the corresponding percentage of squares in each with hares were as follows: lowland with intensive agriculture 51%, lowland and upland with less intensive agriculture 73%, and upland and mountain with minimal agriculture 88%.

If one equates her third category with the blanket bog around Glenamoy, such findings might at first seem to conflict with those in the Republic. However, there was almost certainly some bias in her results, because she found it much easier to spot hare droppings against the vegetation on heath and bog, which typifies the third category, than on good-quality pasture, where they tended to be hidden, even in short grass. The extent to which signs alone betrayed the presence of hares in squares on mountain, compared with all squares, is shown in Table 22. Clearly the relative ease with which droppings can be distinguished there may have resulted in the relatively high overall percentage of squares with hares for upland with minimal agriculture.

Table 22 - Percentages of all 1 km squares surveyed by Dingerkus in which the presence of hares was determined in various ways compared with corresponding results from squares on mountain alone.

Evidence	All	Mountain
Sightings	33	33
Signs only (droppings etc.)	14	42
Additional reports from landowners only	17	8
Absent	36	17

A further relevant factor is that the vegetation on the blanket bog at Glenamoy was dominated by purple moor-grass and bog-rush, both of which probably provide inferior feeding for hares, together with only a little dwarf heather,[367] which is much more nourishing. Dingerkus particularly remarked that many of the uplands in her study were dominated by heather. Mr Paul Hackney, of the Ulster Museum, confirmed to me that some high ground in Northern Ireland is indeed dominated by heather, but that the dominant plants in others are bog-rush and purple moor-grass. The fertility of even heather-dominated moorland varies and may depend on the nutrients from underlying rocks. This in turn may affect hare numbers. Thus on heather moors in Scotland, populations of mountain hares are very variable, averaging $46/km^2$ on fertile moors over alkaline rocks down to $3/km^2$ over acidic rocks.[367] Therefore the abundance of Irish hares on marginal land and upland is also likely to be very variable, and dependent on local conditions.

As a result of many conversations with landowners, Dingerkus was of the opinion that hares had declined in Northern Ireland in the previous 10-20 years. This was also the view of people in a third of the squares she surveyed, and in 11% hares were believed to have become extinct within the same period. If this is correct, it increases the significance of her results. For hares tended to be scarcer in areas where agriculture was most intensive. Modern agricultural practices may affect hare populations adversely. For one thing there tends to be less and less cover, as more and more rough ground is put under the plough and hedges are grubbed up. Many of Dingerkus's sightings were of individuals that she flushed from patches of tall rushes. Furthermore, the switch from hay to silage means that grass is cut earlier in the year and more often. In Northern Ireland in the mid 1950s 2,230 farms were producing silage but by 1994 20,000 were involved - about 68% of working farms. As leverets, unlike young rabbits, commonly rest in forms above ground, they are highly vulnerable during grass cutting. Increased use of pesticides and herbicides may also be bad for hares. Evidence one way or the other is difficult to come by, but there is one recorded instance in Britain of dead brown hares being found in a recently-sprayed field.[80]

What overall conclusions can be reached from all of this work? Hares are probably commoner on agricultural land than on unfarmed uplands, but only where agricultural practices are conducive to their survival. They also prefer to graze on short grass when there is no livestock present. Intensive farming and even intensive recreational use of an area by the general public appears to affect numbers adversely. Given favourable conditions, densities upwards of 50-100/km^2 are not impossible. At Belfast International Airport all such requirements are met, besides negligible human disturbance, so it is no surprise that the animals are so numerous there.

At least one other habitat seems to suit Irish hares well, namely forestry plantations in their early stages, although there are no figures on population densities. Here the animals can be a pest because of their habit of biting off the tops of saplings.[247] In the 1960s there used to be regular shoots to cull hares on forestry properties in Northern Ireland where there were young trees. I attended one such myself.

Both Jeffrey and Wolfe caught hares on their study sites and attached individually coloured-coded ear tags to them, which allowed them to be recognised individually at a distance, and radio collars to some, which were then radio-tracked. Most of the animals were caught by being driven into nets. The type of net used was 100 m long and had a cord running continuously along the top. It was suspended at intervals along a row of forked sticks. When a hare ran into the net, the net unhitched from the fork above, and fell, trapping the beast, which was then placed in a sack to quieten it until it could be dealt with.

IRISH HARE

Jeffrey determined the home ranges by radio-tracking two does (10.6 ha, 14.6 ha) and two bucks (29.9 ha, 29.9 ha) in Wexford;[169] Wolfe successfully tracked four males (average 46.0 ha, minimum 30.6 ha, maximum 70.0 ha) and four females (average 21.5 ha, minimum 11.9 ha, maximum 49.2 ha) on the North Bull.[383] As explained in Chapter 1, home ranges of male mammals are generally bigger than those of females of the same species. When radio-tracking an animal, initially the estimated home range becomes larger and larger as more and more fixes are taken and the animal reveals more and more of its range. Usually, eventually, the area no longer rises with additional fixes - this is clearest when the data are plotted on a graph - and the estimate reaches a maximum, or what mathematicians call an *asymptote:* and this is the true home range. All of the estimates by Jeffrey and Wolfe were obtained from asymptotes. Even though Wolfe obtained more fixes and generally tracked for longer, in some cases up to a year, there were no significant changes in the ranges with extended time. So once a hare had settled on a home range, it stayed there. However, the animals tended to occupy somewhat different parts of it in during the day than in night-time, although the day and night ranges always overlapped, by between 20-84%. Since the ranges were smaller on the Wexford Slobs, it rather looks as though conditions were better for hares there, because an individual must have been able to obtain everything it needed in a smaller area. This is borne out by the greater number of hares per unit area. The home range sizes in both studies were smaller than those for mountain hares and indeed Arctic hares in general elsewhere, doubtless because most of them live in much harsher conditions than prevail in lowland Ireland.[383]

Both Wolfe[382] and Jeffrey[169] watched their hares over long periods to see how they spent their time. This sort of work is extremely tiring and impossible to do accurately continuously for anything like all of the time. Over any given period of observation, therefore, they sampled activity over an extended series of much shorter intervals. The two adopted somewhat different approaches. Wolfe concentrated on a male and female that he had radio-tagged, following them in the open and noting what one or other was doing during an hour and a half either side of midday, midnight, sunset and sunrise. Through the hours of darkness he used an image intensifier or "sniperscope" to watch them. Jeffrey selected a part of her study area where there were nearly always at least half a dozen hares, and she observed a selection of the ear-tagged ones there from a hide from dawn to dusk. Parts of the daily cycle were thus missing from each study. Nevertheless, both sets of results either overlap or dovetail well. It should be borne in mind that, although I generalise below, all of the results are essentially from grassland and might differ on other habitats.

Irish hares apparently spend most of their time resting or feeding and these activities together occupy the majority or their time. Overall about 40% of the time is taken up with crouching or, much less often, lying down and apparently doing nothing. During the day this figure rises to 60% and, in late morning and early afternoon, there is very little activity at all. In contrast, around midnight the animals are usually at rest only 10-20% of the time. So Irish hares are most active at night, although they also keep fairly busy in the early morning and evening.

They feed for around 30% of the time, although this rises to more than 50% around midnight and is usually negligible at midday. The next most time-consuming activities, filling between 5-10% of the time, are, first, moving - walking or running - and, second, remaining stationary but alert with the ears cocked, either sitting up, standing on all fours or erect on the hide legs, looking around. The last is often the posture initially adopted when an animal is disturbed. Other activities, each taking up much less of the animals' time include grooming, sniffing food, scent-marking and refection - hares, as well as rabbits eat their own faeces, and so give the food a second run through the gut to improve its digestion. After the first passage, the faeces are soft, moist and covered with mucous. It is only following the second treatment that they emerge in their familiar form: firm and relatively dry. Scent marking is mainly done with the chin. Both rabbits and hares also have a scent gland there and the secretions from it are rubbed off on various objects in the animal's home range.

Although "madness" is the sort of behaviour that one tends to hear most about, no more than 1% of an adult hare's time is given over to social interaction of any sort, including mating. Most of it is occupied in a much more prosaic way in resting and feeding.[169,382]

In continental Europe the food of the Arctic hare is predominantly the leaves, twigs and bark of shrubs, but such a selection is quite possibly largely ordained by the vegetation in which the animals live there, and especially because of prolonged snow cover, which hides the herb layer and obliges the beasts to browse on shrubs. On moorland in Scotland, heather, again a shrub, is the staple item on the mountain hare's bill of fare, but this too is most likely to be because heather is the dominant plant and snow lies for so long in winter.[345] So what happens in Ireland where snow rarely lasts for long? Does the Irish hare feed mainly on shrubs here too? There have been several studies of the diet by analysis of droppings collected from assorted habitats, usually in every month of the year. The overall findings from most of these are displayed in Table 23.

Table 23 - Summary of the main plants eaten by Irish hares in Connemara, on North Bull Island and on the Wexford Slobs. For an explanation of the terms monocotyledon and dicotyledon, see Chapter 10. + = present.

Food item	Connemara					North Bull	Wexford
	Bog 1	Bog 2	Bog 3	Rough grassland	Machair	Coastal grassland	High quality grassland
Ferns	+	+	-	-	-	-	-
Mosses	-	-	-	-	-	+	-
Rushes	+	+	+	-	-	+	§1
Sedges	36	26	36	8	6	1	-
Grasses	25	45	30	71	63	94	97
*Other monocotyledon	12	10	9	9	6	-	-
Dicotyledon	27	19	24	11	25	4	2

* mainly unidentified and probably predominantly grass or sedge.

§ rushes and sedges combined.

Dr Denis Tangney investigated the diet on five areas of Connemara, Co. Galway, at NUI Galway.[345] Three of these were on bog, one on rough grassland and one on *machair,* the calcium-rich, coastal grassland, usually kept short by grazing, which is typical of much of the western seaboard of Scotland, but which also appears along the west coast of Ireland.

As Bog No.2 was only 0.6 km from the rough grassland, it is entirely possible that the hares there were grazing the grassland as well. This doubtless accounts for the much higher percentage of grass eaten there than on the two other bog sites. Bog sites 1 and 3 probably give a more accurate picture of the diet of the Irish hare on bog. There grass contributed no more than 30% of the food. Readers may remember that the analyses of rabbit droppings in the last chapter (Table 19) showed that grass was always by far the largest element in the rabbit's diet. It is clear from Table 23 that, if necessary, Irish hares can rub along with much less grass. On bog, sedges were at least as important to them, and about a quarter of the food was dicotyledon. Of this, three-quarters is shrub, particularly heather, which was abundantly available on the bogs. In fact botanists had given Tangney's bogs a thorough examination, so the vegetation was well documented, and most of the dominant plants, such as heather, white beak-sedge, bog-cotton, bog-rush (both of which are sedges) and purple moor-grass were all well represented in the droppings, hinting that the hares were not feeding too selectively and could get along well enough by browsing most of the plants on the bog. Of course this is only to be expected of a species which, elsewhere in the world, in the persona of the Arctic hare, can survive perfectly well on tundra. The animals on the bogs ate rather more heather during the winter, when it would have provided living tissue at a time when many herbs would have supplied only dead material above ground. On the rough grassland the hares also ate heather from the surrounding bog during the wintertime, presumably for the same reason. In summer there, however, when fresh grass was is steady supply, they apparently lost interest in heather completely.

On the rough grassland the hares consumed much more grass than on bog, as they also did on machair. On the latter dicotyledons were just as important as on bog, but here they were mostly herbs, which, because they die back in winter, were then taken more sparingly. The only shrub available was thyme and, as you might expect, because it is a source of fresh tissue throughout the winter, the hares browsed it more often then. Overall thyme contributed 6% of the diet. Obviously the adaptable Irish hare is quite happy to add this shrub to its bill of fare, although rabbits dislike it. In Siobhan Duffy's analyses of rabbit faeces in the last chapter, thyme comprised less than 0.1% of the food, although it was a common plant at Mweelin and only somewhat less so at Cur.

As numbers of hares on the Wexford Slobs were relatively high, and those on North Bull Island believed to be constrained by disturbance rather than fodder, one must regard these as superior habitat for Irish hares. The menu on both was almost exclusively grass.[169,385,386] So, although the hares can subsist on the poor quality forage on bog, they do better on good quality grassland. Indeed at Wexford no less than 48% of the diet was composed of a single species - perennial rye grass.

This is particularly nutritious and easy to digest, and has a high yield. It is probably the most valuable grass for grazing livestock and for hay, especially on rich heavy soils on lowlands, as on the Wexford Slobs. It does rather less well on infertile soils. It is the most widely sown grass in Britain and Ireland in the formation of new pastures.[159,281] Evidently the hares in Wexford knew what was good for them. They appear to be entirely at home on carefully managed grassland, provided they are undisturbed and have enough cover. If the seeding on the treated bog at Glenamoy was mainly of perennial rye grass, this would go far to explain the high densities of hares there. As in Whelan's study near Celbridge, the hares at Wexford did not bother much with cereals, and barley comprised only 1% of the diet.

As this book goes to press, Dingerkus has published her substantial studies of the diet of hares in Northern Ireland.[81] She gathered and analysed monthly samples of droppings over 2 years at three main sites, besides making smaller collections at 13 others. Her work puts the icing on the cake regarding our knowledge of the diet of the Irish hare. While her results furnish some useful additional information, they also broadly agree with those from the other studies. Dingerkus too showed that grass predominated in the diet except on upland and mountain, where dicotyledons (especially heather) and sedges are together usually more important. Just as on bog in Connemara, on high ground in Northern Ireland the hares grazed and browsed a wide variety of plants. Two of her main sites were on fairly good quality lowland and here the diet was less varied and included 10% and 17% of perennial rye grass. Dingerkus was inclined to believe that the Irish hare requires a varied diet, and that one of the reasons for the putative decline in numbers in Northern Ireland might be that intensively farmed land there does not provide sufficient variety. In particular she mentions the present tendency to sow a ryegrass and clover mix on pasture. Notwithstanding, in view of the large population of hares on the Wexford Wildfowl Reserve, and that almost half of their food is perennial rye grass, such reasoning is open to question.

Apart from man and dogs, the fox is the Irish hare's chief predator. Both badger (Chapter 3) and pine marten (Chapter 5) also take it occasionally, and, in former times, the golden eagle.[350] It is highly probable that leverets, exposed above ground as they usually are, are far more likely to end up as dinners for predators than the adults, except when the latter are infirm or sick. When I was researching on foxes in Northern Ireland in the 1960s, it was not unusual to find the remains of young hares among the cadavers that had accumulated outside earths housing cubs.[111] In order to investigate predation by foxes on hares on North Bull Island, Wolfe compared the fine structure of the fur of rabbits and hares in some detail under the microscope and eventually came up with a way of differentiating the two. Out of a total of 53 fox scats he picked up on North Bull Island, eight (15%) contained remains of hare, but most of these may have been of leverets.[384] He saw a fox pass within 10 m of grazing adult hares at night 12 times. In every instance the hares stayed put and merely stopped feeding. Only once did he see a fox run at a group of hares, which immediately made off. But the fox seemed half-hearted about it - Reynard can move like lightning once he has crept close enough to rabbits - and may merely have been trying to see whether any of the hares was weak enough to justify more serious effort.[382]

We know very little about the parasites and diseases of the Irish hare. Like the rabbit, it sometimes suffers from coccidiosis, but no details are available.[11] As discussed in Chapter 2, those species of mammal which do not have a nest or a regular bedding place usually have no fleas specific to them. Since the hare's form is temporary, hares, unlike rabbits, therefore, do not suffer from fleas to anything like the same extent, and those which are picked up normally parasitise other animals. The rodent flea *Ctenophthalmus nobilis* has been recorded once on an Irish hare and rabbit fleas from time to time.[322] As the latter insects are almost certainly the vectors of myxomatosis in Ireland, they have probably also been responsible for occasionally spreading the disease to hares. Myxomatosis has been reported in Ireland in hares by veterinary surgeons in Cos Cavan and Kildare, the findings being published as "Myxomatosis in the common hare - *Lepus europaeus*". But the common hare in Ireland is the Irish hare and I suspect that it was, in fact, the latter which they examined.[54,375] Deane recorded a further instance.[63]

On 28 January 1955 it was reported that a number of hares were dying from the disease in the Middletown area of Co. Armagh and one sent to the Veterinary Research Establishment, Surbiton, Surrey, confirmed that the Irish form of the variable hare had become infected with myxomatosis, the first instance of a hare contracting the disease in the British Isles....Reports of hares dying from the disease have been received from Co. Tyrone but so far only this one instance has been confirmed.

Whatever the reliability of identification of hares by vets in the laboratory, I am convinced that the number of people who can be depended upon to distinguish brown and Irish hares in Ireland in the field is extremely few. This raises the hoary question of whether there have, in fact, been any brown hares in Ireland at all for many years.

When hares are coursed, the brown hare is said to be faster than the Irish. On the other hand, the latter is supposed to have superior acceleration and to twist and turn more readily to escape the greyhounds. It was probably therefore largely in order to introduce variety into coursing, that hares from Ireland were often introduced to Britain and *vive versa* in the second part of the nineteenth century. A few of the latter imports were of mountain hares from Scotland, but as these and Irish hares are one and the same species and could therefore interbreed producing fertile offspring, the mountain genes would have been swamped with Irish within a few generations. There were also several instances of brown hares being brought in, and the records are summarised in Table 24. These were accumulated by Barrett-Hamilton through correspondence, along with other alleged instances which he eventually concluded were bogus.[20,21]

Most of the brown hares seem to have died out, although they certainly lasted for at least a few years in several places. Some were damaged in transit and this cannot have prolonged their survival or made for successful breeding. The first consignment sent to Powerscourt, Co. Wicklow, of about 40 animals, was apparently transported rather like sardines, ten to a crate, and all arrived blind, Lord Powerscourt making the melancholy observation that "They ran up against trees and fences....and they all died". The second batch, again of 40, each of which was accommodated separately for the journey, did well and bred, so that 2 years afterwards "200 were killed in a day". Several of Barrett-Hamilton's correspondents particularly remarked that the aliens did not mix with the native Irish hares, the two species always forming separate groups, although an instance of interbreeding between the two species has recently been reliably reported in Sweden.[351]

Even though the brown hare does better on lowland than the mountain hare in Britain, initially at any rate the immigrants may have suffered in competition with the established Irish natives, and the only district in which the species appears to have survived well into the twentieth century was at Strabane. When the brown hares were turned

Table 24 - A summary of recorded introductions of brown hares into Ireland in the latter nineteenth century.

Location	Year	Fate
Co. Armagh		
Lurgan	1860s	Still around in 1910.
Co. Cork		
Fermoy	c1858	Probably survived for 20 years then died out.
Ballyhooley	c1848	Did well but were killed off by man.
Castlemartyr	c1852	May have survived for some time.
Castlemartyr	1890s	Probable introduction. No information on fate.
Trabolgan	1882	Died out.
Co. Down		
Copeland Island	1860s	Died out.
Co. Fermanagh		
Cleenish Island	?	"Nearly extinct" by 1896.
Co. Galway		
Salruck	?	Probable introduction. Probably present in 1891.
Co. Tyrone		
Strabane	1876-77	May have survived until present.
Baronscourt	c1876	Probable introduction only.
Co. Wicklow		
Powerscourt	c1865-66	Mainly died out by 1895.

down there, the Irish hare was described as having become extinct locally owing to overhunting by harriers. So the foreigners would have had things more or less to themselves for some time. In the 1930s people in Cos Tyrone and 'Derry wrote to Moffat informing him that the brown hare was locally the common species there. So it was then reputedly well entrenched in mid Ulster.[231] But Moffat's information was acquired solely through such correspondence, and therefore at second hand.

In January 1981 I visited Lifford, Co. Donegal, which is just across the River Foyle from Strabane, to look at the 70 or so hares that were being held in a paddock there by the East Donegal Coursing Club for a subsequent meeting. A good number of these showed whitening on the flanks. The animals had been collected over east Donegal and west

BROWN HARE

Tyrone. Although I was not able to pick out any brown hares, Mr Eamon McColgan, the Club Secretary, told me that five of the beasts were what were known locally as "thrush hares": they had longer ears and a speckled coat, resembling a thrush's breast. These were very probably brown hares, the speckling being due to the dark underfur where it showed between the longer guard hairs - a feature characteristic of the species.[108] This would appear to indicate that brown hares had survived in mid Ulster into the 1980s, although, judging by the relative numbers of brown and Irish hares in the paddock, the latter was the commoner species; but I must repeat that I did not pick out the "thrush hares" myself.

The matter has recently been settled by Karina Dingerkus, who tells me that she positively identified some brown hares that were being held by a coursing club. These had most probably been caught in Co. Tyrone.

Deane asserted that around 1950 brown hares were released in the northern part of Co. Down, but his information was hearsay.[70] I have heard of at least one other supposed introduction since then, and the situation may be complicated because it is alleged that hares may occasionally be moved from one part of Ireland to another for coursing. Those hares that survive at a coursing meeting - usually the majority - are, to the best of my knowledge and belief, released afterwards.

In 1993 the Ulster Wildlife Trust surveyed the presence of the brown hare in Northern Ireland and recorded it as patchily spread over wide areas in all six counties.[13] However, Dingerkus, in her hare survey of Northern Ireland in 1994-96 noted only "three unconfirmed reports" of brown hares out of a total of 105 individual records. Dingerkus was a hare specialist and probably has had more experience of identifying the animals in the field in Northern Ireland than anyone living. I am therefore more inclined to accept her version of the situation. During the Trust's survey it was unknown that the upper surface of the Irish hare's tail could be dark, and those relying upon this to distinguish the two species, as would be natural if the hare was making off with its back to the observer, might thus easily misidentify Irish hares as brown hares.

The only other recent relevant published information - and that negative - comes from Cos Waterford and Cork. There Pat Smiddy, a most rigorous and experienced field mammalogist, stated categorically that of the many hundreds of hares he has seen over some 20 years in these two counties, none were brown hares.[333]

To sum up, there has been only one absolutely reliable record of brown hares in Ireland in the last 100 years and this is almost certainly from Co. Tyrone. There is additional supportive evidence of their presence in mid Ulster, but probably very locally. Records from elsewhere remain dubious.

Chapter 12. Man's Messmates - Rats and House Mice

Then we saw *them,* an' a rale tarrible sight it was to us standin'
out on the ice in mid-lake. Rats, rats be the thousand, comin'
towards us in a great army stretchin' away each side an' deep
as the Shannon. The ice was crowded with them, a black,
running field that changed its shape every minute with
widenin' an' closin', hoppin' along as if mad to get at us.
"The shadow on the lake". Joseph Lyster. *Journal of the Old
Athlone Society* 1974-75.

There are two species of rat in Ireland: the brown or common rat *Rattus
norvegicus* and the black or ship rat *Rattus rattus.* The words brown
and black are slightly misleading, for, as will be explained below, some
black rats are brown and some brown rats black. Consequently, recently
many zoologists have favoured the alternative terms common and ship.
Nevertheless, I have preferred the names pertaining to colour because
they are at present still the ones most often used in Ireland.

The usual colouration of the black rat is dark grey-black above and
pale grey beneath. However, there are two widely recognised variants:
brown with a creamy white belly, and brown with a grey belly, both of
which have been recorded in Ireland,[61,231] although authenticated records
of such varieties are scarce. For the vast majority of people who
encounter a rat have little interest in such nice detail; dead or alive, they
simply want rid of it.

Brown rats are mostly grey brown above with a paler grey underside,
but black forms sometimes turn up. A black variety with a white patch
on its chest was recorded as present in several parts of Ireland by
William Thompson and in 1837 he pronounced it a distinct species, the
"Irish rat" *Mus hibernicus.* The existence of such a uniquely Hibernian
rodent species was more or less accepted until toward the close of the
nineteenth century, when its true nature was acknowledged. In the
meantime a formidable - and useless - literature had accumulated on its
precise relationships. Incidences of white and buff coloured forms have
also been reported in Ireland.[21,350]

The one precise, instantly recognisable and infallible feature in distinguishing between *Rattus norvegicus* and *Rattus rattus* is the length of the tail. The black rat's tail is always at least as long, and often longer, than its head and body. The brown rat's is always shorter than its head and body. The brown's is also stouter and paler beneath, that of the black rat being uniformly dark. Brown rats have a shaggier coat, and the eyes and ears are relatively smaller. The brown rat's ears are furry, whereas those of the black rat are almost hairless. The black rat is also smaller, slimmer and altogether a more elegant creature. The brown rat's appearance is that of a bruiser, which is no defamation of its character. Few unmutilated rats have been weighed in Ireland. Barrett-Hamilton weighed 20 brown rats in Co. Wexford: the average was 392 g and the largest was just over 500 g,[21] which is around the general size for adults recorded in Britain. The maximum verified weight there of 794 g[151] may well be approached occasionally by Irish specimens. As far as I know there are no recorded weights for black rats in Ireland. Elsewhere it is unusual for adults to exceed 200 g.

There are two species of mouse in Ireland, the field mouse *Apodemus sylvaticus,* sometimes called the wood mouse, and the house mouse *Mus domesticus* (formerly called *Mus musculus*). Almost all field mice live out of doors, although visits to buildings are by no means unknown. The house mouse, on the other hand, is essentially a close associate of man - a point that we shall return to at length later. The house mouse is grey above and slightly lighter beneath, which is usually enough to distinguish it from the field mouse, which has a rich wood-brown back and white belly. However, as we shall see, there is a great deal of variation in the colour of house mice in the wild and the range of colours bred by mouse fanciers in captivity is astounding. The largest wild individuals that I weighed, in a total of 146 from a barn in Co. Limerick, were around 20 g, with a maximum of 23 g. For more technically minded readers, I should add that, although I am treating the house mouse here as a single species, modern ideas are that there are, in fact, about eight closely related species, some of which hybridise. Nevertheless, even of these, *Mus domesticus* is by far the most common and widespread throughout the world.[37,151]

Rodents have a large pair of incisors at the front of both their upper and lower jaws. These grow continuously throughout life. So, even if a rodent were not gnawing food, or gnawing something so as to get at food, or gnawing off scraps of material to build a nest, it would have to gnaw simply to keep its incisors from growing so long as to prevent their owner from feeding at all. Rodents are compulsive gnawers and their very name derives from the Latin verb *rodo* = I gnaw.

Like many other rodents, rats and house mice have relatively poor sight but excellent senses of hearing, smell, taste and touch. They can find their way in pitch darkness on a combination of touch, the scent marks previously deposited via their scent glands, urine and faeces, and a heightened sense of *kinaesthesis,* or muscle awareness. The latter enables them to memorise the muscular movements needed to travel along a familiar path. This means that one can escape from a potential predator at great speed, if necessary in absolute darkness, using the "prerecorded" movements in its brain. Internationally acclaimed instrumental soloists, who play long and technically complex concertos from memory, must acquire a comparable facility.

Whereas in Ireland one species of mammal, namely the rabbit, was formerly cultivated in the wild by man (Chapter 10), rats and house mice have succeeded in completely turning the tables and living mainly off humans. They are what are known technically as *kleptoparasites,* a term applied to animals which make at least part of their living by stealing foodstuffs from others. But the actual items that rats and house mice pilfer is only part of the problem. They usually spoil far more food than they devour, either by fouling it with faeces and urine, or through the spillage caused by their gnawing open bags or boxes. They also damage both the fabric and fittings of buildings in their endless quest for the food from man's table. Moreover, their small size and remarkable fecundity have enabled them not only to enter premises with relative ease, but also to build up populations swiftly, and, furthermore, often for them to recover quickly after campaigns of extermination. Because rats and particularly mice are so small, and so have an extremely large ratio of surface area to volume, in order to maintain their body temperature they need to eat far more than larger animals in proportion to their size. They are also customary nest-builders, a nest providing thermal insulation when they are at rest.

While the health hazards caused by rats and house mice are indisputable, particularly in regard to bacteria and food poisoning, in Ireland there are few *infectious* diseases which can be said with confidence to be spread solely by rats. Leptospirosis, also known as Weil's disease or spirochaetal jaundice, is an exception and can nearly always be attributed to the agency of the brown rat. The bacteria which cause this sickness live in the rat's kidneys, but cause the rodent no trouble. They are shed in the urine and may multiply in water. They probably enter the human body through abrasions on the skin or when someone drinks infected water. They then multiply in the blood. The disease in man is characterised at first by fever, muscular pains, diarrhoea and vomiting followed, in a week or so, by jaundice as a result of liver damage. Eventually the bacteria pass to the kidneys, where they may cause death through kidney failure. Leptospirosis is an occupational hazard of sewer workers, farm hands and others whose work may bring them into contact with contaminated water.[353] Freshwater bathers and canoeists are also at risk. Figures for cases in Northern Ireland in recent years are 1996-97 0, 1997-98 1, 1998-99 5; and for the Republic 1997 8, 1998 12, 1999 6.

Another rat-borne disease, which can affect most mammals, is trichinosis, caused by the tiny parasitic worm *Trichinella spiralis*. The adult worm lives in the intestine but the larvae move through its wall, travel in the blood vessels to various parts of the body and then bore into the muscles where they form cysts, in the process causing intense pain and sometimes the death of the. host. Infection occurs when uncooked meat containing the cysts is eaten. The commonest route to man is via uncooked pork, the pigs having become infected through eating rats, which are probably the main reservoir in Ireland .[57,142]

It is hardly surprising that rats are almost universally hated and are also feared by most people, except by certain laboratory technicians, who sometimes develop an affection for white rats. These are an albino strain of *Rattus norvegicus*, which some people also keep as pets. But even wild rats are not always without friends, and Moffat, while fully accepting the pest status of the brown, confessed that

> I once cultivated for some time the friendship of a rat of this species that inhabited a hole in the wall of an out-house, and which, when I tapped with my knuckle at the entrance to its home, would quickly appear, put out its hands, and take a small piece of toast, which it lost no time in conveying to its mouth.[225]

The house mouse is almost equally as abhorrent as rats to many people, and with good reason. It is a restless, erratic and peripatetic feeder, nibbling a little here and then moving on to snack again somewhere else, defaecating frequently and continually dribbling urine, and so polluting far more food that it devours. Premises long infested with house mice bear characteristic miniature landmarks: tiny pillars of dust bonded together with mouse urine, which proclaim the sites at which the beasts habitually relieve themselves. Mouse urine also has a characteristic, unpleasant, musty odour, which may be evident even in moderate infestations.

However, some people will tolerate the companionship of a few mice, small size, nursery tales and cartoon films all having their pernicious influence. Moffat, who was a shy bachelor, poignantly observed of the house mouse that "in small numbers....its company is often a welcome relief to loneliness".[226] Once, when I called to see my Departmental Head at his home in Galway, I found him at dinner and was invited to stay for dessert. The talk turned to mice. Upon his unconcernedly remarking that he had some lodgers of this sort, one, perfectly on cue, emerged from the ceiling, shinned down a pipe with some dexterity - it was then that I first fully appreciated the importance of the tail in maintaining balance during such feats - and disappeared behind the refrigerator.

The original stamping ground of the house mouse was probably in the dry steppes of Iran and Turkestan, and it is still to be found there, living in the open on seeds of wild grasses.[128] As such provender is widely scattered, the animal became adapted to gleaning a little at a time, first here and then there, and, of course, it has retained the habit to this day, even in the midst of plenty. With the cultivation of cereal crops in the Neolithic, the mouse found easier pickings by associating itself with humankind, an attachment which has proved of lasting benefit to the rodent. Man has been unable to shake off his undesirable little messmate, and in company with him, it has spread throughout most of the world.

The house mouse is unobtrusive, nimble, versatile and hardy, and is well able to coexist with man, save in the face of the most rigorous control measures. It can squeeze through an opening as tiny as 6 mm in diameter and does not need to drink, an ability doubtless developed as an adaptation to the dry districts where it evolved. Thereafter, as it dispersed around the world with man, it colonised a wide range of habitats, from coral atolls in the Pacific to near Antarctic conditions in tussock grass on South Georgia.[30] It has, however, an Achilles heel - in the wild it is a poor competitor with many other rodents of comparable size. Where there are such competitors, the house mouse cannot survive entirely independent of man.

In Ireland there have been numerous live-trapping studies of populations of small rodents in woodland, both deciduous and coniferous, but the rodents captured have nearly all been field mice (and bank voles in such parts of Ireland where they occur). House mice in the traps have been an extreme rarity, and even then almost always where there was an inhabited building nearby. Trapping in several other habitats, other than farmland, has generally been no more productive of *Mus*.[117]

The largest batch of data on house mice on agricultural land was obtained during studies of the spread of the bank vole in Ireland. This small rodent, which lives only in the wild and has a particular preference for heavy ground cover, was first discovered in Ireland in Co. Kerry in 1964; it had probably been introduced not many years beforehand. On various occasions from 1969 up to 1982, Chris Smal and I, individually, trapped over a wide area to monitor its subsequent spread, to most of Cos Cork, Kerry and Limerick and parts of Tipperary and Clare. Because the traps, which were usually of the conventional break-back type, had to be set in many different places each day, and on the morrow lifted and put down somewhere else in the light of the catch, we did almost all of the trapping without the permission of landowners. I simply stopped the car, made certain that I was unobserved, nipped through the hedge or fence and laid my traps. Woodland, copse and scrub would have been the best places for the voles, but these are scarce over much of rural Ireland and thus unavailable near most of the places at which we wanted to trap. So we made do with overgrown banks, thick hedges and, more rarely, cover in fields instead. Because the work was unauthorised, and we did not want our traps to be interfered with, it was expedient to keep well away from dwellings and farm buildings as much as possible. Altogether we trapped at 715 different places.

The results from this work give some indication of the prevalence of house mice on agricultural land. We caught a grand total of 1,169 field mice, 24 house mice, 489 bank voles, 53 pygmy shrews and four young brown rats. Unfortunately we did not always record the kind of habitat that we had trapped in, but at five of the sites where we took house mice, surviving notes show that there were buildings within 100 m, which could, of course, have increased the chances of catching them. When the catches for these sites were subtracted, the figures were 1,154 field mice, 13 house mice, 481 bank voles, 53 pygmy shrews and four young brown rats. So less than 1% of the total catch was house mouse, which would seem to indicate that the rodent is scarce on farmland. Incidentally, the figures for shrews and brown rats do not reflect their relative abundances. Shrews eat insects and would not normally be interested in the chunks of coconut that we used as bait; perhaps those that were caught had been attracted to insects which had landed on it. And most rats at large are far too big to be caught in a mouse trap.[117]

In 1985 Mr Martin Dowie and Professor Ian Montgomery, of Queen's University Belfast, live-trapped small rodents, both in summer and in winter, in the hedges in an area of 25 km[2] of good pastoral farmland in the Craigantlet district of northern Co. Down.[232] They also set traps rather less often in cover adjacent to the farms and in the farm buildings themselves, mainly in battery and deep litter hen houses, meal stores, tack rooms and barley lofts. In addition they trapped in the farm buildings again in the winters of 1990, 1991 and 1992. Their results, as numbers of field mice and house mice caught in the farms themselves, adjacent to the farms and in hedges are presented in Table 25.

Table 25 - Numbers of field mice and house mice trapped in an area of farmland in north Co. Down: in farm buildings, adjacent to farm buildings and in hedge-away from farms, in winter and summer.

	Inside farm buildings		Adjacent to farm buildings		Away from farm buildings	
	Field mouse	House mouse	Field mouse	House mouse	Field mouse	House mouse
Winter	3	33	11	-	308	-
Summer	4	11	16	10	362	39

The much greater overall captures of rodents away from farm buildings simply reflects the greater trapping effort there. However, the results were quite clear-cut in regard to where the house mice lived in the different seasons. In winter they were only present in buildings. None whatsoever were caught outside. Moreover 39% of them were fecund, verifying that indoors in Ireland the species continues to breed throughout the year. House mice always predominated over field mice in the buildings. In summer house mice were also fairly common adjacent to buildings and occurred in hedges in smaller numbers; most of the rodents caught there were field mice. It must, however, be pointed out that the ratio of field mouse to house mouse in hedges in summer, of 1:0.11, is much greater than the ratio of 1:0.01 that Smal and I recorded in south-west Ireland, although our trapping did extend over most of the year, including winter. Notwithstanding, the relative scarcity of house mice away from buildings, despite the fact they are common in the wild in parts of the world where the climate is much harsher than in Ireland - but where there is no competition from other rodents - confirms that the house mouse competes unfavourably with the field mouse over most of rural Ireland. It is greatly dependent upon man to survive. It should perhaps be added that the work at Craigantlet is not conclusive evidence that there are no house mice whatsoever abroad in the field in winter. I have trapped them on the verge of the M1 Motorway in the suburbs of Belfast in November and December and, as will be seen later, Irish owls manage to catch them during the winter, so there must be some available outside buildings.

Curiously, on Ireland's offshore islands, house mice seem to be commoner away from human habitation than on the mainland, and this may be something to do with the great hardiness of the species. Although the field mouse is to be found in Ireland almost anywhere that there is cover and the ground is not built up or waterlogged,[111] it is clear that here, and elsewhere over its geographical range, it is most at home in woodland. Our offshore islands are exposed and provide less in the way of seed - the staple diet of field mice, particularly seeds from trees and shrubs - than on the mainland. Generally populations of field mice everywhere are highest in late autumn and early winter. Although breeding usually ceases at the end of October or in November, there is plenty of seed available; the young born in autumn therefore survive better and the population peaks at the close of the breeding season. Numbers then fall through spring and summer until about July, even though the mice commence breeding about April. Few of the young survive in early summer because seed is in short supply. In August-September numbers begin to recover with the fresh crop of seed.[111] On at least some Irish islands, field mice are very scarce indeed in the summer months.[175] I spent a most frustrating summer in 1963 camping and trapping for field mice on islands off the west coast, including stays of a week or so on some during which I caught only one individual, or sometimes none at all. This is probably because the field mouse finds Irish islands a particularly tough environment.

In the circumstances, it is not surprising to find that on Ireland's offshore islands house mice are evidently relatively commoner away from buildings than on the mainland. Table 26 provides the numbers of

Table 26 - Numbers of field mice and house mice mostly caught away from inhabited buildings during trapping studies on various Irish islands.

County & island	Field mouse	House mouse	Habitat	Months
Co. Antrim				
Rathlin	17	4	Available cover	May-Jul
	19	3	Available cover	Sep
Co. Cork				
Bere	1	1	Available cover	Aug
Cape Clear	3	3	?	Aug
Sherkin	1	5	Available cover	Aug
Co. Donegal				
Aranmore	0	3	Available cover and wooded glen	Jul
Tory	17	21	Walls, derelict building, hay ricks, fields	Sep-Oct
Co. Galway				
Gorumna	5	-	Available cover	Jan
Inishbofin	1	6	Available cover	Jul
	11	1	Available cover	Jan
Inisheer	10	3	Available cover	Nov
Inishman	15	2	Available cover	Nov
Inishmore	22	5	All house mice caught near houses	Oct
Co. Kerry				
Great Blasket*	10	-	Deserted village and fields	Jul
	8	-	Available cover especially walls	Jul
Valencia	2	-	Woodland	Aug
Co. Mayo				
Achill	2	-	Woodland and hedges	Jul
	14	-	Woodland	Jan
Clare	6	2	Available cover	Dec
Inishturk	13	4	Available cover	Dec
Iniskea North*	7	-	Available cover	Apr
Iniskea South*	37	-	Mostly walls and derelict houses	Various
Mullet Peninsula	36	74	Rough pasture, orchard, coastal moorland	Aug-Sep

*Uninhabited for many years before trapping.

both species obtained during trapping away from buildings on various islands. Where trapping was in woodland, as on Valencia and Achill, field mice would have been very much on favourable ground, and no house mice were caught. Taking the rest of the data as a whole, but omitting the Iniskeas and Great Blasket, the totals are 177 field mouse and 137 house mice, a ratio 1.3:1 If we ignore the Mullet Peninsula, where by far the largest number of animals were caught and which, although it can only be reached via a bridge, is not usually considered an island, the respective figures are still 141 and 63, a ratio of only a little over two to one.[117] It also appears that house mice are particularly common in the open in summer on islands, reflecting somewhat Montgomery and Dowie's findings on the mainland.

The Iniskeas and Great Blasket had been long evacuated when work in Table 26 was undertaken, and no house mice were caught there. There were certainly house mice on Great Blasket until at least shortly after the human population departed. We do not know for certain whether there were ever any on the Iniskeas, but both were inhabited for many years and there was regular traffic with the mainland, so this was highly probable. It therefore seems that although the house mouse competes rather better with the field mouse on Irish islands than on the mainland, it still requires the presence of humans to survive long-term if there are field mice in residence. When man leaves, the house mouse is doomed.

In their study in north Down, Montgomery and Dowie found that the percentage of farms that contained house mice, and the density of house mouse populations in those which did, altered markedly from year to year. The size of house mouse infestations in buildings in general is very variable, and depends on the extent to which mice are controlled, the amount of food available (which can change dramatically over the year) and the chance arrival of a pregnant female, or a minimum of one mouse of each sex, to start the infestation off. We can envisage house mice in mainland Ireland as a whole as existing in a series of relatively dense populations in and around buildings, with comparatively few animals in between. It must be something of a sweepstake for a mouse leaving, or being ejected, from one population to make its way across unexplored and highly unfavourable terrain, where it might very probably starve or be eaten, to a suitable new home. The risk is enormous but the gain equally so if it ends up in uninfested premises. If the mouse is a pregnant female, or there are only a few individuals in its new home including at least one of the opposite sex, the benefits in biological terms - of passing on its genes to further generations - are potentially enormous. If man accidentally transports a few mice from an infested building to another which is mouseless, so much the better.

Since most house mouse populations in a building or group of buildings are therefore probably derived from a few ancestors, there is likely to have been a great deal of inbreeding. If most populations are more or less isolated and are indeed derived from a few founders which trickle in from other buildings in the district, we could further envisage house mice on a map of Ireland mostly as a series of concentrations. Each of these would contain mostly closely related individuals, and the further apart any two populations were, the more distantly they would be likely to be related. This is the theory anyway, but to what extent does it reflect the actual situation?

This very point was investigated by one of my students, Dr Tony Ryan, who compared the *mitochondrial DNA* of house mice from various parts of Ireland at University College Dublin. The Head of the Department of Zoology there, Professor Eamonn Duke, agreed to oversee the laboratory work if I would supply the house mice. The latter proved more difficult that I had imagined, but I was able to catch the blighters in quantity at an animal food store and a poultry farm in Co. Galway. One of the staff in the former obligingly used his forklift truck to shift palettes bearing bags of animal food, and we then dived for the mice as they scurried out. I live-trapped in the poultry farm. Poultry farms often provide excellent bed and board for house mice, but not around the hens themselves, which, with a sharp jab of the beak, tend to make short work of any mouse fool enough to venture too near. I also caught a lot of house mice in the buildings of the Wexford Wildfowl Reserve on the North Slob, where the stores of cereals for the geese there proved a magnet for rodents. I obtained a few individuals from several other places: in Co. Galway and Co. Clare and from eight different sites around Belfast and northern Co. Down, including batches from four farms in Professor Montgomery's study area. These he most generously caught and donated to the project.

DNA, as you dear reader probably already know, is the material on which the genetic code is carried in each of the cells of our bodies. Most of the DNA in a cell is carried in a structure called the *nucleus,* and half of it originally comes from our mother and half from our father, the contributions from both parents uniting when a sperm fertilises an egg. However there is additional DNA in other tiny bodies in the cell, called *mitochondria*. These are the power plants which burn food and so supply the energy for all the cell's needs. Cells which require a lot of energy, such as those in muscle, have a lot of mitochondria. Because only the DNA in the nucleus of the sperm is transferred to the egg at conception, we get all our mitochondrial DNA from our mother. It passes unchanged, without any contribution from the father, to the next generation. Mitochondrial DNA therefore only changes very slowly from generation to generation, and only by *mutation* - by extremely slow spontaneous change which occurs over many generations. Because of this, and because it is simpler than that in the nucleus, it is much easier to study, although extracting and decoding it is still an arduous and complex process.

The results of Ryan's work showed clearly that generally the further two house mouse populations were apart, the greater the differences in their mitochondrial DNA tended to be, which lends considerable weight to the theory above.[293] Moreover, the mice from around Belfast and north Down, even although they came from eight different places, showed much stronger similarities in their mitochondrial DNA than to that in the mice elsewhere in Ireland.

The results from Montgomery's four farms were particularly instructive. The mitochondrial DNA showed that mice from farm 1 were more closely related to those in farm 2 than to those in farms 3 and 4. Correspondingly, the mice in farm 3 were more closely related to those in farm 4 than those in 1 and 2. Why was this, when all were in the same general area and the farms all approximately the same distance apart? The explanation must be that the two pairs of farms were divided by some topographical feature which a house mouse rarely crossed. In fact a tarred road some 5 m wide separates the two pairs of farms. A migrating house mouse, when faced with the choice of crossing the road, where it would be completely exposed, or proceeding along the hedges on its verges, is arguably more likely to choose the latter option. So the road probably hindered the interchange between mice on either side.

Inbreeding in a house mouse population, the consequence of it having been derived from a few founders, may explain why unusual coat colours are prevalent in some populations. The most celebrated example of this in Ireland has been the house mice of North Bull Island, Co. Dublin. The Bull Island appeared as a result of the accumulation of sand following the construction in 1823 of a breakwater to the north of the Port of Dublin, the "North Bull Wall". In October 1895, H. Lyster

Jameson, whom we have already met in Chapter 7, was taking a walk through the dunes there. He happened to catch a glimpse of some mice of a peculiarly pale colour which appeared to him to harmonise remarkably with the sand over which they were running. His curiosity whetted, he laid traps and eventually secured 36 house mice. The shading of their upper parts varied, at one extreme, from the usual hue of the house mouse to, at the other, a pale sandy colour. He argued that, when the latter individuals went abroad on sand, their coloration could have acted as a protective camouflage against predatory birds, all which hunt primarily by sight. When I was on the island in January 1965, there was no shortage of such fowl: I saw both kestrels and short-eared owls there, and no doubt several species of gull would kill mice given the opportunity.

Jameson found that the mice were living in burrows in the sand, those which he dug out extending, sometimes with side branches, for 60-90 cm underground. He unearthed three nests, two made entirely of marram grass and the third of straw, paper, feathers and "tidal refuse". He found house mice to be very common on the island, the dunes being "riddled with their burrows", and he saw them several times at the high tide line, where they had presumably been foraging. As he made it clear that there were no field mice on the North Bull at the time, the beasts had doubtless settled into a relatively untroubled life in the wild, free of competition.[168]

In 1931 Eugene O'Mahony (Chapter 9), who had developed an interest in the fauna of the North Bull, penned a paper on its mammals.[253] To his surprise, he was unable to trap any house mice. But, by this stage, field mice were common on the Bull and competition may have put paid to the house mouse living in its former numbers in the open. In 1929 Moffat had already remarked that house mice were "better known on other parts of the Bull than among the sand-hills" and indeed doubted whether there were still any on the dunes at all. He queried, in his characteristically gentle but incisive way, what use a pale coat would have been to a mouse anywhere on the island but on sand.[226]

O'Mahony, although he failed to catch house mice on the Bull in 1931, mentioned having obtained specimens "not at all unlike" Jameson's sandy mice on the nearby mainland in the Dublin suburbs of Sutton and Kilbarrack. He could be quite sure of the similarity, as Jameson had deposited skins of his animals in Dublin's Natural History Museum, where O'Mahony was employed at the time. As the mice in Sutton and Kilbarrack could presumably have derived no advantage from a sandy pelage, and it is possible that individuals from these, or other contiguous districts, were founders of the population on the Bull, the appearance of sandy mice on the island in the first place may have been fortuitous.

In 1935 O'Mahony reported at last catching house mice on the North Bull: "three on the dunes and one in the coast guard's cottage" and all of the sandy type.[255]

There was no further work on the animals until the early 1960s when Messrs Fergus O'Gorman and Azzo von Rezzori trapped "some thirty" there. They did not publish their findings in detail but recorded "quite a significant difference in the colouration" from Jameson's mice, the animals being "whiter underneath and browner on the back and not sandy".[251] This seems to be saying that, by this time, "genuine" Jameson's mice were no more.

Originally the island was accessible to pedestrians and motor traffic via a narrow wooden bridge connecting the Bull Wall to the mainland. In the late 1960s a causeway carrying a road was built joining the middle of the island to the mainland. This could have facilitated further invasions of house mice which could have further diluted any remaining sandy genes in the original population.

The story of the North Bull mouse is a fragmentary one. We are unlikely ever to learn how long the sandy mice existed there and of whether their colouration was of any real advantage. Perhaps too much has been made of Jameson's discovery. At the close of the nineteenth century, the concept of natural selection was still novel, and sandy-coloured mice living on sand seemed an almost perfect instance of it. Subsequently some people recklessly pronounced Jameson's mice a new subspecies, or even a new species, both totally unjustified. A species, by definition, is *reproductively isolated* - it cannot interbreed with another species and produce fertile offspring. There was no reason to believe that the animals did not interbreed with other house mice, and the spectrum of colours that Jameson found suggests that they did. A subspecies can successfully interbreed with other members of the species to which it belongs, but members of it are sharply distinct from the others in some way, and usually form a separate breeding group, often by being isolated geographically. Today zoologists are very wary of naming subspecies at all. The fact that Jameson's sandy mice were not clear cut from the others, and were not isolated from them in any way, but formed the extreme of a continuum of variation, meant that even the use of a subspecies was unjustified.

HOUSE MOUSE

O'Mahony's exposure to Jameson's mice may have stimulated his interest in the colour variations of house mice in general. At any rate, later in 1935, on the basis of skins of house mice caught on the east coast, he came up with what he considered the arrival of an exotic subspecies of house mouse in Ireland. These animals were sandy-brown above, although less so than Jameson's mice, white beneath and had white feet. After comparison of his skins with others from outside Ireland - he even travelled to the Natural History Museum in London to make comparisons - he concluded that his "subspecies" was identical to one in Egypt, *Mus musculus orientalis,* as distinct from the common form in Ireland.[254]

In the late 1960s it occurred to me that this whole business was distinctly fishy. How could this subspecies remain distinct if it interbred with the ordinary house mice all around it? O'Mahony "found nothing to suggest interbreeding", but how often does one see wild house mice copulating? Although, of course, they must, and often! Moreover the resemblance to the Egyptian form was no proof of origin. If there were sandy-coloured mice on North Bull Island and in the suburbs of Sutton and Kilbarrack, what was so remarkable about somewhat similar kinds elsewhere on the east coast?

In 1970 I examined O'Mahony's *orientalis* skins in the Natural History Museum in Dublin and found them as he had described. However, there were others, some with *domesticus* colouring on the back and *orientalis* undersurfaces, and *vice versa.* A few *domesticus* specimens had white hair on the feet and one, which had not, was otherwise like *orientalis.* It was obvious that O'Mahony had considered only material typical of the two "subspecies" and ignored all the intermediates.[98] There were not two races at all. When I looked over the only house mouse skins in the Ulster Museum at the time, all from Douglas Deane's back garden in Newtonbreda in the Belfast suburbs, there were both *"domesticus"* and *"orientalis"* types and intermediates. O'Mahony had simply been carried away. In 1937 he described a further two subspecies in Ireland, both just as invalid as *orientalis.*[256]

The matter of colour variations in house mice in Ireland goes back a long time in Irish natural history. William Thompson noted that numerous specimens sent to him from "stack yards" in the North of Ireland "were larger, lighter in colour and more handsome than those found in houses".[350] Even more exotic colours turn up from time to time. There are pale orange and orange-grey skins in the Natural History Museum in Dublin; O'Mahony mentioned a completely black animal;[256] and semi-albinos were recorded from a corn-stack at Katesbridge, Co. Down, which were described as "biscuit-coloured" with "brilliant red eyes".[157,339]

One phenomenon in regard to house mice, which I have never encountered myself, nor indeed have been told of by anyone in Ireland who has, seems to have been common in Moffat's time.[226]

> Most of us have, no doubt, been occasionally entertained by "singing mice", and been struck with their tameness when pouring out their little melody - a melody often listened to, and rightly enough, with pathetic interest. It seems to be now beyond doubt that the song is an involuntary performance, due to some disease or derangement in the respiratory organs of the little singer.

Neither house mice nor rats are indigenous to Ireland and there has been much speculation as to the times of arrival of the mouse and the black rat. In Britain house mice were certainly present in the Iron Age (over 2,000 years before present) and there are many authenticated instances from Roman times. The black rat, like the house mouse, was originally an Asian rodent, hailing from India and further east. There is now good evidence that it had also reached Britain by Roman times, the earliest recorded specimen being from a Roman well in York, which had been filled in about 110 A.D. Remains from the third century have also been excavated at a similar locale at Fenchurch Street in London, from another London site dated as fourth century, and from yet another at Wroxeter dating from the fifth.[15,387]

If the house mouse and black rat were widespread and common in Britain in the early centuries A.D. it would not be at all surprising if both, as incorrigible camp-followers of man, had made their way to Ireland within a few years. House mice were almost certainly abundant in Ireland by the seventh or eighth century, as surviving contemporary legal material particularly stresses the importance of cats in keeping mice away from barns.[174] However, it is by no means certain that black rats were ubiquitous in Britain in Roman times. They might only have been present in a few urban centres. There is a great deal of negative archaeological evidence to suggest that they may have become exceedingly rare or extinct in Northern Europe during the dark ages (500-700 A.D.). This could have been as a consequence of the withdrawal of the Romans, with their urban way of life and central heating, for black rats are susceptible to cold. Furthermore, as has been demonstrated by studies of tree-rings in Ireland, the European climate deteriorated from about 400-800 A.D. The decrease in trade over this period would also have reduced the rodents' opportunities for travel. Reinfestation of Northern Europe, from about the ninth century onward, may have occurred through a combination of an improvement in climate and an increase in sea voyages, among others by the Arabs, Vikings and Franks. From the eleventh century, during the Medieval period, not only did travel and trade further expand, but the human population increased. There is extensive archaeological evidence of black rats in Britain during the Middle Ages.[15] It was therefore quite possibly not until the latter part of the first millennium A.D. that the black rat first made its home in Ireland.

Bones of black rats have been recovered from an early Christian site at Rathmullan in Co. Down[174] and a skull was excavated at Waterford dating from the mid-twelfth century.[204] It seems that black rats certainly became well established in Ireland in Medieval times. In early Irish manuscripts there appears to be only one word for a domestic rodent (*luch* = mouse). The distinction *luch beag* (= small mouse) and *luch frangcach* (= French mouse) - presumably referring to house mouse and black rat respectively - only emerges in post-Norman documents.[174]

Rather less reliable information comes from two literary sources. Giraldus Cambrensis (Chapter 5), writing in Latin of Ireland at the close of the twelfth century, speaks of the abundance of mice of both the smaller and larger kinds, and the latter can only have been rats. However, he also asserts that in the sixth century St Yvorus, Bishop of Ferns, cursed rats and expelled them from the south-east of Ireland, probably because they had gnawed his books.[224] "None afterwards bred there, or could exist if they were introduced". There is also an illustration in the eighth century *Book of Kells* said to be of a cat catching a rat, although both animals are so curiously portrayed as to correspond to no known species, unless the imagination is allowed fullest play. Anyway, even allowing the appearance of a rat in the *Book of Kells* is no proof that it was actually present in Ireland at the time.

The black rat's flea *Xenopsylla cheopis* is the vector of plague and was no doubt responsible for the outbreaks of the "black death" in Ireland in the fourteenth century.

The black rat continued as the only rat of Ireland until the eighteenth century, when it was displaced by the brown. Consequently, very little is known of its biology here. When it occurs in Britain, unlike the brown rat, it rarely burrows or swims and is a more agile climber - not to say that the brown is a poor hand at this - with a tendency today to occur more in the upper stories of buildings. Its replacement by the brown rat in many temperate parts of the world is possibly primarily due to the latter's greater ability to withstand cold. In the tropics the black is well able to co-exist with the brown.[151]

There are intriguing hints that, even before the brown rat had colonised Ireland, the black was absent from some districts. If so, this may have been because it was less tolerant of the Irish climate. O'Flaherty in his *A Choreographical Description of West or H'Ir Connaught* (1684) states that West Connaught "admits no rats to live anywhere except the Isles of Aran and the district west of the liberties of Galway" and in John Evelyn's diary for 15 November 1661 there is a curious statement, made to him on that day by the Duke of Ormonde, that there were in Ireland "not many rats till of late, and that but in one county".[231] A further convincing indication of local absence comes in a letter of 1708 from Thomas Wadman, of Donegal Town, to Dr Thomas Molyneaux, then one of Ireland's foremost scientists.[9] Wadham had "made diligent inquiry about Ratts" in his district and concluded that

....there is considerable tract of Land about Donegall [Town] wherin there is not one Ratt, tho' Ballyshannon....and Killybegs....have enough to send Colonies to adjacent Countries. This is the more strange because Donegall is a sea-port town....but it is not the only town of Donegall that is blessed with the absence of these vermin, the whole parish of Drumholm, and two or three more of the adjacent Parishes, pertake of the same happiness....

The brown rat's ancestral home appears to have been in the steppes of central Asia. Early in the eighteenth century for some reason the beast expanded its range into Europe and thereafter, in company with man, over much of the globe. The Jacobites, and others who disapproved of the house of Hanover, claimed that it had entered England in the baggage of George I. It spread rapidly and largely displaced the black rat.

The earliest date for its appearance in Ireland is given by Rutty in his *Essay toward a Natural History of the County of Dublin* (1772).[292] Since the event was then recent, and must have caused more than a little alarm, his account is perceptive.

BLACK RAT

....here commonly called, the *Norway* Rat, which first began to infest these parts, about the year 1722, and which, tho' it has devoured our pease, grain, &c. has at the same time, in a great measure, rid us of other rats [i.e. the black rat] and partly destroyed the Frogs. It also feeds on fish, nor does it wholly spare Birds, Rails having been devoured by it.

The *Norway* Rat is amphibious, and burrows in the ground, under the water, yet it lives dry as the Beaver, and tho' it dives, it cannot live long under water, its claws being only a little webbed at the bottom, whereas the claws of the common rat [i.e. the black rat] are entirely cloven, and adapted for climbing, not swimming. It has been eaten.

An account in *Walsh's Impartial Newsletter* of 1729 points up the sensational aspect of the debut of the brown rat in Dublin.[233]

This morning we have an account from Merrion that a parcel of those outlandish Marramounts which are called Mountain Rats, who are now here, grow very common; that they walk in droves and do a great deal of mischief - the writer describes how they ate a woman and nurse child in Merrion. People killed several who are as big as Katts and Rabbits. This part of the country is infested with them. Likewise we hear from Rathfarnham that the like vermin destroyed a little girl in the Field; they are to be seen like Rabbits, and are so impudent that they suck the cows - nay abundance of them are to be seen in Fleet Street.

By 1744 the brown rat was probably the common species over much of Ireland, as a statute of that year, offering bounties for killing several animals, includes "water rats, commonly called 'Norway Rats'" but does not mention the black rat, which may therefore have then been of much less significance.[231] Oliver Goldsmith, who was born in 1728 in Co. Longford, remarked in 1774 of the black rat, in his *History of Animated Nature,* that it

....is almost totally extirpated by the great rat....It is become so scarce, that I do not remember ever to have seen one.

If the brown rat did indeed come to Ireland in 1722, then it must have spread and ousted the black rat swiftly over much of the land for Goldsmith never to have seen it. He believed that the brown had first arrived in the British Isles in Ireland in ships that "traded in provisions to Gibraltar".[145]

The extent to which the black rat hung on in Ireland thereafter is difficult to piece together, chiefly because alleged identifications cannot always be depended upon, there having been a tendency to determine species on colour alone. There have been a number of reliable sporadic records and the animal persisted in a few ports up until the latter part of the twentieth century, almost certainly as the result of repeated re-invasion from ships. This explains the alternative name of "ship rat", even though brown rats commonly stowaway too.

In the first half of the nineteenth century, William Thompson received several reports of black rats, but considered only two such reliable: from Portglenone, Co. Antrim, and Cork City. He also obtained actual specimens from Tallaght, Co. Dublin, and Glenravel, Co. Antrim.[350] Subsequently there were records of a litter at Levitstown, Co. Kildare, in 1876, of an individual on the Department of Agriculture's steamer *Helga,* which was believed to have come aboard in 'Derry in 1903, of a colony in a corn-store in Dungarvan, Co. Waterford, in 1911, and a specimen from Baltinglass, Co. Wicklow in 1923.[21,231]

Douglas Deane, writing in 1952, doubted whether the black rat had ever become extinct in the City of Belfast, the Pest Control Officer for the port area having informed him that it had been resident there for at least the previous 80 years (i.e. since 1872).[61] Few people in Ireland living outside Belfast appreciate the vast extent of the harbour there and consequently the potential for settlement of rats from ships. Annual figures of black rats destroyed on docks, quays, wharves and warehouses in Belfast from 1929 to 1937 ranged from 72 to 302 and averaged 202.[213] Deane stated that in 1948 he was able to obtain as many specimens as he wished from the docks and in 1949 received a number from a colony in a factory within a mile (1.6 km) of them.[61] In 1961 the black rat was still reported as "always present somewhere" in Belfast Harbour.[26] With advances in shipping, especially the transport of edibles in bulk containers and rapid discharge, there is now much less of a rodent problem. Mr Terry Nevin, of the Port Health Division, Belfast City Council, told me that there have been no colonies of black rats in the Belfast Harbour area in recent years, although they are occasionally found aboard ships calling there.

Deane also asserted in 1952 that, elsewhere in Northern Ireland, the species was present in 'Derry and Newry and he also had two further probable records for 1950, which had been brought to his attention by the British Ratin Company: at Lurgan, Co. Armagh, in a food factory, and at Tyrella, Co. Down, near a café. He suggested that, as lorries from Belfast docks travelled frequently to both localities, they had probably accidentally transported the rats there too.[61] Other inland infestations may have arisen in a similar way. Black rats were certainly still being reported by the port authorities in 'Derry in 1951[25] and 1961.[26] They were again recorded in the 'Derry area by Rentokil Ltd. in the 1970s and a few were caught in a piggery there as late as 1987.[354] Messrs Randall Crawford, who are responsible for pest control in the port of 'Derry as I write, have come across no black rats there in recent years.

From 1932 to 1937, 71,580 rats were recorded as having been destroyed on the docks, quays and wharves at Dublin, and it is stated that "about 80% belonged to the brown variety", which implies that the black rat was still to be found there in quantity.[213] This is not surprising in view of the continuation of the animal in Belfast for so long thereafter. However, I have been unable to obtain much more information on the frequency of the black rats in Dublin, until recent times. Mr Peter Mooney, of the Dublin Port Company, tells me that they are now a rarity, although there was a sighting a few years ago.

The most recent reliable record of a black rat in Cork City is for 1976.[322]

The colony which probably persisted longest into the twentieth century is on Lambay Island, Co. Dublin. In 1920 there had been a probable record by the Hon. Cecil Baring, the owner, which was followed in 1935 by a specimen which Moffat identified.[231] In the 1960s Fergus O'Gorman told me of seeing them there and in 1985 remains of both black and brown rats were recovered from the pellets of short-eared owls wintering on the island.[361] In the late 1990s Mr Dominic Berridge informed me that there were still black rats on Lambay.

The brown rat - and for the remainder of this chapter this is the species of rat with which we are concerned - is widespread in Ireland but is usually dependent on man to augment its food supply. But because of the sophisticated techniques of control now available, the species is very much less numerous than heretofore. One indication of the abundance of rats in rural Ireland in Victorian Times - perhaps a unique one - may be found in a notebook preserved in the Public

Records Office of Northern Ireland. This records the payments against the rats killed, at one penny (0.4p) each, in Lissadell House and its outbuildings in Co. Sligo, from 20 June 1846 to 30 December 1847.[278] Over this period receipts totalled £32 13s 5d, representing no fewer than 7,841 rats!

Today the brown rat would abound in urban areas if efforts to control and exclude it were relaxed. It is at home in water, as Rutty pointed out, and is particularly associated with sewers and waterways in cities. In the country it is sometimes known as the "water rat" but should not be confused with Ratty in the *Wind in the Willows,* who was a water vole *Arvicola terrestris,* a species which does not occur in Ireland. Rats, incidentally, unlike house mice, must have water to drink. Mice can make do without it if need be but will drink if it is available.

The rat often inhabits farms and is at home in the open in agricultural land, particularly during the summer. In Britain it has been demonstrated that many rats move from farm land in autumn into farm buildings. This is associated with the harvesting of crops, which means less food and cover in the field, and with a drop in temperature and generally poorer weather. There is a corresponding movement out of buildings onto the land in spring and summer.[161] The latter is probably a more gradual, measured and less risky affair than the autumn migration, which is forced upon the animals. They are then moving over unfamiliar territory and are very much open to predation, as we shall see. Mr George Douglas, Pest Control Unit, Belfast City Council, tells me that there is an increase in reports of rat infestations in Belfast with the onset of colder weather in late autumn and Mr Sherwin Curran, of Rentokil Initial Group, agrees.

Unless there is suitable provender at hand, rats are generally absent from moorland and forest.

Rats turn predator with ease and warrens may be infested with them. The dealings of rats with rabbits, especially those not fully grown, were well described by a naturalist living near Ballina, Co. Mayo, in 1886, who was at first puzzled to come across so many young rabbits dead or wounded, with their hindquarters paralysed, or shrieking beneath bushes until

> ....one day I heard a young Rabbit screaming under some briars, and on approaching cautiously....I saw a great Rat at the mouth of a hole, having hold of a young Rabbit across the loins, trying to drag it back into the hole; but on seeing me

the Rat retreated....leaving the unfortunate Rabbit in such a helpless state that I was obliged to kill it....since then I have frequently rescued young Rabbits from Rats, and have been much surprised at the size of the Rabbits attacked, and at the great strength shown by the Rats in holding their victims. About two years ago....I saw an old mangy-looking Rat having fast hold of a half-grown Rabbit across the loins, and struggle as he did most violently, he could not gain an inch on the Rat, which held on like a bulldog, with his feet firmly placed against a bunch of grass....[365]

As far as can be determined, brown rats live on all Ireland's offshore islands inhabited by man, with the exception of Tory Island, off Co. Donegal. Rats too live entirely independent of man on some uninhabited islands, especially where there are ground-nesting birds, such as on the Saltees, Co. Wexford,[282] and on the islands in some loughs, like Strangford, Co. Down.[67] Sheep Island, off Co. Antrim, is uninhabited but holds large colonies of sea birds. Rats lived there too until they were exterminated by the National Trust.[280] Waterfowl nesting on the ground around several of Ireland's inland lakes also provide a living for colonies of the rodents, both young birds and eggs being equally acceptable fare. On Lough Funshinagh, Co. Roscommon, for instance, where eggshells punctured by the rodents' teeth were "lying in heaps", rats were reckoned to be by far the worst destroyers of eggs.[160] Outside the nesting season, on the other hand, pickings for the rodents must be slimmer.

The rat is a versatile and resourceful feeder, readily utilising practically all man's foodstuffs and at times showing little fear of man himself. Dr Tim Sharrock wrote of living on his own in a house on Cape Clear Island, Co. Cork, where the rats ran over his sleeping bag and crept onto the table as he ate his dinner. After chasing them out of the door with a broom, he returned to find others gathered around his plate helping themselves.[299] Thompson described how rats regularly stole food from a ferret hutch, only a metre from where its inmates slept.[350]

Outdoors rats will feed on crops, particularly root vegetables, raid bird tables, prey on anything they can catch and even pilfer from fruit trees. Thompson mentioned a pear tree growing against a house at Holywood, Co. Down, with a well beaten path leading from it to a rat hole. The householders remarked that "about a bushel" [36 litres] of the fruit had been purloined by the greedy rodents during the previous summer.[350] Nor need the fruit be as luscious as pears. In the *Irish Naturalists' Journal* of 1951, are two descriptions from Co. Dublin of rats munching hawthorn berries in the uppermost parts of hedges,[35,290] one

> ....at a place where the road runs through a cutting about ten feet [3 m] deep, with hedges about six feet [2 m] high on top of the banks, I heard a peculiar crunching noise coming from the hedge. On investigation, I found four or five rats in the top of the hedge, busily eating haws.

On occasion the animals will even gobble up flowers. One was seen in a garden in Co. Dublin

> ....sitting on its haunches breakfasting from a clump of pinks about to burst into bloom. Bud after bud was nipped off and held squirrel-like in the fore paws to be comfortably nibbled. On looking at the clump later I found over 100 buds had been eaten....[212]

The opportunism of rats is well illustrated by the observations of a Belfast zoologist who drove regularly from July to November 1974 along the M1 Motorway from Balmoral, in the Belfast suburbs, to Moira, Co. Down. There he often saw rats, apparently quite accustomed to the traffic, either on the hard shoulder or on the central reservation which, in those days, was of earth well colonised with vegetation. There they probably fed on dead birds and grain spilled from lorries, many of which were open-topped at that time. The rodents were rather more plentiful on the verges of the outbound carriageway, the reason doubtless being that lorries travelling from Belfast were more likely to carry a load of grain, whereas those returning were usually empty. The main hitch to such a life might have been a shortage of water and it is pertinent that the largest number of rats he saw at any time - 14 near Lisburn - was shortly after a shower, and that the animals were drinking from puddles.[165]

Perhaps the most curious description of rats' enterprise in trying to obtain food, but one in which the animals often overreached themselves, was recorded in the folklore of Clare Island, Co. Mayo. I have heard a similar tale at Portavogie, Co. Down, and there is also a supposed instance from the island of Dunstaffnage in Scotland.

However, the evidence is short of certainty and the truth of the matter is anyone's guess. Limpets *Patella vulgata* were said to be able to catch rats. Not by outrunning them, but as described by one of the islanders thus.[53]

> Of a soft evening them bornyacks [limpets] do rise up and go travelling a bit over the rocks, and the rat'll come up and slip in her tongue to lick the meat, and the bornyack'll clap down its shell and catch her by the tongue. And I tell you the power of man couldn't loose the hold of the bornyack, and the rat is caught there till the tide comes up and drowns her. I seen them myself hanging by the tongue and they dead.

In view of the rat's manifest economic importance, it may seem surprising that there has been only one serious investigation of the composition of wild rat populations in Ireland. This is because rats are seldom to the taste of people who have chosen to research our wild mammals. The work in question was undertaken by Dr Fidelma Butler at University College Dublin, who studied populations in two pig farms, 8 km apart, in Co. Kildare.[40] The pigs were fed on swill, so there were ample victuals for the rodents in the form of bread, potatoes, offal and what have you. Disused machinery and rubbish heaps scattered around the buildings furnished more than adequate refuges. Butler caught a sample of the animals each month over a full year in break-back and other lethal traps, totalling 366, with about equal numbers of each sex.

She found rats in breeding condition throughout the year, as is usual for populations with excess food in and around buildings. Judging by the numbers trapped, the populations in both piggeries were highest in summer and in late autumn. Butler weighed the bodies "clean", in other words after she had removed the guts. So the percentage weight distribution in Table 27, of the 345 animals which were sufficiently undamaged to be weighed, underestimates full body weight. Generally the larger, and therefore older, rats were in the best condition, as demonstrated by the amount of fat inside the abdomen: 44% of rats of 300 g or more had significant fat deposits, but only 4% of animals of less than 100 g. No females below 100 g were in breeding condition, but about half of those from 100-199 g, although few of these were pregnant. On the other hand, most females of 300 g or more were or had been pregnant. No males of less than 100 g were fecund and only a minority of those weighing 100-199 g; over 90% of those over 300 g were.

Butler was able to check 348 rats for wounds, which she did by both examining the tail and by skinning each animal, turning the pelt inside out and scrutinising it for wear and tear, which is mostly inflicted during internecine strife. Few rats under 200 g, in other words those which were mostly sexually immature or just matured, had been wounded and then the damage was confined mostly to the tail. Most males over 200 g had suffered some battle scars, often on the back as well as on the tail, whereas only about three-quarters of females had. The most battered animals of all were the biggest males.

Table 27 - Percentages of 345 rats from two pig farms in Co. Kildare in various clean weight classes.

Clean weight class	Percentage
<100 g	44
100-199 g	22
200-299 g	15
300-399 g	12
400-499 g	7

Since there were plenty of comestibles throughout the year, the size of these rat populations was not being limited by available food. All rats examined were in a comparatively healthy state and none had an excessive load of parasites. Although there were cats on both farms and a dog on one, they did little to earn their living and seemed largely uninterested in rats. It is also virtually certain that the numbers of rats being born greatly outstripped the numbers of adults dying from old age alone. The rate at which young were being produced must indeed have been impressive: about half of the females were in breeding condition; a female rat should be able to produce young every 3-6 weeks under favourable conditions;[151] and Butler's beasts had litters of 9.0 on average. It can only be concluded that the populations were being limited by space, and by conflict between one rat and another, resulting either directly in deaths, or in ejection of individuals from the piggeries to the much less favourable ground outside.

Even though there was plenty of food, the young rats evidently did not eat sufficient to lay down fat, which suggests that they were low in the social scale, harried by their elders and perhaps often prevented from feeding uninterrupted. Nevertheless, their subservient state was probably achieved mainly without recourse to serious violence, as few were wounded and then only slightly and predominantly on the tail. With sexual maturity there was an increase in wounding and males especially often showed scars of combat, a significant proportion probably sustained in disputes over access to females. On several occasions Butler actually saw large males attack smaller ones and chase them away from females. However, the females did not get off Scot free, and probably led only a slightly less competitive and stressful life. Few rats can actually have survived to breed successfully. Rats are evidently as tough on each other as they appear to be to the world in general. It may perhaps be worth adding here that one distinguished Irish naturalist twice noted a large rat kill a smaller one and observed that when a rat is caught by the leg in a break-back trap it will be "set upon by others, killed and demolished, all but the skin".[23]

Apart from *Leptospira* and *Trichinella,* we do not known much about the diseases or internal parasites of rats and mice in Ireland. The common fleas of the brown rat have already been dealt with in Chapter 2. It also plays host to a species of louse *Polyplax spinulosa.*[257] The house mouse has a flea all of its own, *Leptopsylla segnis,* which rarely parasitises anything else.[322]

Rodents, almost anywhere they occur in the whole world over, are a significant source of food for carnivorous mammals. Despite this, there is no instance on record, either through direct observation, or from analysis of stomach contents or faeces, of house mice having been eaten by any of Ireland's wild mammals, save the stoat.[99] This, yet again, underlines the scarcity of house mice away from human habitations. Even domestic cats, hunting abroad, probably secure only a few. The only information from this quarter comes from near Oughterard, Co. Galway, where one naturalist carefully logged the corpses brought home by Duchess and Blodwyn, two well fed, female, neutered pets, from August 1989 to December 1992.[374] The prey included no house mice whatsoever but otherwise comprised three frogs, 146 small birds of 16 different species including a brood of six chicks, two larger birds of different species which had been dead when the cats found them, two pipistrelle bats, 191 field mice, 93 pygmy shrews and 59 rats. The cats procured no rats from June, July or August in 1989-91, although they delivered a single specimen on each of these months in 1992. Perhaps there was too much cover for the rats in summer. The cats caught rats most often in autumn and early winter when, as discussed above, rats in the field run short of food and cover and may be moving over unfamiliar ground and hence into buildings, and so are especially vulnerable. Incidentally, the above astonishing catalogue gives some idea of the wanton destruction of wildlife perpetrated by some pussies.

It will already be evident from earlier chapters that the mink, stoat and otter all kill rats, that martens may catch them occasionally, and may have slain black rats regularly in former times. The fox is an important predator in this regard. Of 340 stomach contents from adult foxes and 163 from cubs which I examined in one study in north-east Ireland, respectively 17% and 16% contained rat.[97] Consumption peaked in November or December indicating again the vulnerability of rats in the field in late autumn and early winter. In a much smaller study after myxomatosis had reduced rabbit numbers, I found rats in 52% of the scats and stomach contents that I examined.[93]

One of the more unusual Irish mammals said to have killed a rat in Ireland is the hedgehog. At the end of the nineteenth century a tame hedgehog was kept in a yard infested with rats. One day when one of the latter was feeding from a dog's unfinished dinner, the hedgehog was observed to creep stealthily up behind it and then, with a sudden rush, to seize the rat by the back. The aggressor then quickly partially curled itself, presumably to discourage any effective resistance, and then proceeded to eat the unfortunate rodent.[269] Hedgehogs apparently proceed in a somewhat similar way when attacking snakes.[279]

In Ireland, the great black-backed gull occasionally kills rats[38] and the kestrel eats both rats and house mice in small numbers.[101,114] Herons *Ardea cinerea* also sometimes catch rats. In 1986-87 one ornithologist repeatedly saw up to three herons at a time foraging along the verges of a road running between Athy, Co. Kildare, and Stradbally, Co. Laois. There were always rats there at the same time, feeding on the grain and sugar beet which had fallen from passing lorries. The birds were seen to kill the rodents three times and a heron which had expired after being hit by a vehicle contained mammalian hair, mainly underfur, which is difficult to identify, and a few free guard hairs; I judged the latter to be almost certainly from a rat.[140] Caroline Shiel described to me how she watched a heron swallow a rat on the banks of the Garvoge River in Sligo Town. The rodent went in head first and the whole process took about 15 minutes.

The predators *par exellance* of rats and house mice in Ireland are owls, and we know a lot about their diets, largely because of the comparative ease with which they can be determined by analysing pellets.

The short-eared owl is essentially a winter migrant in Ireland. Elsewhere it has a tendency to be nomadic and to settle for a time in places where rodents are particularly plentiful, especially lemmings and field voles (*Microtus* species), both of which are absent from Ireland. Probably for this reason the diet in Ireland varies very much with locality. The owl is again probably making use of local concentrations of prey. At one time and place or another it has been found to be feeding here predominantly on bank voles,[171] birds,[4,362] shrews[328] and, at several places, on rats.[94,361] Rats are locally important because the terrain on which short-eared owls take up residence is often coastal or wet. Whereas the field mouse is the most important prey for other owls in Ireland, it tends to be scarce where the land is waterlogged and is strictly nocturnal, whereas the short-eared owl is largely diurnal. In Ireland short-eared owls catch house mice only occasionally.

Table 28 - Prey of barn and long-eared owls inside and outside the bank vole's range in Ireland as revealed by some pellet analyses. Figures are percentages of total items of vertebrate prey. + = present.

Owl species and inside/ outside range	Prey								Total items of prey
	Pygmy shrew	Bats	Bank vole	Field mouse	House mouse	Brown rat	Birds	Frog	
Barn owl									
Inside range	12	1	19	54	4	7	2	1	5932
Outside range	14	+	-	61	9	8	6	1	1945
Long-eared owl									
Inside range	2	-	7	72	4	3	12	-	1123
Outside range	1	-	-	82	7	4	6	-	2847

Representative data on the diets of long-eared *Asio otus* and barn owls *Tyto alba* in Ireland are provided in Table 28.[55,95,158,326] Rats and house mice together make up over 10% of the preys caught by both birds. Such predation is clearly somewhat reduced where bank voles are available. The barn owl roosts mainly in ruins or old buildings and is more closely associated with man than the long-eared, so it tends to take more of both rodents. Barn owls living in urban areas literally go to town on rats and house mice. Indeed, this may be the reason why they are able to live in towns at all. For instance, of 279 prey items taken by owls living in a church in Mitchelstown, Co. Cork, 34 (12%) were house mice and 118 (42%) were rats.[326] Corresponding figures in 377 prey items in pellets from a roost in the city centre of Waterford were 132 (35%) and 68 (18%) respectively.[360] In both places therefore these rodents together accounted for more than half of the items taken.

Data on the diets of both species of owl in Ireland have often shown a rise in the proportion of rat in the diets in late autumn and early winter.[56,95,119,158] This, yet again, emphasises the vulnerability of rats in the open during this season.

Rats are even more important to owls than the foregoing suggests because a rat provides a bigger meal to an owl than a mouse, bank vole or shrew. On the other hand it seems unlikely that a barn owl, weighing some 280-450 g, or a long-eared owl of 200-390 g[214] could possibly dispatch a full-grown rat of 500 g.

This problem was addressed by Dr Pat Morris, an English zoologist. He established a laboratory colony of wild rats and then killed 89 representing a wide range of sizes. The lower jaw in all mammals is composed of one bone on either side, the dentary, left and right dentaries being mirror images of each other. Morris removed the dentaries from the carcasses, cleaned them and measured their lengths. He then plotted length of dentary against body-weight of rat on graph paper and fitted a line to the data. Given the dentary length of any rat, he was thus able to read off its approximate body weight. As rat dentaries survive in owl pellets, it is therefore now a simple matter to calculate the sizes of the rats eaten.[238] Subsequently Chris Smal and I worked up similar graphs for field mice, house mice and bank voles.[118] The average weight of a rat caught by a barn owl in Ireland is about two-and-a-half times that of a mouse or vole. The average for 259 rats in pellets from various roosts was 50 g[118] and the corresponding figure for another 985 in pellets collected from roosts in Co. Cork was 57 g, with the largest rat of all estimated at only 160 g; few exceeded 100 g.[56] There is little comparable data for the long-eared owl. Those available - an average of 52 g and maximum of 177 g for only 21 rats - suggest that the rats eaten are of much the same size as those caught by the barn owl. As

BROWN RAT

none of Fidelma Butler's rats below 100 g were fecund,[40] owls must rarely kill rats in breeding condition, and since mortality among young rats is high even under optimal conditions, it is questionable whether Irish owls ever actually *control* rat populations anywhere in Ireland.

Recently it has proved possible to link the proportion of rats and house mice in the diets of Irish barn owls with some aspects of land use in the countryside. This research would have been impossible without the enthusiasm and intense efforts of Messrs David Cooke and Tony Nagle in Cork City who, in 1992-94 along with Pat Smiddy, regularly collected owl pellets from barn owl roosts all over Co. Cork, which I then analysed.[56] The county, incidentally, exhibits a great variety of land use, from the rich agricultural region in the east to much poorer ground in the west.

The roosts were situated in a total of 29 different District Electoral Divisions and data on land use for each of these were available from the 1991 Census of Agriculture, in particular for areas under cereals, sugar-beet, potatoes, other crops, hay, silage, pasture, rough grazing and woodland. It was a simple matter to calculate the corresponding percentage of land under each of these in each Division.

Before trying to relate these to the prey in the pellets, I first excluded the results for bats, birds and frogs. The owls took few bats, so they were not worth bothering about. Although there were somewhat more frogs, the owls caught these mainly in late winter and early spring, and therefore most probably around the frogs' breeding sites. There was no information in the Census of Agriculture whatsoever on ponds and the like, so I ignored frogs. The barn owl hunts mainly at night and therefore while most other birds are roosting. The owl usually catches other birds by beating the outer branches of a roost with its wing tips, disturbing the occupants and catching those which fly out. Most of the birds I could identify in the pellets were house sparrows *Passer domesticus,* which roost together and are therefore ideal candidates to have been picked off like this. None of the information in the Census can give any indication of potential communal roosting sites in a Division, so birds are irrelevant. I therefore ended up with data only on mammalian prey - field mice, house mice, rats, bank voles and pygmy shrews - of which there were 9,203 individual items in all.

I looked at the proportion of each prey in the diets of the owls in each Division and tried to see whether it was correlated with the proportion of any of the various classes of land use listed above. Each prey can be examined against all the different classes simultaneously using a statistical technique known as stepwise multiple regression. Two interesting correlations emerged: the larger the percentage of the Division under potatoes, then the greater the percentage of rat in the pellets; and the greater the percentage of "other crops" in the Division, the greater the percentage of house mouse in the pellets. What are the explanations?

One might argue that potatoes provide food for rats, but presumably sugar-beet and cereals are equally acceptable. Chris Smal, who has had a wealth of experience of Irish wildlife on agricultural ground, particularly during the Badger and Habitat Survey (Chapter 3), came up with a probable explanation. Rats are particularly prevalent around potato fields in winter, because many potatoes remain unharvested. Modern mechanised lifting leaves the smallest ones in the ground and farm hands usually throw back any that are damaged, green or of irregular shape. Supermarkets today, and ultimately the housewife, are fussy about the appearance of vegetables. Ultimately the rats benefit as do the barn owls.

The association of house mice with "other crops" was particularly marked and thus all the more baffling. Surely cereals should be the most acceptable food for house mice? More puzzling still was that the area under "other crops" in the Divisions ranged only from less than 0.1% to only 3.9% and, as we already know, house mice are scarce on farmland anyway. "Other crops" comprised peas, beans, oilseed rape, turnips, kale, cabbage, other vegetables and fruit, but the largest component (37%) was fodder beet, which should be of restricted interest to house mice. But wait: a farmer who grows fodder beet uses it primarily to feed dairy cattle, which do not produce the best milk yield on grazing alone. Besides fodder beet, the feed is also supplemented by cereal-based concentrates, which are, of course, much more to the taste of house mice. Teagasc (the Agriculture and Food Development Authority) confirmed to me that there was likely to be a strong correlation between the amount of fodder beet grown on a farm and the quantity of concentrates stored there. Where there are more concentrates there are likely to be more house mice in the farm and its vicinity, and barn owls often hunt near farms. So the correlation between house mice and "other crops" although strong, is almost certainly an indirect one.

One predator of rats in Ireland that might not at first spring to mind is a fish, namely the pike. Pike are voracious and will eat just about anything which they can subdue; and rats, of course, often swim. In

Ireland it is often said that pike will swallow rats and indeed this has led to a prejudice against eating the fish here. There is no doubt that such predation does occur and a postgraduate at NUI Galway recovered rats from pike stomachs more than once.

Methods of controlling house mice and rats in Ireland have been multifarious, ranging from the supernatural to the technologically advanced poison baits of today, designed to kill rodents exclusively, the end product of protracted research.

The procedure of "rhyming rats to death", which has a long history in Europe, was once popular in Ireland and was referred to from time to time in the literature of the seventeenth century. For example in Ben Johnson's *Poetaster* there appears the line "Rhime them to death as they do Irish rats".[21]

Another remedy of the supernatural kind is confined to Tory Island, Co. Donegal. Surprisingly, there are no rats there and they are said to have been banished by St Columbkille, who devolved the power of dealing with them to his first convert on the island, a member of the Dugan family. Thereafter it has been passed down to the head of the family from generation to generation, the active agent being a miraculous clay. Thomas Mason, who was having rodent problems in his house in Dublin, gives some space to it in his book *The Islands of Ireland* (1936). On being told of an instance of the efficacy of the clay by a man in Donegal, who emphasised that he was "not superstitious", Mason travelled to the island himself. There he obtained, from the then head of the Dugan family, a paper bag of the clay, which was, he explained, always supplied free of charge to those who did not ask for it in a frivolous spirit. However, when Mason reached home the rodents had left of their own accord.[211] I have spoken by telephone to the post office on Tory and was assured that there are still no rats on the island and that the tradition continues.

There have been innumerable designs of rat and mouse trap, many of which take the rodents alive, either in a cage or box. Some are of the multiple-capture type, others catch the animals one at a time, each having to be removed and disposed of before the trip mechanism can be reset. Such devices are enjoying a degree of popularity at present among those too squeamish to kill house mice outright. Such persons prefer what they believe to be the more humane alternative of releasing the animals unharmed in the countryside. There, in entirely unfamiliar surroundings, the mouse's chances of survival are slim and it will most probably perish anyway, but in abject terror, or whatever passes for terror among mice. Of course no one extends such benevolence to a rat, which there is no reason for believing has any less tender feelings.

Break-back traps are familiar to everyone but there is a variety of equally deadly alternatives in which the victim is flattened or bisected. The gin trap has long been illegal in Ireland for catching rats, or indeed anything else, but there are others with steel jaws that are just as effective - the Imbra, Fenn and Juby in particular - which kill the animal outright by scissor action. These must be set in rats' tunnels, or in artificial ones, or otherwise screened so that other mammals and birds are not polished off accidentally.

Sticky traps, which act in the same way as bird lime can be effective, but are messy and far from humane. I well remember my father using this system successfully to rid his garage premises of a small infestation.

There are also a few gadgets on the market other than traps for dealing with rodents, including some which produce ultrasonic sounds to scare off the pests. The efficacy of such devices is open to question.

Out of doors, rodents may also be gassed, usually by tipping powder into their holes and then closing them over. The powder gives off hydrogen cyanide gas in contact with water or damp soil.

Cats are unpredictable creatures and whereas some are incorrigible assassins of anything small and furry or feathered, instanced by Duchess and Blodwyn above, other felines are unaccountably indolent, as apparently were those in the piggeries in Co. Kildare. It is often said that cats can keep rats down but cannot get them down. In other words on a farm already free of rats, cats can dispatch the occasional fresh immigrant, which is at a disadvantage in not yet being fully acquainted with its surroundings. On the other hand, cats are unable to deal with a well-established infestation, where the rats know their ground intimately, have established holes or other refuges for themselves, and are breeding at full steam.

Some terriers make superb ratters, provided that the quarry is in the open, which generally necessitates flushing it out in some way.

The most exotic animals used to control rats in Ireland have probably been mongooses (species unrecorded), which were introduced into the port areas of Belfast and 'Derry to "keep the rats away from the sides of bacon", the last one having been killed in 1954.[66] At one time rats lived behind the wainscot of the dining-hall in my old school, Campbell College, Belfast. A pair of mongooses was purchased and introduced into the rats' runs. They survived for more than a year and were sometimes seen of an evening in the dining-hall when it was empty and peaceful. The rats vanished.[47] It is just as well that none of the mongooses released in Ireland survived to breed. Those that have been naturalised in other parts of the world to control rats or snakes have often done their work well. Unfortunately, they have also preyed on harmless members of the local faunas, with dire results, and sometimes even on kittens, puppies and the young of farm animals.

Poison is unquestionably the only efficient method of dealing with sizeable rodent infestations and of routine prevention in commercial premises and sewers. The central problem here is to insure that the rodent eats enough poison to kill it and that man, his pets and domestic animals are not put at risk. The behaviour of rats and house mice is highly pertinent. Although rats are enterprising beasts and will eventually learn to exploit virtually every source of sustenance available, they are initially highly distrustful of anything new in their environment. This means that, to begin with, a rat will treat any new source of food, such as poisoned bait, with deep suspicion and, even when it gets around to eating some, will do so only in small quantities, which are often insufficient to kill it. However it will almost certainly feel sick. With its suspicions thus confirmed, it will thereafter avoid the bait. The house mouse also suffers from neophobia, although to a lesser degree. In addition it may fail to ingest a fatal dose because of its diffuse and erratic feeding habits. The efficient poisoning of mice requires many baiting stations scattered around the animals' range.

Up until the middle of the twentieth century all rodenticides were acute: both highly toxic and quick-acting. The baits contained a high percentage of poisonous compounds, which were mostly simple chemically, and so were cheap to produce. However, this meant that they were non-proprietary and that the manufacturers thus had little incentive to develop them further or to provide technical back-up. Furthermore, acute poisons were also lethal to man and other animals, and there were few antidotes. Even where the latter existed, the speed with which the poison acted meant that there was rarely sufficient time to administer them. Needless to say, a rat which nibbled a sub-lethal dose would certainly be ill for a time and thereafter would avoid the bait. Acute poisons have included zinc phosphide, sodium monofluroacetate, thallium sulphate, arsenic and red squill, the last being a plant extract.

Modern rodenticides[37,151,353] which were first used in 1949, are mostly anticoagulants and act by preventing the clotting of blood. When an animal's tissues are damaged, either internally or on the surface of the body, blood usually flows for a time and then clots. The clot prevents excessive haemorrhage and is part of the normal process of body repair. Various factors in the blood necessary for clotting are produced in the liver by a chemical process involving vitamin K. The latter is chemically bound to other compounds during the production of these clotting factors and is then released at the end of the process, so that it is used over and over again. Anticoagulant rodenticides block this release and so, when there is no more free vitamin K left, the clotting factors are no longer produced. When the existing factors in the blood are used up, which usually takes some days, the blood will no longer clot. In the meantime the rodent feels perfectly well and continues about its business, and to partake of the bait. It then suffers a massive haemorrhage and dies.

Warfarin was the most commonly used of the early anticoagulant poisons and the others were chemically closely related to it. Not only was there no difficulty in the rodents ingesting a sufficient dose but, in the concentrations used in the bait (around 0.025%), larger animals needed to eat an enormous amount of it for it to be fatal. In addition the delay in the action of the poison meant that there was time to administer an antidote and the latter was readily available - a large dose of vitamin K.

Resistance to warfarin was first discovered in 1958 near Glasgow and subsequently in various parts of England and then in Denmark. This led to the development of so-called second-generation anticoagulants in the 1970s, including difenacoum, brodifacoum and bromadiolone, which have proved effective against warfarin-resistant rodents. These normally kill most rats and house mice within a few days and are present in extremely low concentrations in baits. As with warfarin, an animal rather bigger than a rat would have to eat a great deal of such baits for them to have a serious effect. As far as I can determine, warfarin resistance has not yet been confirmed in Ireland. Nevertheless, most professional poisoning now mainly involves the use of second-generation anticoagulants, perhaps because rodent poisons are manufactured in Britain. The baits often contain other compounds to enhance the effect of the poison and also bitrex, a compound perfectly palatable to rats and mice, but with an almost unimaginably nasty taste to humans. Baits are also usually dyed bright blue or green. This means that, should a child put any in its mouth, the nature of the substance will be immediately clear to medical personnel by the staining about the lips. With the presence of bitrex, the delay in the effect and the ready availability of an antidote, as far as I am aware, no person, child or adult, in Ireland has died through the accidental consumption of an anticoagulant rodenticide.

Most people think very little about rats and mice if they see no signs of them, but a constant war is waged against them by local government agencies and many private companies. Insights into the former were kindly provided for me by Mr George Douglas (Pest Control Unit, Belfast City Council) and into the latter by Mr Sherwin Curran (Rentokil, Northern Ireland).

Belfast's Municipal Pest Control Unit, which at present has five officers working full-time "in the field", covers the whole of the electoral area of Belfast and serves local government responsibilities, such as schools, council canteens, library boards, Housing Association homes, The Zoo, Botanic Gardens and other properties owned by the City Council, as well as - on request - private homes and very small businesses, such as a combined shop and house. For legal reasons the Unit does not normally service commercial properties, although this sometimes leads to a very nice demarcation, as, for example, when Northern Ireland Railways engage a private firm to deal with rodents beside a railway line, whilst the city's Pest Control Unit is eradicating them in the houses backing onto it. The operatives use mainly brodifacoum and difenacoum baits but occasionally also warfarin.

A major part of the preventative work is the routine baiting of sewers. Brown rats are, of course, very much at home around water, and in a major port such as Belfast, with most of the centre low lying, it is a rule of thumb that about 80% or rats dwell in the sewer system. This furnishes both home and highway, from which individuals may emerge to take up residence in any inviting accommodation. The sewer baits are of whole or cut wheat impregnated with difenacoum, and the operatives provide it in paper bags, each holding "two ice cream scoops". A manhole cover is lifted and the bag placed on the shelf which lies above the level of the water in the sewer beneath. Because of the hazard from gas, operatives may not enter the sewer without breathing apparatus and, in any case, this is unnecessary and would waste time. Extendible tubes are available to allow the bag to be dropped onto the shelf, but even the opening and shutting of these would occupy time needlessly. The procedure at most manholes is simply to chuck the bag through the manhole onto the shelf. This accounts for the use of a paper bag, which rarely bounces off into the water when it lands, as a plastic bag would.

Throwing the bags is an acquired skill but, even if an occasional one falls into the sewer itself, the procedure is cheaper, quicker and more efficient than using the tubes.

The extent of such baiting depends on the district. In some suburbs sewers might only be serviced every 3 years, whereas in central Belfast, especially in thoroughfares with many restaurants, the manholes might be baited once a month. Some idea of the extent of such operations may be had from the fact that, in the financial year 1999-2000, there were 20,399 sewer baitings, a remarkable figure when one considers the labour involved in setting out traffic cones and warning notices and in lifting manhole covers. Routine baiting also extends to many council properties.

Above ground, a particulate bait, such as wheat, is often unsuitable because of its tendency to scatter, and thus has the potential of finding its way into places where it might contaminate food. Baits in the form of pastes, gels, pellets and wax blocks are the alternatives and these are usually placed in containers of some kind, both to keep them in one place and also, for mice at least, to provide some sort of cover for the animals when they are feeding. Gels are extruded using a specially-designed gun. Operatives often break up wax blocks to expose the edible components within, rodents showing limited interest in the wax itself. A further ingenious way of persuading mice to take the bait is to provide it smeared on wicks in a plastic tube, which is then placed along a mouse run. The poison is thus transferred to the mouse's coat, and the animal ingests it whilst grooming.

In 1999-2000 the Unit dealt with 1,825 actual infestations of rats and 1,475 of mice (not to mention other pests), the relative proportions varying from year to year. Infestations are nearly always of one or other species although, from time to time in large buildings, both may be present. The first visit, to lay bait, must be followed up by at least one more, after a week or so, during which excess bait is removed or, if all of it has been taken, it is replenished. Because of their feeding habits, mice are somewhat harder to eradicate than rats and require more baiting stations. Occupiers do not like poison to be laid in the open near their homes, because of the potential hazard to children, dogs and cats. With rats in gardens or around the outsides of houses, baiting sewers is often a successful alternative.

Rodents often gain entry to premises through drains, usually by breaking through a bend in a plastic pipe, where they can best gain initial purchase for gnawing. Such entry points can be detected using an investigative drain test (220 tests in 1999-2000). A generator puffs smoke down the pipe and any leaks are betrayed by its escape through the floor.

Rentokil Initial Group currently services standard annual contracts for routine rodent control in 2,500 premises in Northern Ireland, including many of those used in food production and storage, besides hotels and restaurants. About 80% of contract work is for mice and 20% for rats and most of it is essentially preventative. Once again, operatives rarely find that both species occur together, although sometimes there may be rats living outside a building and mice inside. In a further 500 or so instances, the Group is called out to deal with actual infestations, a few of which are in private houses but most in commercial premises. It is not unusual to find that owners of the latter have already tried to exterminate the rodents themselves, fearing to call in the professionals because they believed (wrongly) that rats and mice will be taken as irrefutable evidence of a lack of hygiene, whereas the rodents have no such prejudices.

Rentokil manufactures its own baits in Britain, the active ingredient most often being bromadiolone, although difenacoum is also used. Baits are commonly gels and pastes, or the poison is incorporated into barley or lard. Most baits are held in plastic trays or boxes and, when the latter are used outdoors, they are securely fastened to fencing or to the ground to keep them out of the way of non-target species. Rentokil also uses dusts, which cling to the fur and are swallowed during grooming. With a heavy initial mouse infestation, operatives sometimes use alpha chloralose, with appropriate precautions. This is an acute poison which lowers body temperature and kills quickly. A further armament in Rentokil's war against rodents is the liquid bait. In the midst of plentiful food, rodents will sometimes take insufficient notice of a bait. However, water is usually in short supply in food stores. Rats must drink and house mice will drink if water is available. So a liquid poison may sometimes prove the best means of eradication.

Both Messrs Douglas and Curran described the usual signs of the presence of rats and mice, including those of feeding, gnawing damage, droppings, nests in product - which sometimes has to be stripped or opened to find the nests - and, in heavy infestations, urine pillars (from house mice) and smears. Because the fur of both rats and mice is oily, its repeated rubbing against walls or woodwork beside runs leaves characteristic smears, which, of course, extend higher above the runs of rats than those of mice.

Rats and mice sometimes gain entry to properties via a hole in the fabric. The electricity cable into most houses is laid in a trench and enters through a hole, which is sometimes actually broken through the wall, and is afterwards buried. The earth around the cable is soft and a rodent can burrow along this without much difficulty and through the hole. Mr Douglas described a case in Belfast of rats in residence in 16 new houses in a row of 24, the animals having gained access through the meter box in every one. If there are any holes big enough to admit a rodent around the emergence of the soil pipe from the upper story of the dwelling, these too may be exploited. A mouse invaded my bedroom one night in just this way in one of my digs in Galway. I had a wash-hand basin in the room and the soil pipe passed to the outside close to my bed head, so I actually heard the beast entering. (The noise made by a mouse as it progresses cautiously when all else is silent - a faint rustling or scratching - is well summed up by the onomatopoeic verb currently used of it in some parts of Northern Ireland: to *fissle*.) Both mice and rats are expert at scrambling up in the space between the wall and soil pipe, perhaps with back wedged against one and feet against the other.

Rodents sometimes chew insulation on electric wiring and this can cause a short, which may, incidentally, prematurely terminate the career or the animal involved. On rare occasions in Belfast, this has led to a fire. A representative of the Royal Sun Alliance Insurance Group took the trouble to explain to me that, whereas their insurance policies will not pay out on the ravages wrought by rats and mice, consequent damage is normally insured. Thus, if a mouse chews the wiring in your house, you will have to pay to replace it yourself, although resultant fire-damage would usually be covered.

All Irish ports take precautions against rats, especially those coming ashore from visiting ships. However, according to my informants, it is only in Belfast - by far the largest port in Ireland - that "deratting" regulations apply, as detailed to me by Mr Terry Nevin (Port Health Division, Belfast City Council). The World Health Organisation in Geneva has formulated international health regulations regarding the control of rodents on ships and has designated certain ports throughout the world as qualified to issue "deratting certificates". Belfast employs a full-time officer for the purpose. Thus every foreign-going vessel docking in Belfast is obliged to hold a valid certificate, which runs for 6 months and must then be renewed. If the certificate has expired, or if it is close to expiry and the master so requests, the ship is inspected for rats and, if any are discovered, they must be exterminated at the owner's expense before a new certificate is issued. In the meantime the gangway must be withdrawn during the hours of darkness and rodent guards fitted to the mooring ropes. These either look rather like bin lids or are conical, with the pointed end facing towards the quay. Even if the master holds a valid certificate, the officer is entitled to inspect merely on suspicion.

Some vessel's crews are shocked to find that they have had rats as shipmates, having never had cause to suspect it. The largest infestation in recent years in the port of Belfast was on a Russian craft, where 40-50 bodies were recovered after control measures. In one Rumanian boat the rats gnawed through a cable in the engine room and started a fire. A ship's engine room is an extremely noisy place. This clearly demonstrates that, despite the innately suspicious nature of rats and the caution with which they approach any novel situation, in the end they may adapt to almost anything.

Botanical Names of Plants Mentioned in the Text

Apple, Crab *Malus sylvestris*
Ash *Fraxinus excelsior*
Barley *Hordeum* species
Bean, broad *Vicia faba*
Beech *Fagus sylvatica*
Beet, Fodder or Sugar *Beta vulgaris*
Bilberry *Vaccinium myrtillus*
Birch *Betula* species
Blackberry/Bramble *Rubus fruticosus*
Blackthorn *Prunus spinosa*
Bog-cotton *Eriophorum* species
Bog-rush *Schoenus nigricans*
Bracken *Pteridium aquilinum*
Cabbage *Brassica oleracea*
Carrot *Daucus carota*
Cherry *Prunus avium*
Clover *Trifolium* species
Coconut *Cocos nucifera*
Elder *Sambucus nigra*
Fescue *Festuca* species
Gooseberry *Ribes uva-crispa*
Hawthorn *Crataegus monogyna*
Hazel *Corylus avellana*
Heather *Calluna vulgaris*
Ivy *Hedera helix*

Lime *Tilia europaea*
Lodgepole pine *Pinus contorta*
Marram grass *Ammophila arenaria*
Matt grass *Nardus stricta*
Nettle *Urtica dioica*
Oak *Quercus* species
Oilseed rape *Brassica napus*
Pea, Garden *Pisum sativum*
Pear *Pyrus communis*
Perennial rye grass *Lolium perenne*
Pink *Dianthus* species
Plum *Prunus domestica*
Potato *Solanum tuberosum*
Purple moor-grass *Molinia caerulea*
Raspberry *Rubus idaeus*
Rhododendron *Rhododendron ponticum*
Ribbed sedge *Carex binervis*
Rowan *Sorbus aucuparia*
Sea buckthorn *Hippophae rhamnoides*
Sitka spruce *Picea sitchensis*
Sow-thistle *Sonchus oleraceus*
Strawberry *Fragaria vesca*
Sycamore *Acer pseudoplatanus*
Thyme *Thymus drucei*
Turnip *Brassica rapa*
Willow *Salix* species
White beak-sedge *Rhynchospora alba*
Yellow flag/Iris *Iris pseudacorus*

References

Figures given in parenthesis, before and after volume numbers, refer respectively to series and part. The latter is included only wherever parts are individually paginated.

1. Alcock, N.H. 1899 The natural history of Irish bats. *Irish Naturalist* **8**: 169-175.

2. Alcock, N.H. & Moffat, C.B. 1901 The natural history of Irish bats. *Irish Naturalist* **10**: 241-251.

3. Allen, P., Forsyth, I., Hale, P. & Rogers, S. 2000 Bats in Northern Ireland. *Irish Naturalists' Journal* **25** (supplement): 1-31.

4. Andrews. D.J. 1992 The diet of wintering short-eared owls on Strangford Lough, Co. Down. *Irish Birds* **4**: 549-554.

5. Anon 1835 *Ordnance Survey Memoirs.* Mss in Library of the Royal Irish Academy. Now published as Day, A. & McWilliams, P. (Editors) 1991 *Ordnance Survey Memoirs. Volume 11.* Institute of Irish Studies and the Royal Irish Academy. Belfast.

6. Anon 1861 *General alphabetical index to the townlands and towns, parishes and baronies of Ireland.* HMSO. Dublin.

7. Anon 1867 *Calendar of the Carew manuscripts.* Longmans, Green. London.

8. Anon 1870 Large otter. *The Field* **35**: 273.

9. Anon 1951 A Donegal naturalist of the early 17th century. *Donegal Annual* **2**: 329-330.

10. Anon 1967 Annual conference 1967. *Bulletin of the Mammal Society of the British Isles* **28**: 2.

11. Anon 1975 Hare seminar held at Clonmel. *Sporting Press* 11 September.

12. Anon 1985 Last word. £250 fine for badger baiting. *Field and Countryside* **2**(6): 70.

13. Anon 1993 *Mammal survey of 5 species.* Ulster Wildlife Trust and Forest Service, Department of Agriculture of Northern Ireland. Belfast.

14. "Aquarius" 1896 Hares going to ground. *The Field* **87**: 185.

15. Armitage, P.L. 1994 Unwelcome companions: ancient rats reviewed. *Antiquity* **68**: 231-240.

16. Avery, M.I. 1985 Winter activity of pipistrelle bats. *Journal of Animal Ecology* **54**: 721-738.

17. Baring, C. 1907 Contributions to the natural history of Lambay Island. Mammals. *Irish Naturalist* **16**: 19-23.

18. Barrett-Hamilton, G.E.H. 1891 Leaping powers of the Irish hare (*Lepus variabilis*). *Zoologist* (3) **15**: 60-61.

19.   Barrett-Hamilton, G.E.H. 1894 The marten in Ireland. *Zoologist* (3) **18**: 134-142, 187.

20.   Barrett-Hamilton, G.E.H. 1898 Notes on the introduction of brown hares into Ireland. *Irish Naturalist* **7**: 69-76.

21.   Barrett-Hamilton, G.E.H. & Hinton, M.A.C. 1910-21 *A history of British mammals.* Gurney & Jackson. London.

22.   Barrington, R.M. 1874 The hairy-armed bat (*Scotophilus leisleri*) in Ireland. *Zoologist* (2) **9**: 4071-4074.

23.   Barrington, R.M. 1875 Rat killing its own species. *Zoologist* (2) **10**: 4662-4663.

24.   Beebee, T. 1985 *Frogs & Toads.* Whittet Books. London.

25.   Bentley, E.W. 1959 The distribution and status of *Rattus rattus* L. in the United Kingdom in 1951 and 1956. *Journal of Animal Ecology* **28**: 299-308

26.   Bentley, E.W. 1964 A further loss of ground by *Rattus rattus* L. during 1956-61. *Journal of Animal Ecology* **33**: 371-373.

27.   Berrow, S. 1990 Feeding observations on mink *Mustela vison* Schreber at Lough Hyne, Co. Cork. *Irish Naturalists' Journal* **23**: 284.

28.   Berrow, S. 1992 The diet of coastal breeding ravens in Co. Cork. *Irish Birds* **4**: 555-558.

29.   Berrow, S. 1993 A further record of a stoat carrying a fish. *Irish Naturalists' Journal* **24**: 345.

30.   Berry, R.J. 1981 Town mouse, country mouse: adaptation and adaptability in *Mus domesticus* (*Mus musculus domesticus*). *Mammal Review* **11**: 91-136.

31.   Boyle, K. & Whelan, J. 1990 Changes in the diet of the badger *Meles meles* L. from autumn to winter. *Irish Naturalists' Journal* **23**: 199-202.

32.   Breathnach, S. & Fairley, J.S. 1993 The diet of otters *Lutra lutra* (L.) in the Clare River system. *Biology and Environment* **93B**: 151-158.

33.   Brooke, A.B. 1870 Natural history of Wicklow and Kerry. *Zoologist* (2) **5**: 2281-2285.

34.   Brown, R. 1990 *Strangford Lough.* Institute of Irish Studies, Queen's University Belfast.

35.   Brunker, J.P. 1951 Rats eating hawthorn berries. *Irish Naturalists' Journal* **10**: 137.

36.   Brunton, T. 1874 Birds observed at Glenarm Castle. *Zoologist* (2) **9**: 3829-3830.

37.   Buckle, A.P. & Smith, R.H. 1994 (Editors) *Rodent pests and their control.* Cab International. Wallingford.

38.   Buckley, N.J. 1990 Diet and feeding ecology of great black-backed gulls (*Larus marinus*) at a southern Irish breeding colony. *Journal of Zoology* **222**: 363-373.

357

39. Butler, F.T. 1994 Arthropod and helminth parasites from rabbits *Oryctolagus cuniculus* in south-west Ireland. *Irish Naturalists' Journal* **24**: 392-395.

40. Butler, F.T. & Whelan, J. 1994 Population structure and reproduction in brown rats (*Rattus norvegicus*) from pig farms, Co. Kildare, Ireland. *Journal of Zoology* **233**: 277-291.

41. Cabot, D. 1999 *Ireland.* HarperCollins. London.

42. Canny, N. 1989 Early modern Ireland c.1500-1700. In Foster, R. (Editor) *The Oxford illustrated history of Ireland.* 104-160. Oxford University Press. Oxford.

43. Carruthers, T. 1998 *Kerry, a natural history.* Collins Press. Cork.

44. Chanin, P. 1983 Observations on two populations of feral mink in Devon, U.K. *Mammalia* **47**: 463-476.

45. Chanin, P. 1985 *The natural history of otters.* Croom Helm. Beckenham.

46. Chapman, P.J. & Chapman, L.L. 1982 *Otter survey of Ireland.* Vincent Wildlife Trust. London.

47. Chase, C.D. 1948 *The natural history of Campbell College and Cabin Hill.* Privately published. Belfast.

48. Clarke, B., O'Connell, J., O'Corry-Crowe, G. & Ronan, M. 1993 *Final report on evacuation of Abbotstown badger sett for Dublin County Council Parks Division.* Unpublished report. Department of Zoology, University College Dublin.

49. Clarke, R. & Scott, D. 1994 Breeding season diet of the merlin in County Antrim. *Irish Birds* **5**: 205-206.

50. Coburn, B. 1978 After 20 years, a pine marten is recorded in County Down. *Belfast Telegraph* 20 May.

51. Cocks, A.H. 1897 Pine marten in Ireland. *Zoologist* (4) **1**: 269-270.

52. Cogan, R. 1999 *Commuting routes used by the lesser horseshoe bat Rhinolophus hipposideros at a roost in Co. Clare.* BSc Thesis. National University of Ireland Galway.

53. Colgan, N. 1911 Clare Island Survey. Gaelic plant and animal names and associated folklore. *Proceedings of the Royal Irish Academy* **31** (4): 1-30.

54. Collins, J.J. 1955 Myxomatosis in the common hare - *Lepus europaeus* - II. *Irish Veterinary Journal* **9**: 268-269.

55. Cooke, D., Nagle, A. & Fairley, J.S. 1995 The diet of long-eared owls within the range of the bank vole in Co. Cork. *Irish Birds* **5**: 305-307.

56. Cooke, D., Nagle, A., Smiddy, P., Fairley, J. & Ó Muircheartaigh, I. 1996 The diet of the barn owl *Tyto alba* in County Cork in relation to land use. *Biology and Environment* **96B**: 97-111.

57. Corridan, J.P. & O'Meara, J. 1968 Trichinosis in Cork. *Irish Journal of Medical Science* **1**: 109-113.

58. Curtis, E. 1932 *Calendar of Ormond deeds 1172-1350.* Stationery Office. Dublin.

59. Day, M.G. 1968 Food of British stoats (*Mustela erminea*) and weasels (*Mustela nivalis*). *Journal of Zoology* **155**: 485-497.

60. Deane, C.D. 1951 Hare breeding beneath a hut. *Irish Naturalists' Journal* **10**: 197.

61. Deane, C.D. 1952 The "black" rat in the north of Ireland. *Irish Naturalists' Journal* **10**: 296-298.

62. Deane, C.D. 1952 Pine martens in Counties Tyrone and Down. *Irish Naturalists' Journal* **10**: 303.

63. Deane, C.D. 1955 Note on myxomatosis in hares. *Bulletin of the Mammal Society of the British Isles* **3**: 20.

64. Deane, C.D. 1955 Myxo empties the burrows - but what next? *Belfast Telegraph* 5 April.

65. Deane, C.D. 1962 Irish golden eagles and a link with Scotland. *British Birds* **55**: 272-274.

66. Deane, C.D. 1964 Introduced mammals in Ireland. *Bulletin of the Mammal Society of the British Isles* **21**: 2.

67. Deane, C.D. 1971 Mammals of Strangford Lough. In Anon (Editor) *Strangford Lough.* 22-23. National Trust. Belfast.

68. Deane, C.D. 1973 Still no respite for the badger. *Belfast News Letter* 4 December.

69. Deane, C.D. 1974 A rare type of man - and his friend. *Belfast News Letter* 12 August.

70. Deane, C.D. 1974 Airport hares have taken flight. *Belfast News Letter* 19 October.

71. Deane, C.D. 1974 On the wild island kingdom of Great Saltee. *Belfast News Letter* 7 December.

72. Deane, C.D. 1975 Upside-down world of the hairy armed bat. *Belfast News Letter* 8 February.

73. Deane, C.D. 1976 Let's be fair to the hare. *Belfast News Letter* 20 March.

74. Deane, C.D. 1983 *The Ulster countryside.* Century Books. Belfast.

75. Deane, C.D. & O'Gorman, F. 1969 The spread of feral mink in Ireland. *Irish Naturalists' Journal* **16**: 198-202.

76. de Buitléar, E. 1993 *Ireland's wild countryside.* Boxtree. London.

77. Denny, G.O. & Wilesmith, J.W. 1999 Bovine tuberculosis in Northern Ireland: a case-control study of herd risk factors. *Veterinary Record* **144**: 305-310.

78. Dent, H.C. 1890 Hares swimming. *The Field* **75**: 893.

79. Devane, D.J. 1984 Stoats capturing fish. *Irish Naturalists' Journal* **21**: 365.

80. Dingerkus, S.K. & Montgomery, W.I. 1997 The distribution of the Irish hare

(*Lepus timidus hibernicus*) in Northern Ireland and its relationship to land classification. *Gibier Faune Sauvage* **14**: 325-334.

81. Dingerkus, S.K. & Montgomery, W.I. 2001 The diet and landclass affinities of the Irish hare *Lepus timidus hibernicus*. *Journal of Zoology* **253**: 233-240.

82. Dolan, L.A. 1993 Badgers and bovine tuberculosis in Ireland: a review. In Hayden, T.J. (Editor) *The badger*. 108-116. Royal Irish Academy. Dublin.

83. Dolan, L.A. 1998 Tuberculin testing data for clear and restricted herds in the east Offaly Badger Research Project for the period 1988 to 1997. In *Tuberculosis Investigation Unit, University College Dublin. Selected papers 1997*. 6-13. Tuberculosis Investigation Unit, University College Dublin.

84. Dolan, L.A., Eves, J.A. & Bray, D. 1995 East Offaly Badger Research Project (EOP): interim report for the period, January 1989 to December 1994. In *Tuberculosis Investigation Unit, University College Dublin. Selected papers 1994*. 1-3. Tuberculosis Investigation Unit, University College Dublin.

85. Dolan, L.A. & Lynch, K. 1992 Badgers and bovine tuberculosis. In *Tuberculosis Investigation Unit, University College Dublin. Selected papers 1990-1991*. 31-34. Tuberculosis Investigation Unit, University College Dublin.

86. Doran, G. 1976 Some observations on general and nesting behaviour of the hen harrier. *Irish Naturalists' Journal* **18**: 261-264.

87. Duffy, S.G., Fairley, J.S. & O'Donnell, G. 1996 Food of rabbits *Oryctolagus cuniculus* on upland grasslands in Connemara. *Biology and Environment* **96B**: 69-75.

88. Dunstone, N. 1993 *The mink*. Poyser. London.

89. Entwistle, A.C., Racey, P.A. & Speakman, J.R. 2000 Social and population structure of a gleaning bat, *Plecotus auritus*. *Journal of Zoology* **252**: 11-17.

90. Erlinge, S. 1968 Food studies on captive otters *Lutra lutra* L. *Oikos* **19**: 259-270.

91. Eves, J. 1993 The East Offaly Badger Research Project. In Hayden, T.J. (Editor) *The badger*. 166-173. Royal Irish Academy. Dublin.

92. Fahie, C.J. 1887 Rabbit taking to the water. *The Field* **70**: 486.

93. Fairley, J.S. 1966 An indication of the food of the fox after myxomatosis. *Irish Naturalists' Journal* **15**: 149-151.

94. Fairley, J.S. 1966 An indication of the food of the short-eared owl in Ireland. *British Birds* **59**: 307-308.

95. Fairley, J.S. 1967 Food of long-eared owls in north-east Ireland. *British Birds* **60**: 130-135.

96. Fairley, J.S. 1967 An indication of the food of the badger in north-east Ireland. *Irish Naturalists' Journal* **15**: 267-269.

97. Fairley, J.S. 1970 The food, reproduction, form, growth and development of

the fox *Vulpes vulpes* (L.) in north-east Ireland. *Proceedings of the Royal Irish Academy* **69B**: 103-137.

98. Fairley, J.S. 1971 A critical reappraisal of the status in Ireland of the eastern house mouse *Mus musculus orientalis* Cretzmar. *Irish Naturalists' Journal* **17**: 2-5.

99. Fairley, J.S. 1971 New data on the Irish stoat. *Irish Naturalists' Journal* **17**: 49-57.

100. Fairley, J.S. 1972 Food of otters (*Lutra lutra*) from Co. Galway, Ireland, and notes on other aspects of their biology. *Journal of Zoology* **166**: 469-474.

101. Fairley, J.S. 1973 Kestrel pellets from a winter roost. *Irish Naturalists' Journal* **17**: 407-409.

102. Fairley, J.S. 1973 Horse-shoe bats in Co. Clare. *Irish Naturalists' Journal* **17**: 424.

103. Fairley, J.S. 1974 Notes on the winter breeding of hares in the west of Ireland. *Irish Naturalists' Journal* **18**: 17-19.

104. Fairley, J.S. 1975 Summer food of pine martens in Co. Clare. *Irish Naturalists' Journal* **18**: 257-258.

105. Fairley, J.S. 1980 The muskrat in Ireland. *Irish Naturalists' Journal* **20**: 405-411.

106. Fairley, J.S. 1980 Observations on a collection of feral Irish mink *Mustela vison* Schreber. *Proceedings of the Royal Irish Academy* **80B**: 79-90.

107. Fairley, J.S. 1981 A north-south cline in the size of the Irish stoat. *Proceedings of the Royal Irish Academy* **81B**: 5-10.

108. Fairley, J.S. 1981 Species of hares in the north-west. *Irish Naturalists' Journal* **20**: 351.

109. Fairley, J.S. 1983 Exports of wild mammal skins from Ireland in the eighteenth century. *Irish Naturalists' Journal* **21**: 75-79.

110. Fairley, J.S. 1984 Otters feeding on breeding frogs. *Irish Naturalists' Journal* **21**: 372.

111. Fairley, J.S. 1984 *An Irish beast book*. 2nd Edition. Blackstaff Press. Belfast.

112. Fairley, J.S. & Clark, F.L. 1972 Notes on pipistrelle bats *Pipistrellus pipistrellus* Schreber from a colony in Co. Galway. *Irish Naturalists' Journal* **17**: 190-193.

113. Fairley, J.S. & McCarthy, T.K. 1985 Do otters prey on breeding natterjack toads? *Irish Naturalists' Journal* **21**: 544-545.

114. Fairley, J.S. & McLean, A. 1965 Notes on the summer food of the kestrel in northern Ireland. *British Birds* **58**: 145-148.

115. Fairley, J.S. & Murdoch, B. 1989 Summer food of otters in the lakes of Killarney. *Irish Naturalists' Journal* **23**: 38-41.

116. Fairley, J.S. & O'Gorman. F. 1974 Food of pine martens in the Burren. *Irish*

Naturalists' Journal **18**: 125.

117. Fairley, J.S. & Smal, C.M. 1987 Feral house mice in Ireland. *Irish Naturalists' Journal* **22**: 284-290.

118. Fairley, J.S. & Smal, C.M. 1988 Correction factors in the analysis of the pellets of the barn owl *Tyto alba* in Ireland. *Proceedings of the Royal Irish Academy.* **88B**: 119-133.

119. Fairley, J.S. & Smal, C.M. 1989 Further observations on the diet of the barn owl in Ireland. *Irish Birds* **4**: 65-68.

120. Fairley, J.S., Tangney, D., Hassett, S., Kirby, N., O'Donnell, G & McAney, K. 1995 *The mammals of the Park. Connemara National Park.* Office of Public Works. Dublin.

121. Fairley, J.S., Ward, D.P. & Smal, C.M. 1987 Correction factors and mink faeces. *Irish Naturalists' Journal* **22**: 324-336.

122. Fairley, J.S. & Wilson, S.C. 1972 Autumn food of otters (*Lutra lutra*) on the Agivey River, County Londonderry, Northern Ireland. *Journal of Zoology* **166**: 468-469.

123. Faris, R.C. 1948 Bats drinking. *Irish Naturalists' Journal* **9**: 123-124.

124. Feehan, J. (Editor) 1983 *Laois: an environmental history.* Ballykilcavan Press. Ballykilcavan, Co. Laois.

125. Fenner, F. & Ross, H. 1994 Myxomatosis. In Thompson, H.V. & King, C.M. (Editors) *The European rabbit.* 205-239. Oxford University Press. Oxford.

126. Feore, S. & Montgomery, W.I. 1999 Habitat effects on the spatial ecology of the European badger (*Meles meles*). *Journal of Zoology* **247**: 537-549.

127. Feore, S., Smal, C.M. & Montgomery, W.I. 1993 Survey of badger setts in Northern Ireland: progress report. In Hayden, T.J. (Editor) *The badger.* 23-25. Royal Irish Academy. Dublin.

128. Fertig, D.S. & Edmonds, V.W. 1969 The physiology of the house mouse. *Scientific American* **221**: 103-110.

129. Finlan, M. 1985 Rabbits eat their way through Inishmore Island. *Irish Times* July 11.

130. Fitzgerald, W. 1916 Richard Eustace, tenant of Gorteenvacan and Ballybyrne in the County Kildare, 1600. *Journal of the Kildare Archaeological Society* **8**: 161-162.

131. Flavin, D.A., Biggane, S.S., Shiel, C.B., Smiddy, P. and Fairley, J.S. 2001 Analysis of the diet of Daubenton's bat *Myotis daubentonii* in Ireland. *Acta Theriologica* **46**: 43-52.

132. Flemyng, W.W. 1897 Pine marten in County Waterford. *Zoologist* (4) **1**: 327.

133. Flux, J.E.C. & Fullagar, P.J. 1992 World distribution of the rabbit *Oryctolagus cuniculus* on islands. *Mammal Review* **22**: 151-205.

134. Flynn, J.E. 1935 The lesser horse-shoe bat (*Rhinolophus hiposideros*) in Co.

Cork. *Irish Naturalists' Journal* **5**: 228-229.

135. Foot, F.J. 1860 Description of Ballyallia Cave, Ennis, with account of the discovery of the lesser horseshoe bat. *Proceedings of the Dublin Natural History Society* **2**: 152-154.

136. Foot, F.J. 1863 Natural history notes on the Mammalia of the west coast of Clare. *Proceedings of the Dublin Natural History Society* **3**: 104-106.

137. Forbes, A.C. 1937 The pine marten in Co. Wexford. *Irish Naturalists' Journal* **6**: 301.

138. Forster, R. & Fairley, J.S. 1975 Further data on barn owls feeding on bank voles, and a new county record of the lesser horse-shoe bat. *Irish Naturalists' Journal* **18**: 251-252.

139. Fowler, P.A. & Racey, P.A. 1988 Overwintering strategies of the badger, *Meles meles,* at 57°N. *Journal of Zoology* **214**: 635-651.

140. Fox, J.B. 1989 Grey herons taking roadside rats. *Irish Birds* **4**: 70-71.

141. Frost, W.E. 1939 River Liffey Survey II. The food consumed by the brown trout (*Salmo trutta* L.) in acid and alkaline water. *Proceedings of the Royal Irish Academy* **45B**: 139-206.

142. Furnell, M.J.G. 1957 The incidence of trichinosis in Ireland. *Annual Report of the Medical Research Council in Ireland* **1957**: 43-44.

143. Garvey, T. 1935 The musk rat in Saorstat Éireann *Journal of the Department of Agriculture of the Republic of Ireland* **33**: 189-195.

144. Gethin, R.G. 1936 Pine marten visitor. *Irish Naturalists' Journal* **6**: 145-146.

145. Goldsmith, O. 1774 *An history of the earth and animated nature.* Nourse. London.

146. Gormally, M.J. & Fairley, J.S. 1982 Food of otters *Lutra lutra* in a freshwater lough and an adjacent brackish lough in the west of Ireland. *Journal of Zoology* **197**: 313-321.

147. Griffin, J.M., Quigley, F., Towey, K.P., Costello, E., Hammond, R.F., McGrath, G. & Sleeman, D.P. (1999) The four-area badger study: progress report 1998. In *Tuberculosis Investigation Unit, University College Dublin. Selected papers 1998.* 1-9. Tuberculosis Investigation Unit, University College Dublin.

148. Gunning, B.E.S. & Pate, J.S. 1954 Golden eagle in Co. Antrim. *Irish Naturalists' Journal* **11**: 208.

149. Hammer, O. 1940 Biological and ecological investigations on flies associated with pasturing cattle and their excrement. *Videnskabelige Meddelelser fra Dansk Naturhistorisk Forening i Kjøbenhavn* **105**: 1-257.

150. Hannon, C., Berrow, S.D. & Newton, S.F. 1997 The status and distribution of breeding Sandwich *Sterna sandvicensis,* roseate *S. dougallii,* common *S. hirundo,* Arctic *S. paradisea* and little terns *S. albifrons* in Ireland. *Irish*

Birds **6**: 1-22.

151. Harris, S. (Editor) In press *The handbook of British mammals.* 4th Edition. Blackwells. Oxford.

152. Harting, J.E. 1873 Martens robbing hives. *The Field* **41**: 478.

153. Harting, J.E. 1883 *Essays on sport and natural history.* Cox. London.

154. Harting, J.E. 1894 The marten in Ireland. *Zoologist* (3) **18**: 100-107, 222.

155. Hayes, R.J. (Editor) 1965 *Manuscript sources for the history of Irish civilization.* Irish National Library. Dublin.

156. Hayward, P.J. (Editor) Undated *Mullet 1970-72.* University of Reading.

157. Heron, E.D. 1929 Unusually coloured wild mice. *Irish Naturalists' Journal* **2**: 212.

158. Hillis, P., Fairley, J.S., Smal, C.M. & Archer, P. 1988 The diet of the long-eared owl in Ireland. *Irish Birds* **3**: 581-588.

159. Hubbard, C.E. 1954 *Grasses.* Penguin. London.

160. Humphreys, G.R. 1978 Ireland's former premier breeding haunt of aquatic birds. *Irish Birds* **1**: 171-178.

161. Huson, L.W. & Rennison, B.D. 1981 Seasonal variation of Norway rat *Rattus norvegicus* infestations of agricultural premises. *Journal of Zoology* **194**: 257-289.

162. Hutchinson, C.D. 1979 *Ireland's wetlands and their birds.* Irish Wildbird Conservancy. Dublin.

163. Hutchinson, C. 1989 *Birds in Ireland.* Poyser. Calton.

164. Il-Fituri, A.I. & Hayden, T.J. 1993 Craniometrics and age determination of the Eurasian badger. In Hayden, T.J. (Editor) *The badger.* 58-63. Royal Irish Academy. Dublin.

165. Irwin, A.G. 1975 Motorway rats. *Irish Naturalists' Journal* **18**: 257.

166. Jameson, H.L. 1894 Irish bats: how to collect specimens with a view to the investigation of their distribution. *Irish Naturalist* **3**: 67-71.

167. Jameson, H.L. 1897 The bats of Ireland. *Irish Naturalist* **6**: 34-43.

168. Jameson, H.L. 1898 On a probable case of protective coloration in the house-mouse (*Mus musculus* Linn.). *Journal of the Linnean Society* **26**: 456-473.

169. Jeffrey, R.J. 1997 *Aspects of the ecology and behaviour of the Irish hare, Lepus timidus hibernicus (Bell, 1837) on lowland farmland.* PhD Thesis. Trinity College Dublin.

170. Jenkins, D. & Burrows, G.O. 1980 Ecology of otters in northern Scotland: III. The use of faeces as indicators of otter *Lutra lutra* density and distribution. *Journal of Animal Ecology* **49**: 735-754.

171. Jones, E. 1979 Breeding of the short-eared owl in south west Ireland. *Irish Birds* **1**: 377-380.

172. Jones, G. & Rayner, J.M.V. 1988 Flight performance, foraging tactics and

echolocation in free-living Daubenton's bats *Myotis daubentoni* (Chiroptera: Vespertilionidae). *Journal of Zoology* **215**: 113-132.

173. Joyce, P.W. 1910 *The origin and history of Irish names of places.* Longmans, Green. London.

174. Kelly, F. 1998 *Early Irish farming.* Institute for Advanced Studies. Dublin.

175. Kelly, P.A., Mahon, G.A.T. & Fairley, J.S. 1982 An analysis of morphological variation in the fieldmouse *Apodemus sylvaticus* (L.) on some Irish islands. *Proceedings of the Royal Irish Academy* **82B**: 39-51.

176. Kelsall, J.E. 1887 The distribution in great Britain of the lesser horse-shoe bat. *Zoologist* (3) **11**: 89-93.

177. Kinahan, J.R. 1860 Mammalogica Hibernica: Part 1. - Sub-Class, Lissencephala; Order Cheiroptera, Insectivoridae; - or, a general review of the history and distribution of bats in Ireland; with remarks on Mr Foot's discovery in Clare of the lesser horseshoe bat, a species hitherto unrecorded in Ireland. *Proceedings of the Dublin Natural History Society* **2**: 154-170.

178. Kinahan, J.R. 1863 Three days among the bats of Clare. *Proceedings of the Dublin Natural History Society* **3**: 94-99.

179. King, C. 1989 *The natural history of stoats and weasels.* Christopher Helm. London.

180. King, F. 1983 Barn owl catching prey on the wing. *Irish Birds* **2**: 345-346.

181. King, W. 1859 On the occurrence in Galway of the lesser horse-shoe bat (*Rhinolophus hipposideros*). *Proceedings of the Dublin Natural History Society* **1**: 264-268.

182. Kingston, S., O'Connell, M. and Fairley, J.S. 1999 Diet of otters *Lutra lutra* on Inishmore, Aran Islands, west coast of Ireland. *Biology and Environment* **99B**: 173-182.

183. Kinney, G.M. 1998 *The prevalence and impact of internal parasites on the wild rabbit,* Oryctolagus cuniculus (L.). PhD Thesis. Queens University Belfast. 1997. Abstract in *Aslib Index to Theses* **47**: 1139.

184. Kirkham, G. 1981 Economic diversification in a marginal economy: a case study. In Roebuck, P. (Editor) *Plantation to partition: essays in honour of J.L. McCracken.* 64-81. Blackstaff Press. Belfast.

185. Korky, J.K. & Webb, R.G. 1999 Resurvey, biogeography and conservation of the natterjack toad *Bufo calamita* Laurenti (Anura; Bufonidae) in the Republic of Ireland. *Bulletin of the Irish Biogeographical Society* **23**: 2-52.

186. Kruuk, H. 1989 *The social badger.* Oxford University Press. Oxford.

187. Kruuk, H. 1995 *Wild otters.* Oxford University Press. Oxford.

188. Kyne, M.J., Kyne, M.J. and Fairley, J.S. 1990 A summer survey of otter sign on Roundstone Bog, south Connemara. *Irish Naturalists' Journal* **23**: 273-276.

189. Kyne, M.J., Smal, C.M. & Fairley, J.S. 1989 The food of otters *Lutra lutra* in the Irish Midlands and a comparison with that of mink *Mustela vison* in the same region. *Proceedings of the Royal Irish Academy* **89B**: 33-46.

190. Lavelle, D. 1976 *Skellig, island outpost of Europe.* O'Brien Press. Dublin.

191. "Lepus hibernicus" 1891 The Irish hare. *Irish Sportsman* **22**: 444.

192. Lever, C. 1977 *The naturalized animals of the British Isles.* Hutchinson. London.

193. Longfield, A.K. 1929 *Anglo-Irish trade in the sixteenth century.* Routledge. London.

194. Lunnon, R. 1996 Otter (*Lutra lutra* L.) distribution in Ireland. In Reynolds, J.D. (Editor) *The conservation of aquatic ecosystems.* 111-116. Royal Irish Academy. Dublin.

195. McAney, C. & Fairley, J. 1990 Activity of Leisler's bats *Nyctalus leisleri* (Kuhl, 1817) at a summer roost in Ireland. *Myotis* **28**: 83-91.

196. McAney, C.M. 1994 The lesser horseshoe bat in Ireland - past, present and future. *Folia Zoologica* **43**: 387-392.

197. McAney, C.M. 1994 *West of Ireland hibernation survey 1994.* Unpublished report.

198. McAney, C.M. & Fairley, J.S. 1983 Horseshoe bat echolocation pulses. *Irish Naturalists' Journal* **22**: 362.

199. McAney, C.M. & Fairley, J.S. 1988 Activity patterns of the lesser horseshoe bat *Rhinolophus hipposideros* at summer roosts. *Journal of Zoology* **216**: 325-338.

200. McAney, C.M. & Fairley, J.S. 1989 Observations at summer roosts of the lesser horseshoe bat in County Clare. *Irish Naturalists' Journal* **23**: 1-6.

201. McAney, C.M. & Fairley, J.S. 1989 Analysis of the diet of the lesser horseshoe bat *Rhinolophus hipposideros* in the West of Ireland. *Journal of Zoology* **217**: 491-498.

202. McAney, K. 1992 *Bats and bridges.* Office of Public Works. Dublin.

203. McAney, K. 1999 Horseshoes, Leisler's and barbastelles - the work goes on. In McAney. K. (Editor) *Second Irish Bat Conference. Abstracts.* Unpaginated. National Parks and Wildlife Service and the Vincent Wildlife Trust. Ballyvaughan.

204. McCormick, F. 1998 The animal bones. In Hurley, M.F., Scully, O.M.B. & McCutcheon, S.W.J. (Editors) *Late Viking age Medieval Waterford.* 819-853. Waterford Corporation. Waterford.

205. McCracken, E. 1971 *The Irish woods since Tudor times.* David & Charles. Newton Abbot.

206. McDonald, D. 1995 *European mammals evolution and behaviour.* HarperCollins. London.

207. McFadden, Y.M.T. & Fairley, J.S. 1984 Food of otters *Lutra lutra* (L.) in an Irish limestone river system with special reference to the crayfish *Austropotamobius pallipes* (Lereboullet). *Journal of Life Sciences of the Royal Dublin Society* **5**: 65-76.

208. McGuire, C. 1998 Survey of lesser horseshoe bats *Rhinolophus hipposideros* (Bechstein) and other bat species in north Co. Clare, Ireland. *Irish Naturalists' Journal* **26**: 43-50.

209. MacLochlainn, C. 1986 Bunny is not funny anymore. *Irish Times* September 15.

210. Mahaffy, R.P. 1907 *Calendar of the state papers relating to Ireland.* HMSO. London.

211. Mason, T.H. 1936 *The islands of Ireland.* Batsford. London.

212. Massy, A.L. 1926 Rat in flower garden. *Irish Naturalists' Journal* **1**: 115.

213. Matheson, C. 1939 A survey of the status of *Rattus rattus* and its subspecies in the sea-ports of Great Britain and Ireland. *Journal of Animal Ecology* **8**: 76-93.

214. Mikkola, H. 1983 *Owls of Europe.* Poyser. Calton.

215. Millais, J.G. 1904 *The mammals of Great Britain and Ireland.* Longmans, Green. London.

216. Moffat, C.B. 1897 Irish bats. *Irish Naturalist* **6**: 135.

217. Moffat, C.B. 1898 *The life and letters of Alexander Goodman More.* Figgis. Dublin.

218. Moffat, C.B. 1900 The habits of the hairy-armed bat *Vesperugo leisleri* Kuhl. *Irish Naturalist* **9**: 235-240.

219. Moffat, C.B. 1904 Bats, hedgehogs and frogs in winter. *Irish Naturalist* **13**: 81-87.

220. Moffat, C.B. 1922 The habits of the long-eared bat. *Irish Naturalist* **31**: 105-111.

221. Moffat, C.B. 1926 The badger. *Irish Naturalists' Journal* **1**: 130-132.

222. Moffat, C.B. 1927 The Irish hare. *Irish Naturalists' Journal* **1**: 271-273.

223. Moffat, C.B. 1928 The rabbit. *Irish Naturalists' Journal* **2**: 28-30.

224. Moffat, C.B. 1928 The black rat. *Irish Naturalists' Journal* **2**: 47-49.

225. Moffat, C.B. 1928 The brown rat. *Irish Naturalists' Journal* **2**: 87-89.

226. Moffat, C.B. 1929 The house mouse. *Irish Naturalists' Journal* **2**: 195-197.

227. Moffat, C.B. 1930 The pipistrelle. *Irish Naturalists' Journal* **3**: 26-29.

228. Moffat, C.B. 1930 The hairy-armed bat. *Irish Naturalists' Journal* **3**: 50-54

229. Moffat, C.B. 1931 The long-eared bat. *Irish Naturalists' Journal* **3**: 182-185.

230. Moffat, C.B. 1932 Daubenton's bat. *Irish Naturalists' Journal* **4**: 26-28.

231. Moffat, C.B. 1938 The mammals of Ireland. *Proceedings of the Royal Irish Academy* **44B**: 61-128.

232. Montgomery, W.I. & Dowie, M. 1993 The distribution of the wood mouse *Apodemus sylvaticus* and the house mouse *Mus domesticus* on farmland in north-east Ireland. *Irish Naturalists' Journal* **24**: 199-203.

233. Moore, C. 1901 Extracts from Dublin newspapers in 1693 and 1729 regarding hawks and rats. *Royal Society of Antiquaries of Ireland Journal* (5) **11**: 179-180.

234. Moriarty. C. 1955 Behaviour of hares on mud flats. *Irish Naturalists' Journal* **11**: 310.

235. Moriarty, C. 1961 Pine marten *Martes martes* (L.) in Co. Wicklow. *Irish Naturalists' Journal* **13**: 239.

236. Moriarty, C. 1978 *Eels: a natural and unnatural history.* David & Charles. Newton Abbot.

237. Moriarty, C. 1997 *Exploring Dublin wildlife, parks, waterways.* Wolfhound Press. Dublin.

238. Morris, P. 1979 Rats in the diet of the barn owl (*Tyto alba*). *Journal of Zoology* **189**: 540-545.

239. Murphy, K.P. & Fairley, J.S. 1985 Food of otters *Lutra lutra* on the south shore of Galway Bay. *Proceedings of the Royal Irish Academy* **85B**: 47-55.

240. Murphy, K.P. & Fairley, J.S. 1985 Food and sprainting sites of otters on the west coast of Ireland. *Irish Naturalists' Journal* **21**: 477-479.

241. Neal, E. & Cheeseman, C. 1996 *Badgers.* Poyser. London.

242. Nelson, B. & Smiddy, P. 1997 Records of the bat bug *Cimex pipistrelli* (Hemiptera: Cimicidae) from Cos Cork and Waterford. *Irish Naturalists' Journal* **25**: 344-345.

243. Ní Lamha, E. 1979 *Provisional distribution atlas of amphibians, reptiles and mammals in Ireland.* An Foras Forbatha. Dublin.

244. Noonan, N.L., Sheane, W.D., Harper, I.R. & Ryan, P.J. 1975 Wildlife as a possible reservoir of bovine tuberculosis. *Irish Veterinary Journal* **29**: 1.

245. Norberg, U.M. & Rayner, J.M.V. 1987 Ecological morphology and flight in bats (Mammalia: Chiroptera): wing adaptations, flight performance, foraging strategy and echolocation. *Philosophical Transactions of the Royal Society of London.* **316**: 335-427.

246. Nowak, R.M. 1991 *Walker's mammals of the world.* 5th Edition. John Hopkins. London.

247. O'Carroll, N. 1984 *The forests of Ireland.* Turoe Press. Dublin.

248. O'Connor, R. & O'Malley, E. 1989 *Badgers and bovine tuberculosis.* Eradication of Animal Disease Board. Dublin.

249. O'Corry-Crowe, G., Hammond, R., Eves, J. & Hayden, T.J. 1996 Effect of reduction in badger density on spatial organisation and activity of badgers (*Meles meles* L.) in relation to farms in central Ireland. *Biology and*

Environment **96B**: 147-158.

250. O'Donnell, G. 1992 Stoat carrying a fish. *Irish Naturalists' Journal* **24**: 174-175.

251. O'Gorman, F. 1977 Mammals of the island. In Jeffrey, D.W. (Editor) *North Bull Island, Dublin Bay - a modern natural history.* 111-113. Royal Dublin Society. Dublin.

252. O'Gorman, F. & Fairley, J.S. 1965 A colony of *Plecotus auritus* from Co. Kilkenny. *Proceedings of the Zoological Society of London* **145**: 154-155.

253. O'Mahony, E. 1931 Notes on the mammals of the North Bull, Dublin Bay. *Irish Naturalists' Journal* **3**: 199-201.

254. O'Mahony 1935 Discovery of a second race of house-mouse in Ireland. *Irish Naturalists' Journal* **5**: 218-219.

255. O'Mahony, E. 1935 The North Bull house mouse. *Irish Naturalists' Journal* **5**: 291.

256. O'Mahony 1937 On some forms of the house mouse *Mus musculus* Linn. in Ireland. *Irish Naturalists' Journal* **6**: 288-290.

257. O'Mahony 1944 A note on the Irish Anoplura Siphunculata. *Entomologist's Monthly Magazine* **80**: 5.

258. O'Rorke, D.V.W. 1956 Pine martens in Co. Mayo. *The Field* **208**: 692.

259. O'Sullivan, B.W., Fairley, J.S. & Stronach, N. 1975 New records of horseshoe bats. *Irish Naturalists' Journal* **18**: 190-191.

260. O'Sullivan, P. 1993 Pine marten. In *Predator Day.* Unpaginated. Unpublished report. Badgerwatch Ireland. Dublin.

261. O'Sullivan, P. 1994 Bats in Ireland. *Irish Naturalists' Journal* **24** (supplement): 1-21.

262. O'Sullivan, P.J. 1983 The distribution of the pine marten (*Martes martes*) in the Republic of Ireland. *Mammal Review* **13**: 39-44.

263. O'Sullivan, W.M. 1991 The distribution of otters *Lutra lutra* within a major Irish River system, the Munster Blackwater catchment, 1988-90. *Irish Naturalists' Journal* **23**: 442-446.

264. O'Sullivan, W.M. 1993 Efficiency and limitations of the standard otter (*Lutra lutra*) survey technique in Ireland. *Biology and Environment* **93B**: 49-53.

265. O'Sullivan, W.M. 1994 Summer diet of otters on part of the River Blackwater catchment. *Irish Naturalists' Journal* **24**: 349-354.

266. O'Sullivan, W.M. 1996 Otter conservation: factors affecting survival, with particular reference to drainage and pollution within an Irish river system. In Reynolds, J.D. (Editor) *The conservation of aquatic ecosystems.* 117-133. Royal Irish Academy. Dublin.

267. Otway, S.G. 1840 Remarkable change of habit in the hare. *Annals and Magazine of Natural History* **5**: 362-363.

268. Pack-Beresford, R. 1928 Hares going to ground. *Irish Naturalists' Journal* **2**: 104.

269. Passingham, G.H. 1895 Rat and hedgehog. *The Field* **86**: 903.

270. Patterson, R. 1894 Notes on the occurrences of the marten (*Martes sylvatica*) in Ulster. *Irish Naturalist* **3**: 106-109.

271. Payne, R. 1589 *A briefe description of Ireland.* Dawson. London. Facsimile Edition 1973. Da Capo Press. New York.

272. Pease, A.E. 1898 *The badger.* Lawrence & Bullen. London.

273. Pentland, G.H. 1902 Strange conduct of a badger. *Irish Naturalist* **11**: 24.

274. Pentland, G.H. 1917 Badgers and hedgehogs. *Irish Naturalist* **26**: 20.

275. Perry, K.W. & Warburton, S.W. 1977 *Birds and flowers of the Saltee Islands.* Privately published. Belfast.

276. Portlock, J.E. 1836 On some peculiar habits of the *Otus brachyotus.* *Proceedings of the Royal Irish Academy* **1**: 52-53.

277. Public Records Office of Northern Ireland. Downshire Estates. Documents under D607/A.

278. Public Records Office of Northern Ireland. Lissadell Estate. Document D/4131/C/11/4.

279. Reeve, N. 1994 *Hedgehogs.* Poyser. London.

280. Ritsema, A. 1999 *Discover the islands of Ireland.* Collins Press. Cork.

281. Robson, M.J., Parsons, A.J. & Williams, T.E. 1989 Herbage production: grasses and legumes. In Holmes, W. (Editor) *Grass, its production and utilization.* 2nd Edition. 7-88. Blackwell. Oxford.

282. Roche, R. & Merne, O. 1977 *Saltees islands of birds and legends.* O'Brien Press. Dublin.

283. Rogers, A.R. 1959 Pine marten *Martes martes* (L.) in Co. Down woods. *Irish Naturalists' Journal* **13**: 42.

284. Russ, J.M., Hutson, A.M., Montgomery, W.I., Racey, P.A. and Speakman, J.R. 2001 The status of Nathusius' pipistrelle (*Pipistrellus nathusii* Keyserling & Blasius, 1839) in the British Isles. *Journal of Zoology* **254**: 91-100.

285. Russ, J.M. & Montgomery, W.I. In press *Biodiversity action plans for bats in Northern Ireland.* Environment and Heritage Service (Northern Ireland) and Queen's University. Belfast.

286. Russ, J.M., O'Neill, J.K. & Montgomery, W.I. 1998 Nathusius' pipistrelle bats (*Pipistrellus nathusii,* Keyserling & Blasius 1839) breeding in Ireland. *Journal of Zoology* **245**: 345-349.

287. Ruttledge, R.F. 1920 The pine marten in Ireland. *Irish Naturalist* **28**: 125-127.

288. Ruttledge, R.F. 1943 Lesser horse-shoe bat (*Rhinolophus hipposideros*) in Co. Mayo. *Irish Naturalists' Journal* **8**: 77.

289. Ruttledge, R.F. 1943 Occurrences of the lesser horse-shoe bat (*Rhinolophus hipposideros*) in Co. Mayo. *Irish Naturalists' Journal* **8**: 261.

290. Ruttledge, R.F. 1951 Rats eating hawthorn berries. *Irish Naturalists' Journal* **10**: 217.

291. Ruttledge, R.F. 1966 *Ireland's birds.* Witherby. London.

292. Rutty, J. 1772 *An essay toward a natural history of the County of Dublin.* Sleator. Dublin.

293. Ryan, A.W., Duke, E.J. & Fairley, J.S. 1993 Polymorphism, localisation and geographical transfer of mitochondrial DNA in *Mus musculus domesticus* (Irish house mice). *Heredity* **70**: 75-81.

294. Rydell, J. & Racey, P.A. 1995 Street lamps and the feeding ecology of insectivorous bats. *Symposia of the Zoological Society of London* **67**: 291-307.

295. Sampson, G.V. 1802 *Statistical survey of the County of Londonderry.* Graisberry & Campbell. Dublin.

296. Schaller, G.B. 1972 *The Serengeti lion.* University of Chicago Press. Chicago.

297. Schober, W. & Grimberger, E. 1989 *A guide to the bats of Britain and Europe.* Hamlyn. London.

298. Scott, D., Clarke, R. & McHaffie, P. 1992 Hen harriers successfully breeding in a tree nest of their own construction. *Irish Birds* **4**: 566-570.

299. Sharrock, J.T.R. 1973 *The natural history of Cape Clear Island.* Poyser. Berkhampstead.

300. Sheail, J. 1971 *Rabbits and their history.* David & Charles. Newton Abbot.

301. Shiel, C. 1999 *Bridge usage by bats in County Leitrim and County Sligo.* Unpublished report. Heritage Council. Kilkenny.

302. Shiel, C.B., Duvergé, P.L., Smiddy, P. & Fairley, J.S. 1998 Analysis of the diet of Leisler's bat (*Nyctalus leisleri*) in Ireland with some comparative analyses from England and Germany. *Journal of Zoology* **246**: 417-425.

303. Shiel, C.B. & Fairley, J.S. 1998 Activity of Leisler's bat *Nyctalus leisleri* (Kuhl) in the field in south-east County Wexford, as revealed by a bat detector. *Biology and Environment* **98B**: 105-112.

304. Shiel, C.B. & Fairley, J.S. 1999 Evening emergence of two colonies of Leisler's bat (*Nyctalus leisleri*) in Ireland. *Journal of Zoology* **247**: 439-447.

305. Shiel, C.B. & Fairley, J.S. 2000 Observations at two nursery roosts of Leisler's bats *Nyctalus leisleri* (Kuhl, 1817) in Ireland. *Myotis* **37**: 41-53.

306. Shiel, C.B., McAney, C.M. & Fairley, J.S. 1991 Analysis of the diet of Natterer's bat *Myotis nattereri* and the common long-eared bat *Plecotus auritus* in the west of Ireland. *Journal of Zoology* **223**: 299-305.

307. Shiel, C., McAney, C., Sullivan, C. & Fairley, J.S. 1997 *Identification of arthropod fragments in bat droppings.* Mammal Society. London.

308. Shiel, C.B., Shiel, R.E. & Fairley, J.S. 1999 Seasonal changes in the foraging behaviour of Leisler's bats (*Nyctalus leisleri*) in Ireland as revealed by radio-telemetry. *Journal of Zoology* **249**: 347-358.

309. Sleeman, D.P. 1987 Records of fleas (Siphonaptera) from Irish stoats. *Irish Naturalists' Journal* **22**: 256-257.

310. Sleeman, D.P. 1988 Recent records of Leisler's bat from the south coast. *Irish Naturalists' Journal* **22**: 416.

311. Sleeman, D.P. 1988 *Skrjabingylus nasicola* (Leukart) (Metastrongyloidae) as a parasite of the Irish stoat. *Irish Naturalists' Journal* **22**: 525-527.

312. Sleeman, D.P. 1988 Irish stoat road casualties. *Irish Naturalists' Journal* **22**: 527-529.

313. Sleeman, D.P. 1989 *Stoats & weasels polecats & martens.* Whittet Books. London.

314. Sleeman, D.P. 1989 Ectoparasites of the Irish stoat. *Medical and Veterinary Entomology* **3**: 213-218.

315. Sleeman, D.P. 1990 Dens of Irish stoats. *Irish Naturalists' Journal* **23**: 202-203.

316. Sleeman, D.P. 1991 Home ranges of Irish stoats. *Irish Naturalists' Journal* **23**: 486-488.

317. Sleeman, D.P. 1992 Long-distance movements in an Irish badger population. In Priede, I.G. & Swift, S.M. (Editors) *Wildlife telemetry remote monitoring and tracking of animals.* 670-676. Ellis Horwood. London.

318. Sleeman, D.P. 1992 Diet of Irish stoats. *Irish Naturalists' Journal* **24**: 151-153.

319. Sleeman, D.P. 1993 Habitats of the Irish stoat. *Irish Naturalists' Journal* **24**: 318-321.

320. Sleeman. D.P. & Kelly, T.C. 1987 Records of fleas (Siphonaptera) from common rats in Ireland. *Irish Naturalists' Journal* **22**: 256-257.

321. Sleeman, D.P. & Mulcahy, M.F. 1993 Behaviour of Irish badgers in relation to bovine tuberculosis. In Hayden, T.J. (Editor) *The badger.* 154-165. Royal Irish Academy. Dublin.

322. Sleeman, D.P., Smiddy, P. & Moore, P. 1996 The fleas of Irish terrestrial mammals: a review. *Irish Naturalists' Journal* **25**: 237-248.

323. Smal, C. 1991 *Feral American mink in Ireland.* Office of Public Works.

324. Smal, C. 1995 *The badger and habitat survey of Ireland.* National Parks and Wildlife Service and the Department of Agriculture. Dublin. ALSO *The badger and habitat survey of Ireland summary report.* Government Publications. Dublin.

325. Smal, C. 2000 *Evacuation of badgers and excavation of badger setts at Carrigoran, Newmarket on Fergus. Co. Clare.* Unpublished report. Clare County Council. Ennis.

326. Smal, C.M. 1987 The diet of the barn owl *Tyto alba* in southern Ireland, with reference to a recently introduced species - the bank vole *Clethrionomys glareolus*. *Bird Study* **34**: 113-125.

327. Smal, C.M. 1988 The American mink *Mustela vison* in Ireland. *Mammal Review* **18**: 201-208.

328. Smal, C.M. 1989 Notes on the diet of short-eared owls in Ireland. *Irish Birds* **4**: 73-75.

329. Smal, C.M. 1990 Den excavation by feral mink *Mustela vison* Schreber in the Irish midlands. *Irish Naturalists' Journal* **23**: 204-205.

330. Smal, C.M. 1991 Population studies on feral American mink *Mustela vison* in Ireland. *Journal of Zoology, London* **224**: 233-249.

331. Smiddy, P. 1991 Bats and bridges. *Irish Naturalists' Journal* **23**: 425-426.

332. Smiddy, P. 1993 The status of the otter in east Cork and west Waterford. *Irish Naturalists' Journal* **24**: 236-240.

333. Smiddy, P. 1994 Hare species in Co. Cork. *Irish Naturalists' Journal* **24**: 417-418.

334. Smiddy, P. 1996 Badgers preying on nestling birds. *Irish Naturalists' Journal* **25**: 224-225.

335. Soulsby, E.J.L. 1982 *Helminths, arthropods and Protozoa of domesticated animals*. 7th Edition. Ballière Tindall. London.

336. Speakman, J.R. 1995 Chiropteran nocturnality. *Symposia of the Zoological Society of London* **67**: 187-201.

337. Stebbings, R.E. 1988 *The conservation of European bats*. Christopher Helm. London.

338. Stebbings, R.E. & Griffith, F. 1986 *Distribution and status of bats in Europe*. Institute of Terrestrial Ecology. Abbots Ripton.

339. Stendall, J.A.S. 1930 Unusually coloured wild mice. *Irish Naturalists' Journal* **3**: 48.

340. Stendall, J.A.S. 1947 Pine marten in Co. Down. *Irish Naturalists' Journal* **9**: 48.

341. Stringer, A. 1714 *The experienc'd huntsman*. Blow. Belfast. New Edition 1977. Blackstaff Press. Belfast.

342. Sullivan, C.M., Shiel, C.B., McAney, C.M. & Fairley, J.S. 1993 Analysis of the diets of Leisler's *Nyctalus leisleri,* Daubenton's *Myotis daubentoni* and pipistrelle *Pipistrellus pipistrellus* bats in Ireland. *Journal of Zoology* **231**: 656-663.

343. Swift, S.M. 1998 *Long-eared bats*. Poyser. London.

344. Tangney, D.E. & Fairley, J.S. 1994 Otter signs and diet in Connemara National Park and its environs. *Irish Naturalists' Journal* **24**: 434-440.

345. Tangney, D., Fairley, J.S. & O'Donnell, G. 1995 Food of Irish hares *Lepus*

373

timidus hibernicus in western Connemara, Ireland. *Acta Theriologica* **40**: 403-413.

346. Tapper, S. (Editor) 1999 *A question of balance.* Game Conservancy Trust. Fordingbridge.
347. Teacher, D. & Gough, K. 1936 Stoats capturing fish. *Irish Naturalists' Journal* **6**: 152.
348. Thompson, H.V. & Worden, A.N. 1956 *The rabbit.* Collins. London.
349. Thompson, R. 1802 *Statistical survey of the County of Meath.* Graisberry & Campbell. Dublin.
350. Thompson, W. 1849-56 *The natural history of Ireland.* Bohn. London.
351. Thulin, C.-G., Jaarola, M. & Tegelström, H. (1997) The occurrence of mountain hare mitochondrial DNA in wild brown hares. *Molecular Ecology* **6**: 463-467.
352. Trout, R.C., Chasey, D. & Sharp, G. 1997 Seroepidemiology of rabbit haemorrhagic disease (RHD) in wild rabbits (*Oryctolagus cuniculus*) in the United Kingdom. *Journal of Zoology* **243**: 846-853.
353. Twigg, G.I. 1975 *The brown rat.* David & Charles. Newton Abbot.
354. Twigg, G.I. 1992 The black rat *Rattus rattus* in the United Kingdom in 1989. *Mammal Review* **22**: 33-42.
355. Ussher, R.J. 1886 Bird life on the Saltees and Keeraghs, Co. Wexford. *Zoologist (3)* **10**: 88-89.
356. Ussher, R.J. 1898 Breeding of the marten in Co. Waterford. *Irish Naturalist* **7**: 171-172.
357. Ussher, R.J. & Warren, R. 1900 *The birds of Ireland.* Gurney & Jackson. London.
358. Vaughan, N. 1997 The diets of British bats. *Mammal Review* **27**: 77-94.
359. Viney, M. & Viney, E. 1999 *A wildlife narrative.* Irish Times. Dublin.
360. Walsh, P.M. 1984 Diet of barn owls at an urban Waterford roost. *Irish Birds* **2**: 437-444.
361. Walsh, P.M. 1988 Black rats *Rattus rattus* (L.) as prey of short-eared owls *Asio flammeus* (Pontopiddan) on Lambay Island, Co. Dublin. *Irish Naturalists' Journal* **22**: 536-537.
362. Walsh, P.M. & Sleeman, D.P. 1988 Avian prey of a wintering short-eared owl population in south-west Ireland. *Irish Birds* **3**: 589-591.
363. Ward, D.P., Smal, C.M. & Fairley, J.S. 1986 The food of mink *Mustela vison* in the Irish midlands. *Proceedings of the Royal Irish Academy* **86B**: 169-182.
364. Warner, P. & O'Sullivan, P. 1982 The food of the pine marten *Martes martes* in Co. Clare. *Transactions of the International Congress of Game Biologists* **14**: 323-330.

365. Warren, R. 1886 Destruction of young rabbits by rats. *Zoologist* (3) **10**: 241-242.

366. Watson, A. & Hewson, R. 1973 Population densities of mountain hares (*Lepus timidus*) on western Scottish and Irish moors and on Scottish hills. *Journal of Zoology* **170**: 151-159.

367. Watson, A., Hewson, R., Jenkins, D. & Parr, R. 1973 Population densities of mountain hares compared with red grouse on Scottish moors. *Oikos* **24**: 225-230.

368. Watson, A. & O'Hare, P.J. 1979 Bird and mammal numbers on untreated and experimentally treated Irish bog. *Oikos* **33**: 97-105.

369. Westropp, T.J. 1905 A survey of the ancient churches in the County of Limerick. *Proceedings of the Royal Irish Academy* **25C**: 327-480.

370. Westropp, T.J. 1914 Fortified headlands and castles on the south coast of Munster. *Proceedings of the Royal Irish Academy* **32C**: 89-124, 188-226, 249-286.

371. Whatmough, J.A. 1995 Grazing on sand dunes: the re-introduction of the rabbit *Oryctolagus cuniculus* L. to Murlough NNR, Co. Down. *Biological Journal of the Linnean Society* **56** (supplement): 39-43.

372. Whelan, J. 1985 The population and distribution of the mountain hare (*Lepus timidus*) on farmland. *Irish Naturalists' Journal* **21**: 532-534.

373. Whelan, R. & Hayden, T.J. 1993 The reproductive cycle of the female badger (*Meles meles* L.) in east Offaly. In Hayden, T.J. (Editor) *The badger*. 64-77. Royal Irish Academy. Dublin.

374. Whilde, A. 1992 The prey of two rural domestic cats. *Irish Naturalists' Journal* **24**: 173-174.

375. Whitty, B.T. 1955 Myxomatosis in the common hare - *Lepus europaeus* - I. *Irish Veterinary Journal* **9**: 267-268.

376. Whyte, J.J. 1876 The marten in the west of Ireland. *The Field* **47**: 638.

377. Wilde, W. 1859 Upon the unmanufactured animal remains belonging to the Academy. *Proceedings of the Royal Irish Academy* **7**: 181-211.

378. Williams, E. 1890 Varieties of the Irish hare. *Zoologist* (3) **14**: 70-71.

379. Wise, M.H. 1980 The use of fish vertebrae in scats for estimating prey size of otters and mink. *Journal of Zoology* **192**: 25-31.

380. Wolfe, A. 1994 Flehmen and scent trail in Irish hares. *Irish Naturalists' Journal* **24**: 471-472.

381. Wolfe, A. 1994 Coat colour in the Irish hare and the status of the brown hare. *Irish Naturalists' Journal* **24**: 472.

382. Wolfe, A. 1995 *A study of the ecology of the Irish mountain hare* (Lepus timidus hibernicus) *with some considerations for its management and that of the rabbit* (Oryctolagus cuniculus) *on North Bull Island, Dublin Bay*. PhD

Thesis. University College Dublin.

383. Wolfe, A. & Hayden, T.J. 1996 Home range sizes of Irish hares on coastal grassland. *Biology and Environment* **96B**: 141-146.

384. Wolfe, A. & Long, A.M. 1997 Distinguishing between the hair fibres of the rabbit and the mountain hare in scats of the red fox. *Journal of Zoology* **242**: 370-375.

385. Wolfe, A., Whelan, J. & Hayden, T.J. 1996 Dietary overlap between the Irish mountain hare *Lepus timidus hibernicus* and the rabbit *Oryctolagus cuniculus* on coastal grassland. *Biology and Environment* **96B**: 89-95.

386. Wolfe, A., Whelan, J. & Hayden, T.J. 1996 The diet of the mountain hare (*Lepus timidus hibernicus*) on coastal grassland. *Journal of Zoology* **240**: 804-810.

387. Yalden, D. 1999 *The history of British mammals.* Poyser. London.

388. Young, A. 1780 *A tour in Ireland.* Bonham. Dublin.

INDEX